GREEN ENERGY ECONOMIES

GREEN ENERGY ECONOMIES

THE SEARCH FOR CLEAN AND RENEWABLE ENERGY

**Energy and
Environmental Policy,
Volume 10**

JOHN BYRNE
YOUNG-DOO WANG
EDITORS

Transaction Publishers
New Brunswick (U.S.A.) and London (U.K.)

Library of Congress Catalog Number: 2013036235
ISBN: 978-1-4128-5375-0
Printed in the United States of America

Library of Congress Cataloging-in-Publication Data

Green energy economies : the search for clean and renewable energy / John Byrne and Young-Doo Wang, editors.
    pages cm. — (Energy and environmental policy ; volume 10)
  Includes bibliographical references and index.
  ISBN 978-1-4128-5375-0 (alk. paper)
  1. Renewable energy sources—Economic aspects. 2. Renewable energy sources—Government policy. 3. Renewable energy sources. I. Byrne, John, 1949-, editor. II. Wang, Young-Doo, editor.
  HD9502.A2G735 2014
  333.79'4—dc23

                                          2013036235

Contents

Acknowledgments

The idea for this book grew out of the 2011 International Conference on Green Energy Economies, held in Washington, DC from July 28 to 29, 2011, which was organized by the Center for Energy and Environmental Policy (CEEP) and co-sponsored by the National Council for Science and Environment and the Korea Energy Economics Institute. Several participants agreed a book on the topic was overdue, and we agreed to recruit leading researchers for this book. We are grateful that so many were able to prepare thoughtful chapters.

This volume would not have been possible without the dedicated efforts of a team of researchers at CEEP, including Dr. Kristen Hughes, former research fellow at CEEP, and Leon Mach and Michaella Song, graduate students at the Center. We would particularly like to recognize Job Taminiau, a Ph.D. candidate at CEEP, for his significant contribution to this project.

1

The Promise of a Green Energy Economy

John Byrne, Young-Doo Wang, Job
Taminiau, and Leon Mach

Why a Green Energy Economy?

The role of the green energy economy in discussions of potential energy futures has steadily grown. The increasing awareness of—and importance given to—this option is exemplified by the United Nations Conference on Sustainable Development (UNSCD), which convened in Rio de Janeiro, Brazil, from June 20 to 22, 2012, where realizing a new economy fueled by green energy was identified as an essential prong in any strategy to alleviate worldwide poverty and promote sustainable development.[1]

The opportunity—and need—to transition to a new economy is highlighted in current economic- and climate-change challenges.[2,3] The worldwide economic problems since 2008 underscore the need for a shift to alternative development pathways. A major case can be made for the green energy economy as a means to move away from current hardships, while providing methods of reducing dependence on finite energy resources as well as mitigating risks and harms associated with conventional energy sources such as climate change. The consequences of fossil-fuel use in terms of environmental and human-health effects are substantial. For instance, Epstein et al.[4] estimate that the life-cycle social costs associated with coal use are between 10 and 28 cents/kWh. The low estimate of 10 cents/kWh corresponds to a cost of $700 billion annually. In short, transitioning away from the use of fossil fuel sources can further increase national income and prosperity by lowering the burden on healthcare services and environmental costs. While the scale and complexity of such a transition are daunting, many benefits can be accrued from a successful shift.

One of the major benefits can be rapid and broad-based job creation. The Energy Information Administration (EIA) forecasts that world energy consumption will increase by 53 percent between 2008 and 2035.[5] In order to provide the necessary services on the basis of such forecasts, the modern energy paradigm focuses on increasing the use of conventional energy sources. Expanding and updating coal generation capacity has typically been the means of meeting the rising energy demand. Within this paradigm, there is an incentive for utilities to sell more, rather than less, electricity. While this approach might satisfy prevailing demands in the short run, it does little for long-term structural economic change or employment prospects. In fact, for every million USD invested in supplying fossil-fuel-based services, only four permanent jobs are created.[6] Given the same level of investment, the number of permanent job increases is estimated to be three times higher when priority is given to photovoltaics (PV) and other renewable energy installations, and grows by a factor of four when resources are dedicated to the use of information and communication technology (ICT) for a smarter electricity grid and more intelligent transport.[7,8,9] Additionally, every million USD invested in energy efficiency and conservation projects creates between twelve and fifteen permanent jobs.[10] Green energy technologies—PV, wind, ICT, energy efficiency, etc.—offer a development pathway that is less fuel-intensive while requiring deployment of more human resources to build the transition. Such a pathway can substantially improve employment opportunities in the near term while enabling the change in energy infrastructure necessary to solve enduring threats of climate change and energy poverty (i.e., the large unmet needs of a large proportion of humanity living in the light shadow of industrialized economies).

Investment in a green energy economy can potentially establish the infrastructure of a new paradigm. While the modern energy paradigm relies on the "more is better" principle of development at the core of decision making, the new paradigm will focus on providing energy service needs while using far less energy. This means a new paradigm in which a narrative of "just sustainability"[11] is at the core of decision making. Using fewer resources is hardly a recipe for social decline. In fact, research has shown that the "negawatt"[12] (i.e., not consuming energy) in most cases costs one-half or less of the US price of retail electricity.[13] Moreover, the high conversion and distribution losses associated with conventional energy production and consumption underscore the costliness and long-term risk of the energy status quo: every unit of end-use electricity saved through efficiency measures avoids production/consumption of three units of primary energy.[14] The infrastructure of a green energy economy not only improves employment opportunities but also provides the same energy services to consumers with fewer resources so that social opportunity expands into the future. This is in marked contrast to the prevailing order whose use of more energy, drawing from non-renewable sources, can only promise higher costs and widening energy poverty. In brief, investing in green energy negawatts and renewable energy as a global strategy not only mitigates climate change (due to the use of fewer energy resources and, as a result, the release of fewer emissions) but also strengthens the resilience of economies throughout the world.

Of course, a transition to a green energy economy necessitates a substantial increase in production and consumption of certain things, for example, solar cell modules, wind turbine nacelles, and fuel cells. Some wonder whether this is realistic in terms of the scale, complexity, and needed level of commitment. A rapid shift from fossil-fuel resources toward a green energy economy can be accomplished on a timescale that is constrained only by the political will to realize the good. As Lester Brown notes, the shifts in engineering, manufacturing, and ways of daily life associated with the communications revolution of the last thirty years redefined the economic, cultural, and political ties of billions of people.[15] The internet and mobile-phone infrastructures are examples of real-time rapid transformation that should warn skeptics who doubt the pace and scope of paradigm shifts.

Several countries recognize the promise of a green energy economy and are creating incentives to ensure that they are manufacturing tomorrow's energy devices. China's development strategy is an example: between 2006 and 2008, world solar cell production increased by 2,400 MW. In the same period, China expanded its share in the production of solar cells from 11 to 60 percent.[16] Considering that the "green shift" will mean the production of thousands of gigawatts of solar and wind energy, China has positioned itself as a key designer and producer of the new era.[17] Notably, countries committed to the production of new technologies will benefit not only their manufacturing industry—and thus, support their domestic job market—but also their national income and export prospects, and their ability to shape further innovation.

Driven by a technological paradigm of "more is better," the modern energy system requires vast amounts of fuel to function. A parallel economic paradigm of "cornucopian"[18] development completes the modern model of growth without limits. But the model has inescapable contradictions, which are especially evident in the contemporary period. One is political: when a society needs fossil-fuel imports to power its success—and the cornucopian impulse destines *all* moderns to eventually need imports—a political paradox is revealed. Smaller countries must sacrifice a measure of endogenous political control, while larger ones, not withstanding declarations of preference for democratic politics, quickly seek energy hegemony. Additionally, modern society—small or large—must embrace an environmental contradiction: the idea of "normal pollution."[19] Treating its natural surroundings as incidental to economic and energy needs is part and parcel of modernization. Spending the profits of economic growth on cleaning up (the premise of the "environmental Kuznets Curve")[20] will not erase the contradiction: as the ability to clean-up improves, the ability to risk more environmental calamity expands.[21] The destiny of such a way of life is ever-expanding political and environmental insecurity.[22]

In contrast, green energy strategies use domestic energy resources, such as solar irradiation, reduce the length of the supply chain and can lower the risk of political dependence, security conflicts, and environmental harm. For this reason, a green energy economy can empower a politics controlled by domestic goals rather than international conflicts.

The "green shift" can also help in mitigating many of the environmental risks that we currently treat as inevitable. For example, energy and carbon obesity are interlocking features of the contemporary pursuit of happiness.[23] Moderns are busily working on "green titan" strategies that can lower greenhouse gas emissions while fueling relentless economic growth,[24] as though the cure for obesity is a greening of the unhealthy condition. If greenwash is to be averted, the new economy needs to enable a genuinely new direction, which we propose is a recovery of the commons foundation of economics. The "green" in the new economy is to be found in the energy commons of all economic activity. We will discuss this challenge below. But first, we would like to examine how and why what could be a solution to our problems has been resisted so strenuously for so many decades.

The Political Economy of Energy Transitions

A green energy economy could offer a considerable contribution to our problems. Its promise has been noted for at least thirty five years, since Amory Lovins mapped the "soft energy path."[25] Considering its known, significant potential compared to the contemporary "hard path"[26] regime, proponents of change understandably regard the choice as obvious and the case for transition as plain. In essence, this position argues that the intrinsic qualities—from job creation, emission reductions, and energy security—are sufficient to engender the needed social change. The premise is that the green energy economy is virtually self-implementing. After waiting so long for a self-implemented soft path, we might finally reconsider the premise.

One factor in, if you will, the "blindness" of new economy advocacy is that it sees change via the lens of "technological niche"[27] in which new techno-logical capabilities and their potential benefits impel change. This perspective neglects the wider context of meso- and macro-level forces of order and change. The larger context, which includes macroeconomic considerations, macro-level political developments, cultural preferences, and the overall exogenous environment, offers a relationship dynamic that needs to be incorporated in any argument for social change.[28,29] The overall structure of relationships between energy producers and consumers, the alignment of political and economic power, and the institutional, legal, and policy frameworks are considerations that can accelerate or inhibit social change, despite potential benefits or disadvantages of such change. In the evaluation of alternative energy futures and the promotion of change toward such futures, the structure of social valuation—the social dynamic that directs actions and establishes goals based upon its evaluation of their merit—is an essential element for analysis. In this vein, the key question that needs to be asked is the following: how does the structure of social valuation direct energy development and evaluate alternative energy futures?

At its core, the valuation of action and the setting of goals or targets are produced by the existing political and economic architecture. This architecture does not simply exist—it is powerful, setting the context in which decisions about energy at every level (individual to national, policy to market, environment to social) are made. This architecture produces and favors certain decision-making criteria that are aligned with the technological "paradigm of governance"[30] revolving around "more is better" and as such, contains an "institutional bias"[31] toward technology, efficiency (measured in the context of the prevailing architecture)[32], and market-based solutions (where "markets" are referring to the ones that exist). Alternative energy futures, therefore, are evaluated in light of—and much more importantly, in *terms* of—the current energy regime. When reflected upon from this vantage point—and especially when reflected against the criteria for decision making set by the existing energy regime—an alternative energy future such as a green energy economy represents a fundamental shift from the current architecture. Currently, powerful political and economic actors that thrive within the contemporary energy regime perceive such social change as a costly threat—both in political and economic terms.[33] As such, when understood as a problem of political economy, the structure of social valuation in place in the contemporary energy regime clarifies that the transition to a green energy economy will not occur simply because of the recognition of its potential social, environmental, or economic benefits.

The structure of valuation that shaped the technological paradigm of "more is better" puts in place a conservative tendency—a "dynamic conservatism"[34]—as it regards change as a threat to its stable state of energy production and use. The key characteristic of this tendency is that alternative futures are reflected upon from the vantage point of the stable state (i.e., the current energy regime). Strategies of change—such as a pathway toward a green energy economy—are, thus, allocated the burden of change while potential costs of inertia are neglected or justified as the proper charges to an alternative for seeking change. Therefore, aspects of proposed strategies of change that are acceptable to the current political and economic paradigm are likely to agree with the fundamental characteristics of the contemporary energy regime. For instance, as noted by Leigh Glover,[35] the renewable energy discourse has transmogrified itself from a narrative of local counter-strategy to one of behemoth corporate enterprise. In fact, any change that attempts to break away from the current dominant energy regime but remains subjected to the same structure of social valuation that shaped the current dominant energy regime in the first place is more likely to be recognized by its similarity to the contemporary structure than by its differences from the structure.

Positioning the demand and need for social change as, in fact, a "consequence" of the stable state rather than a threat that arose independently from it, alternative energy futures should challenge the role of social arbiter currently awarded to the technological paradigm of "more is better" and aim to reshape the social dynamics and its outcome. The challenge represented by a green energy economy needs to distance itself from the technological and institutional landscape of society, presenting a new meta-narrative that reconstructs our problems as the consequences of business as usual, and our alternatives as those that can break away from business as usual. When the challenge is successful, alternative energy futures will be evaluated against a new foundation of social valuation with associated conditions of political economy.

The reconstruction, however, is substantial. Through economic and technological renewal, the contemporary economic and political system aims to sustain and perpetuate itself along the firmly established "more is better" principle. As a result, social values and priorities will work intensively on finding efficient technological solutions that serve to maintain the stable state, and consequently, position energy development on a fixed trajectory. The term "paradox of innovation"[36] captures the problem in practical form: technological development is a double-edged sword of potential sustainability but also un-sustainability. Continued focus on technological development along decision-making criteria that caused our currently unsustainable energy system to be powerful might "lock-in" an undesired development trajectory. The notion that the continued focus on efficiency in the context of the prevailing architecture and technological innovation (rather than structural social change) might re-enforce the fundamental path-dependent dynamics of society rather than alter them,[37] which reinforces the power of the existing energy regime, making a shift to a fundamentally and structurally different direction increasingly more difficult.

The menu of social change often offered by the existing regime is limited to internalization into the market structure of environmental consequences, end-of-pipe technological development, and institutional reforms that maintain the technological paradigm currently employed by modern society. In effect, this menu directs its attention to the outcomes of social dynamics, rather than the widening conflicts of those dynamics, in an effort to maintain and perpetuate an ever-expanding energy production and use system. For instance, the valorization of ecological services through the establishment of a carbon tax or cap and trade mechanism aims to sensitize social dynamics to ecological stress but fails to explicitly address the meta-narrative and ideology that support high-carbon life as an emblem of social and economic success.

A successful challenge to the existing energy political economy will need to reposition the dynamic relationships among social actors so that transformative social change is possible. Transformative institutional change requires ideas, propositions, and actions that aim to establish a new context of political economy. Falling outside the box of contemporary "energy obesity"[38] offered by the modern cornucopian political economy fueled by "abundant energy machines,"[39] such efforts form the initial conceptualization and implementation of a new context. Supported by the realization that energy and carbon decouple from human needs at higher living standards,[40] these efforts offer the possibility of a menu of social change, which recognizes that more is not *necessarily* better.

Accelerating the Transition

Successful energy transitions have taken decades to centuries to complete.[41,42] Considering the implications of climate change, worldwide poverty and inequality, and dependence on finite fossil fuel reserves, we are searching not only for a transition but one that can unfold more quickly than any transition in the past. In this vein, we put forward here a practical proposition already put into action in the United States and which has, subsequently, gained traction in parts of the United States and internationally. We link the paradigm embedded in this approach to a realignment of social dynamics in local energy development. These efforts can be seen as strategies to redefine the context of political economy in which alternative energy futures are evaluated and implemented.

Institutionalization of Sustainability at the Local Level

Twentieth-century energy utilities institutionalized social action to *supply* rather than *match* needs.[43] Leading energy institutions currently measure their success by the amount of cheap energy they can provide. The technological risks, environmental consequences, and social inequity associated with this energy obesity model of abundant energy machines expose the urgent need for change and can be the starting point for redirecting efforts to *match* rather than *supply needs*. An example of a new institutional approach that contests the "more is better" ethos and supplants it with a new context is the Sustainable Energy Utility (SEU), which is already in action in parts of the United States and being put forward in a growing number of communities in and beyond the United States.[44,45]

At its economic core, the SEU uses future savings from investments in grid-use reduction in order to underwrite investment in sustainability. Instead of a mindset of expansion and growth, the SEU incentivizes reductions and savings by establishing the framework for aggregating community desires to use less in low-risk financing for grid-use reductions, offering the estimated energy savings as collateral. The SEU, thus, transforms energy–environment–society relationships by enabling society to use less energy and to deploy renewable energy to meet the remaining energy demand. The SEU positions itself as a practical tool that celebrates—and excels within—the new paradigmatic context of a green energy economy.

The SEU essentially ties the community demand for green energy to its "supply," enabling energy service companies (ESCOs), public agencies, community organizations, and businesses to save energy, install renewable energy, and incorporate long-term considerations into decision making. In order to attract the required financial capital to implement the grid-use-reduction measures, the SEU bundles its projects together and applies its financing (including bond-issuing) authority to scale up a sustainable energy infrastructure. The potential of this mechanism became clear when the Delaware SEU's inaugural bond issue, on August 1, 2011, offered a $67.4 million revenue bond issue and was oversold in two hours.[46] In fact, the rapidity with which investors subscribed to the bond—quickly oversubscribing nineteen of the twenty-three serial bonds—generated a premium in excess of $5 million.[47] The potential of the SEU bond issue mechanism—and the model of this new utility—as a viable strategy to establish a context of energy conservation, and renewable energy development that can support transformative social change became clear. Touching only 4 percent of Delaware's state-owned/managed buildings owed, the offering is generating $148 million in money savings against total costs (including debt service) of $110 million—a 25 percent effective rate of return. The sovereign pledge to pay energy bills, which are *guaranteed* to be over 22 percent lower, earned the SEU's bond issue an AA+ rating from Standard & Poor's. This use of the public replaces the twentieth-century model in which such a promise securitized the investor-owned utility's economic health with a new strategy of prioritizing public purposes—improving ecosystem health, stimulating local economic development, and restoring social governance of the energy sector—and inviting investors to compete to fulfill these promises.

The promise of the SEU has rapidly attracted wider attention. While first implemented in the US state of Delaware, additional SEUs have been established in Washington DC, and variants of the model are at work in Vermont, Connecticut, and Oregon. Sonoma County Water Agency, in California, is pursuing a $30–$50 million sustainable energy bond program for schools, hospitals, and municipal buildings.[48] The concept has been recognized by the White House[49] and the Asian Development Bank,[50] and the implementation of SEUs is being explored internationally.[51,52,53]

A measure of the model's transformative power can be obtained by considering a nationwide initiative to improve performance of the equivalent of 4 percent of the floor area of state public buildings throughout the United States. This would represent, in a *single* transaction (available for use year after year) an investment of more than $25 billion, creating 300,000 jobs, and lowering public sector carbon emissions by as much as 10 percent. A study of its use *only* for federal facilities concludes that the model would outperform the US government's Energy Service Performance Contract Program by a factor of six and save taxpayers $500 million.[54]

A Reorientation of Policy

The institutionalization of sustainability in the form of energy conservation and renewable energy investment is one aspect of a new menu of social change. Other conditions of political economy such as power, policy, and legal frameworks also need to be comprehensively incorporated in strategies of social change. Regarding policy, it is important to note that no "one-size-fits-all" is possible. But growing evidence exists that the deployment of renewable energy in countries that are leading in terms of installations depends on policy support.[55] As such, while the institutionalized approach of the SEU fundamentally reframes the energy–environment–society relationship and thus represents a significant departure from the modern energy paradigm offering substantial potential for the realization of a new green energy economy, several policy mechanisms and ideas need to be considered in any common effort pursuing a sustainability-based future.

Most national and lower levels of government with a modicum of success in installing renewable energy employ a policy mix of various subsidies, specific commitment percentages, and targets, and offer purchase guarantees. Germany, for instance, has a twenty-five-year policy history of promoting photovoltaic use and is now the global market leader in terms of installations with 3.8 GW—representing 4 percent of the global electricity generated from renewable energy.[56] Germany's efforts in the feed-in tariff policy[57,58] coupled with low to no interest financing and subsidy support for installation costs has boosted the country ahead of others, and perhaps not unrelated, coincides with Germany's status as one of the strongest economies in Europe. Leading by example, Germany's efforts are now copied and applied by several other countries.[59]

In the United States, investment tax credits, production tax credits, and accelerated depreciation tax incentives form a national renewable energy policy context. The federal government so far, however, has not established a consistent and effective policy framework for renewable energy and climate policy. As a result, the United States has been the target of criticism from those seeking stronger climate commitments from a leading world economic power. At the same time, despite disappointing federal efforts to date, local and regional policy efforts in the United States show a strong commitment to renewable energy and climate change mitigation and surpass—both in terms of quantitative and qualitative goals and actions—the federal commitment.[60,61,62,63] The American renewable portfolio standard (RPS) and regional cap and trade policy frameworks are examples of policy tools that have been implemented by lower levels of government in the United States in the absence of a larger coordinating federal narrative. Thirty-six states and Washington DC have committed themselves to an alternative energy future, as they have passed RPS legislation enacting the promise to implement increasing shares of renewable energy in the energy mix.[64]

The US bottom-up discourse represents a paradigmatic move away from conventional conceptualizations of effective climate and energy policy that focus on (inter)national articulations and instead, offers a "direct democracy" pathway for civil society to apply its political voice regarding the direction and narrative of US energy and climate policy.[65] Thus, local, participatory, and accountability-based action interjects democratic considerations into the overall decision-making process allowing for a shift toward new ideas and experimentation of new, transformative political propositions. Lower-level experimentation and innovation with these policies—the RPS is only one of the many policies that have been enacted at the state level[66]—offers a dynamic platform for new ideas and propositions that can circumvent the gridlock of the national political economy. Federal-level inertia can thus be overcome by direct articulation of commitments to the green energy economy.

Repositioning the Social Dynamic—The Community as the Building Block of Change

A final example of how the conditions of political economy can be incorporated in the proposal of an alternative energy future and how social dynamics can be realigned to design and reinforce this future is the community solar discourse. Recognizing the limited power of individuals to capitalize on the benefits of solar technology—primarily due to high upfront investment costs—a community solar discourse repositions this social dynamic away from the individual toward community networks. Individual consumers purchasing solar technology from the market face substantial barriers such as information gaps (e.g., where to buy solar technology, which technology is most suited for my needs, etc.) and high initial upfront costs. Through an aggregation of individuals into a community-based effort for the purchasing of solar energy (rather than simply the purchase of technology), the social dynamics between the consumer and producer change. A community-based effort, for instance, is able to capitalize on economies-of-scale thereby negotiating lower prices for the bulk purchase of goods; it can navigate the policy and legal maze more effectively owing to the pooling of knowledge and resources, and it allows for resources to stay within the community rather than having resources flow to large-scale utilities. There are a host of practical mechanisms to accomplish this aim (including power purchase agreements, municipalization of power delivery, etc.). The key factor, though, is the shift from a technical administration of a community's energy destiny to an institutionalization of community governance of its future. The SEU model is designed to express this aspiration.

A community-based SEU essentially shifts the decision-making dynamics away from the political economy of centralized utilities and formulates a new context of community articulated and long-term energy development priorities, objectives, and needs. A greater decentralization in the electricity sector—particularly when performed with renewable energy—through the development of community-owned energy projects allows more active[67] and effective participation in decision making by individuals and community groups.[68] Furthermore, a community-based approach offers many social and economic co-benefits such as local economic development, local job creation, energy infrastructures that can be changed as community objectives change, and community-designated environmental and governance goals (which likewise can change through participation rather than the prevailing model that pits communities against experts in contests of technical acumen, ironically paid for by community members).[69,70,71]

A combined dynamic of institutionalization of sustainability and policy experimentation and innovation offers the potential of a commons-based approach to energy development that recognizes and values long-term ecological considerations that are currently neglected by the conditions of political economy. The social innovation of the SEU offers, through the institutionalization of sustainability, a pathway of actual energy saving beyond the rhetorical recognition now offered with realized savings as a share of sales averaging an embarrassing 1 percent.[72] The paradigmatic shift in energy and climate change policy away from failed top-down policy to bottom-up initiatives allows for civil society engagement in the formulation of the vision of social change and a circumvention of the lobbying power of centralized energy's defenders at the national level.[73] Finally, the community-solar approach is an example of a repositioning of social dynamics—from the individual consumer to the perspective of the community landscape inhabitant—that offers power to outline energy development to the community. The continued expression of—and experimentation with—such new transformative social innovations and ideas offers a pathway of strategic social change that positions livelihood-centered energy and economic development and participatory governance at the center stage in the pursuit of the green energy economy. Such strategic social change can then be positioned as a more appropriate picture of social change—a "social change 2.0"—as it focuses on the implementation of transformative social innovations, a collaborative playing field, the empowerment of people, the transformation of dysfunctional social systems, and the expansion of social innovations at larger levels of scale.[74]

Conclusions

Energy decisions are entangled in large-scale infrastructures, sometimes influencing their evolution, and later, depending upon these infrastructures to pursue different objectives. With age, these infrastructures and the decision making that depends upon them seem inevitable, even essential, and in this way become barriers to, and possibly preclude rapid transformative change. For instance, our current built environment has become difficult to change given policy and economic priorities that emphasize short-term gains, fast construction, and low-cost finance.[75] The inevitability of our built environment, however, is ephemeral when one considers that many new buildings will be constructed, many more will be renovated, and still others will be demolished. This turnover could offer an opportunity for a new era of sustainability, which could, in turn, motivate momentum toward "sustainable cities". The next generation of buildings, then, take into account long-term considerations, minimize energy use, and generate (a portion of) their own energy. This is an exceptional opportunity to change the future spaces where we will live and work—that is, to increase indoor air quality and sunlight exposure while reducing the environmental effects of wasteful fossil and nuclear energy consumption and overuse of water. In effect, a new policy and institutionalized direction for the built environment—and for all other sectors of the economy—can position the principles of sustainability and equity at the core of decision making.

Energy must inevitably be used, even if nothing is wasted. With considerations of climate change and other social and environmental challenges in mind, this energy must increasingly come from renewable energy sources. In addition to energy-saving measures, smart use of these energy sources will substantially decrease the pressure on the environment while providing many social benefits. The prices of coal, oil, and natural gas show—despite concerted efforts to maintain a steady and low price—a high volatility and an overall rising trend.[76] In contrast, renewable energy sources benefit from rapid technological improvements, increased economies of scale, and manufacturing experience. As a result, these technologies experience rapid growth: the PV industry market demand has expanded at an annual rate in excess of 40 percent.[77]

In effect, the new policy context of energy productivity conservation and renewable energy represents a paradigm shift from the "|more is better" principle toward a foundation built on enjoying less. Such a paradigm shift fundamentally re-arranges the energy–environment–society relationship as the policy framework concentrates on efforts to fulfill human needs and wants. Such a paradigm prioritizes the *energy service needs* of society and attempts to fulfill these equitably and sustainably. This chapter has elaborated on the concept of the SEU as a practical tool for actively focusing on *matching energy supply to the needs* of participants rather than *supplying the needs* with ever-expanding energy.[78] The evolution of energy and climate policy in the United States through a bottom-up discourse offers the potential to circumvent entrenched conditions of political economy and allows for the more direct input of civil society to formulate action. Similarly, efforts to build community-scale solar systems could offer a pathway to capitalize on the promise of the green energy economy by repositioning social dynamics from the consumer–producer relationship to the community–producer–user relationship. These initiatives, however, require thinking and action, which departs from an incremental model of cost and benefit built on the premise of the inevitability of the status quo. In brief, paradigm change, not merely improvement in economics of sustainability, is needed.

A green energy paradigm is a means of transformation, enabling society to author a different future. This requires an understanding that nothing is inevitable as well as actions that create new institutions such as SEUs to empower decisions on behalf of a sustainable and equitable future. Rather than perpetuating the oxymoronic pursuit of endless growth with finite fossil resources, the green energy economy requires implementing energy systems that utilize natural and renewable capital and reduce the overall pressure on the environment through a significant dial-back in energy use. For more jobs and economic prosperity in the short run as well as sustained economic and ecological well being in the future, it is time we make energy infrastructural investments and decisions that will provide us with a safer, cleaner, and more equitable way of life—a green energy economy.

Notes

1. United Nations Conference on Sustainable Development (UNCSD), *Report of the United Nations Conference on Sustainable Development*, Conference Report A/CONF.216/16 (New York: United Nations, 2012).

2. Brown, Lester, *Plan B 3.0: Mobilizing to Save Civilization* (New York: W.W. Norton, 2008).

3. McKibben, Bill, *Eaarth: Making a Life on a Tough New Planet* (New York: Times Books, 2010).

4. Epstein, Paul. R. et al., "Full Cost Accounting for the Life Cycle of Coal," In *Ecological Economics Reviews*, ed. Robert Costanza, Karin Limburg en Ida Kubiszewiski. (Annals of the New York Academy of Sciences, 2011): 73–98.

5. Energy Information Administration (EIA), *Annual Energy Outlook 2011—With Projections to 2035* (US Department of Energy, Washington, DC: Energy Information Administration (EIA), 2011).

6. Singh, Virinder, and Jeffrey Fehrs, *The Work That Goes into Renewable Energy*, Research Report (Washington, DC: Renewable Energy Policy Project (REPP), 2001).

7. Ibid.

8. Erhardt-Martinez, Karen, and John A. "Skip" Laitner, *The Size of the US Energy Efficiency Market*, Research report number E083 (Washington, DC: American Council for an Energy-Efficient Economy (ACEEE), 2008).

9. American Solar Energy Society (ASES), *Renewable Energy and Energy Efficiency: Economic Drivers for the 21st Century*, Research Report (Boulder, Colorado: Colorado Printing Company, 2007).

10. Erhardt-Martinez and Laitner, 2008.

11. Agyeman, Julian, Robert Bullard, and Bob Evans, *Just Sustainabilities—Development in an Unequal World* (New York: Earthscan Publications Ltd, 2002).

12. Amory Lovins of the Rocky Mountain Institute coined the term "negawatt" for conserved energy. The term describes one megawatt of electricity conserved for the duration of one hour.

13. Erhardt-Martinez and Laitner, 2008.

14. Randolph, John, and Gilbert M, *Energy for Sustainability: Technology, Planning, Policy* (Washington, DC: Island Press, 2008).

15. Brown, L. R., Brown, L. R., and Earth Policy Institute, *Plan B 3.0: Mobilizing to Save Civilization*. (New York: W.W. Norton, 2008).

16. Prometheus Institute, "25th Annual Data Collection Results: PV Production Explodes in 2008," *Pv News* 28, no. 4 (2009).

17. China, however, is also adding traditional power plant capacity at rapid rates. For instance, since 2007, China's coal production increased from about 57 quadrillion British thermal units (BTUs) to almost 77 quadrillion BTUs, a 34 percent increase, making China the largest coal producer in the world (see the EIA, international energy statistics www.eia.gov). As with any period of significant change, initial signals of how such change will occur are mixed.

18. Byrne, John, and Sin-Jin Yun, "Efficient Global Warming: Contradictions in Liberal Democratic Responses to Global Environmental Problems," *Bulletin of Science, Technology & Society* 19, no. 6 (1999): 493–500.

19. Byrne, John, Leigh Glover, and Cecilia Martinez. "The Production of Unequal Nature," In *Environmental Justice—Discourses in International Political Economy—Energy and Environmental Policy Vol. 8*, ed. John Byrne, Leigh Glover, and Cecilia Martinez (New Brunswick (USA), London (UK): Transaction Publishers, 2002), 261–91.
20. Stern, D. I., "The Rise and Fall of the Environmental Kuznets Curve," *World Development* 32, no. 8 (2004): 1419–39.
21. Beck, U., "Politics of Risk Society," In *Debating the Earth: The Environmental Politics Reader*, ed. J. Dryzek, and D. Schlosberg (Oxford: Oxford University Press, 1998), 587–95.
22. O'Hanlon, Michael, "How much does the United States spend on protecting Persian Gulf oil?," In *Energy security: economics, politics, strategies, and implications*, ed. Carlos Pascual and Jonathan Elkind (Washington, DC: Brookings Institution Press, 2010), 59–72.
23. Byrne, J., Martinez, C., and Ruggero, C, "Relocating Energy in the Social Commons—Ideas for a Sustainable Energy Utility," *Bulletin of Science, Technology, and Society* 29, no. 2 (2009): 81–94.
24. Byrne, J., and Toly, N, "Energy as a Social Project: Recovering a Discourse," In *Transforming Power: Energy, Environment, and Society in Conflict*, ed. J. Byrne, N. Toly, and L. Glover (New Brunswick, NJ: Transaction Publishers, 2006). 1–34.
25. Lovins, A. B., *Soft Energy Paths: Towards a Durable Peace*. (Harper Colophon Books, 1977).
26. Ibid.
27. Berkhout, Frans, Adrian Smith, and Andy Stirling, *Socio-Technological Regimes and Transition Contexts*, Paper No. 106 (Sussex, UK: SPRU Electronic Working Papers Series, 2003).
28. Solomon, Barry D., and Karthik Krishna, "The Coming Sustainable Energy Transition: History, Strategies, and Outlook," *Energy Policy* 39 (2011): 7422–31.
29. See Note 27.
30. Hisschemoller, Matthijs, Ries Bode, and Marleen van de Kerkhof, "What Governs the Transition to a Sustainable Hydrogen Economy? Articulating the Relationship between Technologies and Political Institutions," *Energy Policy* 34 (2006): 1227–35.
31. Ibid.
32. The meaning of efficiency in the context of a prevailing architecture can be stated succinctly: efficiency that is consistent with and reinforces the rationality of that architecture.
33. Byrne, John, and Daniel Rich. "The Solar Energy Transition as a Problem of Political Economy," In *The Solar Energy Transition—Implementation and Policy Implications*, ed. Daniel Rich, Jon M. Veigel, Allen M. Barnett, and John Byrne (Boulder, Colorado: AAAS Selected Symposium, 1983), 163–86.
34. Ibid.
35. Glover, L., "From Love-Ins to Logos: Charting the Demise of Renewable Energy as a Social Movement," In *Transforming Power: Energy, Environment, and Society in Conflict*, ed. John Byrne et al. (New Brunswick, NJ and London: Transaction Publishers, 2006), 249–70.
36. Frances Westley et al., "Tipping toward Sustainability: Emerging Pathways of Transformation," *Royal Swedish Academy of Sciences* 40 (2011): 762–80.
37. Kulkarni, J. S., "A Southern Critique of the Globalist Assumptions about Technology Transfer in Climate Change Treaty Negotiations," *Bulletin of Science Technology Society* 23, no. 4 (2003): 256–64.
38. Tertzakian, Peter, *The End of Energy Obesity: Breaking Today's Energy Addiction for a Prosperous and Secure Tomorrow* (Hoboken, NJ: John Wiley & Sons, 2009).

39. Byrne, J., and Rich, D, "In Search of the Abundant Energy Machine," In *The Politics of Energy Research and Development*, ed. J. Byrne, and D. Rich (New Brunswick, NJ: Transaction, 1986), 141–60.
40. Steinberger, Julia K., and Timmons R. J, "From Constraint to Sufficiency: The Decoupling of Energy and Carbon from Human Needs, 1975–2005," *Ecological Economics* 70 (2010): 425–33.
41. Solomon and Krishna, 2011.
42. Smil, Vaclav, *Energy Transitions: History, Requirements, Prospects* (Santa Barbara, CA: Praeger, 2010).
43. See Note 23.
44. Sustainable Energy Utility Task Force, *The Sustainable Energy Utility: A Delaware First*, Prepared for the Delaware State Legislature (Dover, DE: Delaware General Assembly, Sustainable Energy Utility Task Force, 2007).
45. Houck, Jason, and Wilson Rickerson, "The Sustainable Energy Utility (SEU) Model for Energy Service Delivery," *Bulletin of Science, Technology and Society* 29, no. 2 (2009): 95–107.
46. Delaware Sustainable Energy Utility 8/1/2011 Press Release.
47. Citi, *Delaware Sustainable Energy Utility—Energy Efficiency Revenue Bonds, Series 2011: Post-Pricing Commentary* (New York: Citigroup, 2011).
48. Foundation for Renewable Energy and Environment (FREE), *Sonoma County Efficiency Financing (SCEF) Program*, http://www.scwa.ca.gov/scef/ (accessed October 14, 2012).
49. The White House—Office of the Press Secretary, *We Can't Wait: President Obama Announces Nearly $4 Billion Investment in Energy Upgrades to Public and Private Buildings*, http://www.whitehouse.gov/the-press-office/2011/12/02/we-cant-wait-president-obama-announces-nearly-4-billion-investment-energ (accessed July 31, 2012).
50. Asian Development Bank (ADB), "Communiqué—Special Roundtable to Develop a Regional Plan of Action for Clean Energy Governance, Policy, and Regulation," *Asia-Pacific Dialogue on Clean Energy Governance, Policy, and Regulation* (Manila: Asian Development Bank (ADB), 2011).
51. Yu, Jung-Min, "The Restoration of a Local Energy Regime Amid Trends of Power Liberalization in East Asia: the Seoul Sustainable Energy Utility," *Bulletin of Science, Technology and Society* 29, no. 2 (2009): 124–38.
52. Mathai, Manu V, "Elements of an Alternative to Nuclear Power as a Response to the Energy-Environment Crisis in India: Development as Freedom and a Sustainable Energy Utility," *Bulletin of Science, Technology and Society* 29, no. 2 (2009): 139–50.
53. Agbemabiese, Lawrence, "A Framework for Sustainable Energy Development Beyond the Grid: Meeting the Needs of Rural and Remote Populations," *Bulletin of Science, Technology and Society* 29, no. 2 (2009): 151–58.
54. Schafer, Zach, *The Future of Federal Energy Efficiency Finance: Options and Opportunities for a Federal Sustainable Energy Utility* (Newark, Delaware: Center for Energy and Environmental Policy (CEEP), 2012).
55. Byrne, John, and Lado Kurdgelashvili, "The Role of Policy in PV Industry Growth: Past, Present, and Future," In *Handbook of Photovoltaic Science and Engineering*, ed. A. Luque, and S. Hegedus (Hoboke, NJ: Wiley and Sons, 2011).
56. Solangi, K.H., M. R. Islam, R. Saidur, N. A. Rahim, and H. Fayaz, "A Review on Global Solar Energy Policy," *Renewable and Sustainable Energy Reviews* 15 (2011): 2149–63.
57. Byrne and Kurdgelashvili, 2011.
58. Solangi et al., 2011.

59. Ibid.
60. Byrne, J, K. Hughes, W. Rickerson, L. Kurdgelashvili, "American Policy Conflict in the Greenhouse: Divergent Trends in Federal, Regional, State and Local Green Energy and Climate Change Policy," *Energy Policy* 35 (2007): 4555–73.
61. Rabe, Barry G, *Statehouse and Greenhouse: The Emerging Politics of American Climate Change Policy* (Washington, DC: Brookings Institution Press, 2004).
62. Rabe, Barry G, "Second Generation Climate Policies in the American States: Proliferation, Diffusion, and Regionalization," *Issues in Governance Studies* (August 2006a): 1–9.
63. Rabe, Barry G, "States on Steroids: The Intergovernmental Odyssey of American Climate Policy," *Review of Policy Research* 25, no. 2 (2008): 105–28.
64. "Database of State Incentives for Renewables and Efficiency," Information can be found at: http://www.dsireusa.org/, (accessed July 31, 2012).
65. See Note 60.
66. See Note 60.
67. Sauter, R., and Watson, J, "Strategies for the Deployment of Micro-Generation: Implications for Social Acceptance," *Energy Policy* 35, no. 5 (2007): 2770–79.
68. Walker, G., and Devine-Wright, P, "Community Renewable Energy: What Should It Mean?," *Energy Policy* 36, no. 2 (2008): 497–500.
69. Farrell, J, *Community Solar Power: Obstacles and Opportunities*, Retrieved from: www.newrules.org/energy/publications/community-solar-power-obstacles-and-opportunities, 2010.
70. Walker, G.P., and Cass, N, "Carbon Reduction 'The Public' and Renewable Energy: Engaging with Socio-Technical Configurations," *Area* 39, no. 4 (2007): 458–69.
71. Center for Energy and Environmental Policy (CEEP), *Policies to Support Community Solar Initiatives: Best Practices to Enhance Net Metering*, A Renewable Energy Applications for Delaware Yearly (READY) Project Final Report (Newark, DE: Center for Energy and Environmental Policy (CEEP), University of Delaware, USA, 2012).
72. The graph below presents the cumulative energy savings of several third-party energy efficiency utilities compared to regular utilities. It demonstrates how regular utilities realize much lower energy savings over time. The graph is taken from Byrne, John (2012). Presentation to the 2012 ARPA-E Energy Innovation Summit: The Future of Energy Efficiency Finance. Washington, DC.

Figure 1
Compared Cumulative Energy Savings between Various Energy Utility Models

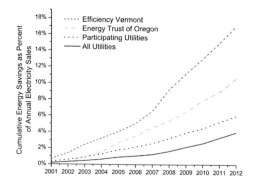

73. One diagnosis of the failed Kyoto process is the inordinate power of centralized energy at national political scales, while local politics is often less prone to takeover by energy elites. See, for example:

    See Note 60 and Sanya, C., and Browne, T. R. (published online). "Innovative US Energy Policy: a review of states' policy experiences," Wiley Interdisciplinary Reviews: Energy and Environment. http://dx.doi.org/10.1002/wene.58.

74. Gershon, David, *Social Change 2.0—A Blueprint for Reinventing Our World* (West Harley, NY: High Point/Chelsea Green, 2009).

75. Hughes, T. P., "Technological Momentum in History: Hydrogenation in Germany 1898–1933," *Past and Present* 44 (August 1969): 106–32.

76. Energy Information Administration (EIA), *Annual Energy Outlook 2011—with projections to 2035*, US Department of Energy (Washington, DC: Energy Information Administration (EIA), 2011b).

77. Byrne, John, et al., *World Solar Energy Review: Technology, Markets, and Policies*, Research Report (Newark, DE: Center for Energy and Environmental Policy (CEEP), 2010).

78. See Note 23.

2

The (New) Thirty Years' War: Fighting for Global Energy Dominance

Michael T. Klare

From 1618 to 1648, Europe was engulfed in an interlocked series of brutal conflicts collectively known as the Thirty Years' War. In basic terms, this extended bloodbath represented a power struggle between the Protestant states of Northern Europe and the Catholic empires of Spain and Austria; at the same time, it constituted a structural contest between an imperial system of governance and the emerging nation-state. Although the fighting did not resolve these struggles in every respect, it did result in a significant shift in power from the southern to the northern states and from imperial to national systems of rule. Indeed, many historians believe that the modern system of nation-states was crystallized in the Treaty of Westphalia of 1648, which finally ended the fighting.

Today, we are embarking upon a new Thirty Years' War—in this case, a global struggle among the principal forms of energy. This contest may not result in as much bloodshed as that of the 1600s, but it will prove every bit as momentous for the future of the planet. This new contest is a succeed-or-perish struggle among all major forms of energy (and the nations and corporations which provide them) over which form of energy will dominate the world's energy supply in the second half of the twenty-first century. The winners in this contest will determine how we live, work, and play in those outlying decades and will profit enormously as a result; the losers will simply disappear, or become niche players at best.

This struggle will last thirty years because that is how long it will take to bring experimental energy systems like hydrogen power, cellulosic ethanol, wave power, algae fuel, advanced nuclear reactors, and others like them from the laboratory stage of development to full-scale industrial development. Some of these systems (or others not yet on the media's radar) will survive this winnowing process, and others will not. At the same time, some existing fuels, such as oil and coal, are likely to witness reduced utility over the next thirty years as a result of both diminished supply and growing concern over carbon emissions.

This will be a *war* because the survival and future profitability of many of the world's most powerful and wealthiest corporations are at risk, and because every major energy producing and consuming nation has a stake in this contest. For giant oil companies like BP, Chevron, ExxonMobil, and Royal Dutch Shell, an eventual shift away from reliance on petroleum will have massive economic consequences, forcing them to adopt new economic models—based on the production of alternative energy products—or risk collapse or absorption by more powerful competitors. At the same time, new companies will arise, some eventually coming to rival the oil giants in wealth and importance.

The fate of nations will also be at stake in this contest because states make bets on one or another set of technologies—or simply cling to their existing energy patterns out of inertia. States that make the right choices will emerge strong and vibrant in the second half of the twenty-first century; states that do not make appropriate choices are likely to decline in wealth and power. Also, because most states view the acquisition of adequate supplies of energy as a matter of national security, struggles over vital resources—oil and natural gas now, perhaps lithium or nickel (used in advanced batteries) later—could lead to armed violence.

When these three decades are over, as with the Treaty of Westphalia in 1648, global society is likely to possess the foundations of a new structural system—this time, one organized around energy production. In the meantime, the struggle for energy resources is guaranteed to grow even more intense for a simple reason: there is no way the existing energy system can satisfy the world's future requirements. It must be replaced or supplemented in a major way by renewable alternatives or the planet will be subject to environmental disaster of a sort hard to imagine today. But no obvious successor to the existing system is in sight—nor will be for some time—so the major contenders will be forced to fight amongst themselves for funding and support until one (or more) emerges as the clear victor. This is the prospect we face for the years now looming ahead.

2011: A Break with the Past

The global energy system has, of course, been evolving for a very long time in response to the development of new fuels and their widespread utilization. Such transitions are often marred by jolts or "shocks" of one sort or another, as demand outpaces supply or problems arise in the distribution of supply. One such "shock" was triggered by the Arab oil embargo of 1973–74 and the accompanying increase in prices mandated by the Organization of Petroleum Exporting Countries; a similar shock arose in 1979, following the Islamic Revolution in Iran. In 2000–01, a severe "energy crisis" erupted in California and elsewhere as a result of efforts by the (now bankrupt) Enron Corporation to manipulate the price of natural gas. The spectacular rise in oil prices in 2008 produced another equally momentous crisis. But while these events unleashed powerful effects, they occurred at times when global energy production was undergoing a sustained growth, so their impact was relatively short-lived. In 2011, however, the global energy system experienced a series of shocks that appear to be of another sort entirely: disruptive events that signal a long-term threat to the supply of energy. In short, 2011 constitutes the onset of a new Thirty Years' War.

Typically, any given block of time incorporates some major developments that affect the global flow of energy. These include both human-induced developments such as political upheavals and wars and natural events such as particularly severe hurricanes and earthquakes. Any one of these developments can cause a temporary disruption in the production or delivery of oil or other primary fuels. What was striking about the first half of 2011 was the simultaneous occurrence of several such developments—both human-induced and natural—producing a severe jolt to the global energy system. Not only were these unusually powerful shocks, but their effects are likely to be long-lasting. In this sense, the events of 2011 pose not only a temporary but also a systemic challenge to the global energy system.[1]

While considering these developments, it is essential to understand that the global economy, as currently organized, cannot tolerate zero growth in energy supply; in order to satisfy the still-colossal energy needs of older industrial powers like the United States and Japan as well as the voracious thirst of rising powers like China, the world's energy supply must grow substantially year after year. To meet the anticipated energy demand in 2035, as calculated by the US Department of Energy (DoE), the world energy output must jump from 505 quadrillion British thermal units (BTUs) in 2008 to 770 quadrillion BTUs by the end of this period, an 53 percent increase.[2] This is an extraordinary increase—the 265 quadrillion BTUs of additional energy that will be needed by 2035 is equivalent to the total energy use in 2008 by the United States, Canada, Japan, and all of Western Europe. Even if the energy industry succeeds in generating some portion of this increase, a failure to achieve the full amount will lead to periodic energy shortages, high prices, and severe economic dislocations.

It is true, of course, that changing economic and technological conditions may alter the world's need for energy. A slowing world economy will result in reduced energy requirements, just as an expanding economy will need more. Increased efficiency, new technologies, or new energy habits (such as the widespread use of public transit and all-electric vehicles) could reduce demand. But the DoE, in its 2011 report on world energy consumption, says that the rate of increase in global energy use is *accelerating*, not slowing down—the result of continuing economic growth in China, India, and other rapidly industrializing nations. In 2010, the DoE indicated that the world energy consumption would grow by 1.4 percent per year over the coming quarter-century; in 2011, it raised its projected rate of increase to 1.6 percent per year. This may seem like a small difference, but over a twenty-five-year period, it amounts to 31 quadrillion BTUs—equivalent to current energy usage by South and Central America combined.[3] Any indication that the energy industry is incapable of satisfying the anticipated future demand, therefore, is a matter of utmost international concern.

Certainly, the world can adjust to a *temporary* disruption in the global supply of energy, as has occurred before (for example, following the Iranian Revolution of 1979 and the 1990 Iraqi invasion of Kuwait). Such events produced a brief reduction in oil deliveries but did not lead to long-term difficulties because Saudi Arabia was able to compensate for the missing oil and it was understood that the global supply picture remained buoyant. However, the events of early 2011 appear to be of a different sort altogether from those of 1979 and 1990, in that they suggest a *permanent* rather than merely temporary disruption in the global supply of energy. For this reason, these events merit close attention. In particular, I will examine three of these key developments: (1) the Arab Spring, and its effects on global oil output; (2) the disaster at Fukushima in Japan, and its impact on the future nuclear power output; and (3) the signs of accelerating climate change, and its myriad effects on energy supply.

The "Arab Spring" in the Middle East

The first and most momentous of the year's energy shocks was the Tunisian revolution and the ensuing "Arab Spring" in the Greater Middle East. It will be many years before the full impact of these epochal events are fully evident, but it is already apparent that the 2011 upheaval will have many lasting consequences. Most obvious, of course, are the political shifts: the ouster of long-serving autocrats in Egypt, Libya, and Tunisia, along with the continuing challenges to the survival of leaders in Bahrain, Syria, and Yemen. But economic shifts are also underway, with more such shifts likely in the future. These include, among others, an altered environment in the oil industry.

It is impossible to overstate the importance of the Middle East to the global oil supply. As indicated in Table 1, the major producers of the Middle East jointly account for more than one-third of the current world petroleum output, or about thirty million barrels per day (mbd). But the role of Middle Eastern production is even greater than this figure would suggest: while producers in most of the rest of the world consume some or all of what they recover, producers in the Middle East export a large share of their output, making this region the world's leading source of *exportable* oil. According to BP, the Middle East consumed only about 9 mbd in 2010, leaving 20 mbd for sale in international markets. This 20 mbd represented the single largest pool of oil available for purchase by oil-importing nations to satisfy their energy requirements.[4]

However significant in the present, Middle Eastern oil is expected to prove even more important in the future. This is so because the Middle East possesses a large share of the world's proven reserves—an estimated 59 percent, according to BP—and because output in many other areas of the world is expected to decline in the years ahead while output in the Middle East is expected to grow. Many of the world's other major producing areas, such as the United States, Russia, Mexico, Venezuela, and Indonesia, have been pumping oil for far longer than most Middle Eastern producers, and their major fields are at a more advanced stage of depletion. Therefore, as time goes on, the Middle East is expected to assume an even more pivotal role in the global output of oil. According to the DoE, combined Middle Eastern output will constitute approximately 40 percent of world output in 2035, compared to 36 percent today.[5]

Thus, the following becomes the key question: How will the Arab Spring affect the future availability of oil from the Middle East?

Table 1
Oil Reserves and Production in the Middle East

	Oil reserves		Actual production, 2010		Projected production, 2035	
	In Gbl	As % of world	In mbd	As % of world	In mbd	As % of world
Algeria	13.5	1.0	1.81	2.0	2.3	2.3
Egypt	4.5	0.3	0.74	0.9	0.8	0.8
Iran	137.0	9.9	4.25	5.2	3.9	3.9
Iraq	115.0	8.3	2.46	3.1	6.3	6.4
Kuwait	101.5	7.3	2.51	3.1	4.0	4.0
Libya	46.4	3.4	1.66	2.0	0.8	0.8
Oman	5.5	0.4	0.87	1.0	0.6	0.6
Qatar	25.9	1.9	1.57	1.7	2.2	2.2
Saudi Arabia	264.5	19.1	10.01	12.0	15.4	15.5
Syria	2.5	0.2	0.39	0.5	0.2	0.2
Tunisia	0.4	#	0.08	0.1	n.a.	n.a.
United Arab Emirates	97.8	7.1	2.85	3.3	3.2	3.2
Yemen	2.7	0.2	0.26	0.3	0.1	0.1
Total	817.2	59.1	29.46	35.9	39.8	40.2

Sources: Reserves and production for 2010: BP. *Statistical Review of World Energy June 2011*, pp. 6, 8. Projected production for 2035: US Dept. of Energy, *International Energy Outlook 2011*, Table E2, pp. 230–31 (excludes unconventional liquids).
Note: Columns may not add due to rounding.
Gbl = billion barrels
mbd = million barrels per day
# = less than 0.1 percent

In the short-term, the impact of the Arab Spring has been relatively modest. Several of the countries that have experienced major upheavals—Egypt, Libya, Oman, Syria, Tunisia, and Yemen—are oil producers, but of these, only Libya can be considered a major producer, and at 1.7 million barrels per day, its output is not that substantial. Even so, the elimination of Libyan output due to the fighting there caused some difficulty for importing countries in Europe (especially Italy, which was a major customer) and prompted the International Energy Agency (IEA) to release significant quantities from emergency stocks for only the third time in its history.[6]

If the tumult had spread to Saudi Arabia, however, the picture would have been very different. Many analysts have noted that if Saudi Arabia experienced a revolution or civil war like Egypt, Libya, and Tunisia, the world would have experienced a catastrophic oil shock. "If something happens in Saudi Arabia, [oil] will go to $200 to $300 [per barrel]," said Sheikh Zaki Yamani, the kingdom's former oil minister, on April 5, 2011. "I don't expect this for the time being, but who would have expected Tunisia?"[7]

Saudi Arabia did, in fact, experience a number of anti-government protests, but none on the scale of those seen in Egypt, Libya, Syria, or Tunisia, and none that threatened the survival of the royal family. But many of the conditions that were present in those other countries—high youth unemployment, rapidly growing populations, insufficient affordable housing, high food prices, and distrust of authority—are also present in Saudi Arabia and thus produced jitters among government officials.[8] With this in mind, King Abdullah promised his subjects $130 billion in new benefits, including $70 billion for 500,000 units of low-cost housing.[9] Some analysts believe that these efforts—combined with a fierce crackdown on the few visible signs of protest—have allowed Abdullah to weather the Arab Spring without noticeable strain; others, however, are not so certain of this. "Saudi Arabia is a time bomb," said Jaafar Al Taie, managing director of Manaar Energy Consulting (which advises foreign oil firms operating in the region). "I don't think that what the King is doing now is sufficient to prevent an uprising," he added.[10]

Even if exaggerated, this assessment is critical because it bears on Saudi Arabia's ability to satisfy the world's future needs for exportable oil. For the world's oil-importing countries to meet their needs, Saudi Arabia, in particular, must substantially increase its output. According to the DoE, Saudi oil output is expected to jump from 10.0 mbd in 2008 to 15.5 mbd in 2035, the biggest increase of *any* oil producer during this period.[11] Increased Saudi output is needed to not only sate global export markets but also provide the "spare capacity" required to offset the temporary loss of production from other suppliers due to war or internal disorder (as, for example, when Libya was shaken by civil strife).

But boosting Saudi output will prove harder than ever before because Saudi Arabia—like so many other major producers—has exhausted its inheritance of "easy oil"—low-sulfur, low-viscosity oil that is concentrated in large reservoirs—and so must rely on hard-to-reach, high-sulfur, highly viscous "tough oil." Saudi Arabia possesses a lot of tough oil, but extracting it will require hundreds of billions of dollars in fresh investment. Indeed, the *Wall Street Journal* recently noted in a front-page article entitled "Facing Up to the End of 'Easy Oil'" that any hope of meeting future world oil requirements rests on the Saudis' willingness to invest these vast sums in the development of their tough-oil deposits.[12] But with all the billions of dollars now being devoted to public housing and the like, doubts naturally arise as to whether the government is prepared to invest like amounts in costly new oil projects.

There was a time when oil could be extracted from Saudi Arabia for very little cost—a dollar or two per barrel. But those days are long over: most of the kingdom's easily accessed fields are now in production and have been for some time, so there is no hope for significantly increasing the output in these fields.[13] This means that simply to maintain production at current levels—let alone achieve any increase in output—the Saudis must develop new fields that will prove far more costly to launch and operate. The development of one such field, Khurais, cost $10 billion, and it required the installation of a complicated water-injection system just to permit extraction of oil.[14] (Normally, water injection is only used after a field has been in production for many years.) Any other field brought into operation will require an investment on this or a larger scale.[15] While there is no doubt that the Saudi leadership can afford such largess, they may face a choice between investments in oil and investments in social stability—and under these circumstances, regime survival is almost certain to win over satisfying the world's thirst for petroleum.

Indeed, this was the conclusion drawn in a recent study by the Energy Forum of the James A. Baker III Institute of Rice University. Future investment in production expansion capacity "is likely to be even more expensive [than in the past] given that the kingdom will have to shift to areas that have more complex geology and require greater technological intervention," the report noted. "But the kingdom is also facing competing priorities with higher spending requirements on social services and defense in light of new regional and internal challenges. . . . The pressures for higher defense and social spending will make it that much harder for the government to justify a massive campaign to expand its oil sector."[16]

A second question also arises: To the degree that Saudi Arabia does make the necessary investments in new oil fields, how much of the resulting output will be available for export, and how much will be diverted for domestic use, to satisfy the needs of local consumers (at discounted prices) and to fuel Saudi industries? Saudi Arabia has one of the fastest-growing populations in the world, and as noted, the significant problem of youth unemployment. According to the most recent projections from the UN Population Division, the kingdom's population is expected to jump from 27.4 million in 2010 to 42.2 million in 2040—an increase of nearly fifteen million people.[17] To create new jobs for the ever-expanding armies of young Saudis—and thus dissuade them from entertaining thoughts of rebellion—the regime is investing many billions of dollars in new industries, many relying on abundant inputs of domestic petroleum.[18]

How all this will play out remains to be seen. Under the best-case scenario, Saudi Arabia will remain stable and continue to invest in new oil production. However, given the regime's mammoth spending on housing and public works in response to the Arab Spring, it is unlikely that such investments will ever reach the level needed to achieve the 15.5 mbd projected by the Energy Information Administration (EIA) for Saudi production in 2035. And even if production does rise above the current levels, this increase will probably be used to satisfy domestic, not export, demands. So even under the most benign scenarios, the global supply of exportable oil is likely to fall short of anticipated requirements in the years ahead; oil will still be available on international markets, but far less than needed, producing a condition of perceived scarcity. Of course, should an Egyptian-style revolution erupt in Saudi Arabia, the situation would worsen further.

Fukushima and its Aftermath

The second major energy shock of 2011 occurred on March 11, when an unexpectedly powerful earthquake and tsunami struck Japan. As a start, nature's two-fisted attack damaged or destroyed a significant proportion of northern Japan's energy infrastructure, including refineries, port facilities, pipelines, power plants, and transmission lines. In addition, of course, it devastated four nuclear plants at Fukushima, resulting, according to the DoE, in the permanent loss of 6,800 megawatts of electric generating capacity.[19] This, in turn, has forced Japan to increase its imports of oil, coal, and natural gas, adding to the pressure on global supplies. With Fukushima Daiichi and other nuclear plants off line, industry analysts calculated that Japanese oil imports would rise by as much as 238,000 barrels per day, and imports of natural gas by 1.2 billion cubic feet per day (mostly in the form of liquefied natural gas, or LNG).[20]

The disaster at Fukushima Daiichi produced a widespread reappraisal of nuclear power in Japan. Until March 2011, Japan had expected to place ever-greater reliance on nuclear power to generate electricity. Plans in place called for the construction of fourteen new reactors over the next quarter-century aimed at raising the nation's reliance on nuclear power to 50 percent, from about 25 percent at present.[21] According to the DoE, this would have boosted Japan's power output from nuclear reactors by 70 percent from 245 to 417 billion kilowatt-hours (see Table 2) between 2008 and 2035. In the aftermath of the disaster, however, Prime Minister Naoto Kan abandoned all plans for further reliance on nuclear power and said the country would have to "start from scratch" in devising a new energy policy for the country.[22]

In the months following his announcement, Kan spoke with growing enthusiasm about his hope that Japan would place ever-greater reliance on "green" energy, such as wind, geothermal, and solar power.[23] However, he was forced

Table 2
World Nuclear Energy Consumption (In billion kilowatt-hours)

	Actual 2008	Projections		
		2015	2025	2035
United States	806	839	877	874
OECD Europe	882	965	1,067	1,136
Japan	245	319	342	417
South Korea	143	183	233	266
Russia	154	197	342	388
China	65	223	585	916
India	13	66	157	211
World, Total	2,602	3,178	4,188	4,916

Source: US Dept. of Energy, *International Energy Outlook 2011*, Table A8, p. 165.
Note that these projections were developed prior to the Fukushima disaster of March 2011.

to announce his intention to resign as prime minister in the face of the government's slow and seemingly ineffectual response to the tsunami and reactor disaster, putting his long-term energy plans in question. His successor, Yoshihiko Noda, has ruled out the construction of new nuclear plants for the time being but has not renounced nuclear energy altogether; accordingly, some analysts believe that nuclear power may someday make a comeback in Japan.[24] For now, however, it appears that Japan will place far less reliance on nuclear power than was previously the case and will not proceed with plans for the construction of new reactors.

It was not only in Japan, moreover, where Fukushima's impact was felt: As the scale of the disaster became publicly known and subsequent investigation revealed significant design flaws and maintenance failures at the plant, opposition to nuclear power arose in other countries around the world. In some cases, this has led to a slowdown in plans to expand reliance on nuclear power, and in others, to a complete rejection of this form of energy.

The earliest and strongest reaction came in Germany. Although Germany had no plans for the construction of new reactors, Chancellor Angela Merkel had earlier announced plans to extend the life of 17 existing reactors an average of twelve years in order to serve as a bridge to a "green" energy future based largely on renewable sources of energy.[25] But when hundreds of thousands of Germans took to the streets in anti-nuclear protests following the disaster at Fukushima, Merkel said that she would immediately close two older reactors and suspend plans to extend the life of the remaining others. Within a few years, she indicated, Germany would have no functioning nuclear reactors.[26]

The Swiss government acted next. On May 25, 2011, it announced that it would abandon plans to build three new nuclear power plants and close the last of its existing plants by 2034, joining the list of countries that appear to have abandoned nuclear power for good.[27]

Other countries have not taken such a dramatic action but have chosen to slow their nuclear plans and/or adopt more stringent safeguards. China, which has ambitious plans to increase its reliance on nuclear power (see Table 2), stopped awarding permits for the construction of new reactors on March 16, 2011, pending a review of safety procedures.[28] As at the Fukushima plant, many of China's reactors lack adequate protective devices and backup systems, so would be vulnerable to a complex emergency as occurred in Japan on March 11. Most observers assume that China eventually will proceed with its nuclear plans, but only after tougher safety requirements have been put in place. The much-publicized crash of a high-speed train at Wenzhou in July 2011, revealing apparent failures to adopt adequate safety procedures, is likely to add to the pressure for stricter oversight of the nuclear power industry.[29] This will not halt the construction of new reactors but is likely to produce substantial delays in their commissioning. China's nuclear energy output will continue to grow but probably not by the amount indicated in Table 2.

In the United States, the effect of Fukushima has been to bolster the ranks of those who oppose nuclear power and increase doubts about the desirability of constructing new reactors. Although US reactors, by and large, are considered far safer than the Fukushima Daiichi facility, concerns have arisen over the ability of American plants to cope with severe earthquakes and other major disasters.[30]

These concerns were given added urgency on August 23, 2011, when a 5.8-magnitude earthquake produced a "state of emergency" at the North Anna nuclear power plant in Central Virginia, causing it to lose electricity and automatically shut down. Although backup generators kicked in, preventing the loss of power used to cool the plant's two reactors—thereby averting a catastrophe like one that occurred at Fukushima—the facility received a jolt greater than its initial design called for, and engineers are still inspecting the plant to determine whether it can be reopened safely. The Nuclear Regulatory Commission has also created a special team to study the quake's effects on North Anna, and in the next two years, it must determine whether a score of nuclear plants in the Eastern United States are earthquake-safe.[31]

All this, in turn, has diminished enthusiasm for building new reactors in the United States and many other countries. Although there is no likelihood that nuclear power will disappear from the world's energy portfolio, once-lofty notions of a "renaissance" in nuclear power seem to have evaporated. In early 2011, prior to the Fukushima disaster, the EIA predicted that net output of the world's power reactors would grow by 89 percent between 2008 and 2035,[32] but this projection now appears wildly optimistic. Included in these projections were increments of 254 billion kilowatt hours in Organization for Economic Cooperation and Development (OECD) Europe and 172 billion in Japan—both of which are now likely to see *decreases* in nuclear power output, not increases. In this sense, too, the first half of 2011 has proved to be a pivotal moment.

The Accelerating Impact of Climate Change

The third major energy development of 2011, less obviously energy-connected than the other two, was a series of persistent, often record, droughts gripping many areas of the planet. Typically, the most immediate and dramatic effect of prolonged drought is a reduction in grain production, leading to ever-higher food prices and increased social turmoil. Intense drought in Australia, China, Russia, and parts of the Middle East, South America, the United States, and parts of northern Europe contributed to the record-breaking prices of many basic foodstuffs—and this, in turn, was a key factor in provoking the political unrest that swept through North Africa and the Middle East. But drought has an energy effect as well. It can reduce the flow of major river systems, leading to a decline in the output of hydroelectric power plants—as has occurred in several drought-stricken regions.

By far the greatest threat to electricity generation arose in China, which experienced one of its worst droughts ever. Rainfall levels from January to April 2011 in the drainage basin of the Yangtze, China's longest and most economically important river, were 40 percent lower than the average of the past fifty years, according to *China Daily*.[33] This produced a significant decline in hydropower and severe electricity shortages throughout much of central China.[34]

To meet their electrical requirements in the face of hydropower shortfalls, the Chinese reportedly burned more coal—one of the few sources of energy that China possesses in some abundance. But domestic mines no longer fully satisfy the country's needs, so China has become a major coal importer. Rising demand combined with inadequate supply led to a spike in coal prices, and with no comparable spurt in electricity rates (which are set by the government), many Chinese utilities chose to ration power rather than buy more expensive coal and operate at a loss. In response, some industries were forced to increase their reliance on diesel-powered backup generators—which in turn boosted China's demand for imported oil, putting yet more pressure on global fuel prices.[35]

The drought in China—and the resulting power crunch—is perhaps the most visible sign of the link between climate change, drought, and energy. As climate change gathers momentum, prolonged droughts of this sort are likely to become more frequent and widespread, posing a threat to hydroelectric facilities in many parts of the world. Of course, not every region will see a decline in river-water flow as a result of climate change. Some areas—particularly Northern Canada, Scandinavia, and Siberia—are likely to experience heavier rainfall in the decades ahead, and so could support increased levels of hydropower activity. However, many countries with ambitious plans to tap hydropower as a source of renewable energy, including China and India, are likely to experience diminished rainfall as a result of climate change.[36] In this sense, the drought on the Yangtze in the spring of 2011 is a significant portent of things to come.

As climate change gathers momentum, it will affect energy in many ways—both on the supply and demand sides of the equation. One expected consequence of climate change, for example, is an increase in the frequency and severity of storm events, such as hurricanes, typhoons, and heavy rains.[37] As oil drilling moves into ever deeper waters of the Gulf of Mexico, it becomes even more vulnerable to hurricane activity, which typically reaches greatest intensity while over the warm waters of the deepwater Gulf. Recent hurricanes, such as the notorious Katrina of August 2005, have caused enormous damage to offshore drilling platforms, and any future increase in hurricane activity will pose an even greater threat to the nation's oil supply as an ever-greater share of domestic crude is derived from such platforms.[38]

Hurricanes and intense flooding could also pose a danger to nuclear power plants, many of which are located in low-lying areas near rivers (to obtain water for cooling systems). In June 2011, for instance, the Fort Calhoun nuclear power plant in Nebraska was submerged under two feet of water owing to flooding of the adjacent Missouri River (caused by a combination of a particularly snowy winter and heavy spring rains). Although the reactor itself was protected against water immersion and the floods receded without causing serious damage, experts were worried about a future replay in which floodwaters destroyed essential electrical fixtures, causing water pumps to stop running and the reactor core to overheat—exactly as occurred at Fukushima. Similar anxieties have been expressed concerning the nuclear plants at Indian Point, N.Y. and Turkey Point, Fla., both of which are located in low-lying areas.[39]

Increased storm activity and rising temperatures will also produce an increased demand for energy, whether for reconstruction activities in the aftermath of disasters or to cool buildings during ever more prolonged heat waves. Just how all of this will play out cannot be foreseen, but the events of 2011 provide a vivid foretaste of the climate/energy challenges to come.

The Global Energy Forecast

By the middle of 2011, the world faced continuing unrest in the Middle East, a grim outlook for nuclear power, and a severe electricity shortage in China. What else was evident on the global energy horizon? Although there were some bright spots—notably an increase in the production of natural gas from shale formations—the big picture was one of unrelenting demand and insufficient supply.

According to projections from the DoE, released on September 19, 2011, both supply and demand will increase by huge amounts over the next quarter-century. To satisfy an estimated 53 percent increase in global energy demand between 2008 and 2035, it predicted, the world oil output (including unconventional sources of supply) will have to grow by 30 percent over this period, coal by 50 percent, natural gas by 53 percent, nuclear power by 88 percent, and hydro-power plus renewables by 113 percent (see Table 3).[40] But for many reasons, including the developments described above, such large increases are not likely to materialize—except, perhaps in the renewables column.

Take oil. A growing number of energy analysts now agree that the era of "easy oil" has ended and that the world must increasingly rely on hard-to-get "tough oil."[41] While it is certainly true that the planet harbors a lot of this stuff—deep underground, far offshore, in problematic geological formations like Canada's tar sands, and in the melting Arctic—extracting and processing tough oil will prove ever more costly and involve great human, and even greater environmental, risk. Such is the world's thirst for oil that a growing amount of tough oil will nonetheless be extracted—even if not, in all likelihood, at a pace and on a scale necessary to replace the disappearance of yesterday's and today's easy oil.[42] With the outbreak of the Arab Spring, moreover, it appears unlikely that Saudi Arabia will invest the colossal amounts needed to exploit its own tough-oil reserves, further reducing the future supply of oil. All this being so, the 30 percent increase in oil output projected by the EIA is not likely to be achieved.

Likewise, the future of coal will rest on increasingly invasive and hazardous production techniques, such as the explosive removal of mountaintops (with all excess rock being dumped into the valleys below).[43] Any increase in the use of coal will also accelerate climate change, since coal emits more carbon dioxide than oil and natural gas. It is possible that the introduction of techniques to separate and bury the carbon in coal before its combustion—so-called carbon separation and storage (CSS)—will make coal's use more attractive, but such techniques

Table 3
World Energy Consumption by Fuel, 2008–2035
(In quadrillion British thermal units)

	Actual, 2008	Projections		
		2015	2025	2035
Oil*	173.0	187.2	207.0	225.2
Natural gas	114.3	127.3	149.4	174.7
Coal	139.0	157.3	179.7	209.1
Nuclear power	27.2	33.2	43.7	51.2
Hydro & renewables	51.3	68.5	91.7	109.5
Total	504.7	573.5	671.5	769.8

*Includes unconventional fuels (Tar sands, heavy oil, shale oil, etc.).

Source: US Dept. of Energy, *International Energy Outlook 2011*, Table A2, p. 159.

are still in the development stage and are likely to prove very expensive, making their widespread utilization unlikely. As the damaging effects of climate change become more pronounced, government officials will be forced—eventually—to impose tough restrictions on the use of coal. Under these circumstances, we should assume that coal consumption, too, will fall short of the increases projected by the EIA.[44]

The only bright spot, experts say, is the growing extraction of natural gas from North American shale rock through the use of hydraulic fracturing, or "hydrofracking." Proponents of shale gas claim it can provide a large share of America's energy needs in the years ahead while actually emitting fewer greenhouse gases than coal and oil (per unit of energy released).[45] Though plentiful, the gas can only be pried loose from underground shale formations through the use of explosives and highly pressurized water mixed with toxic chemicals. As use of the technique has spread, growing numbers of opponents are warning of the threat to public water supplies posed by the use of toxic chemicals in the fracking process.[46]

Concern over the safety of water supplies has prompted lawmakers in a number of US states to impose a moratorium on hydrofracking or to impose new restrictions on the practice, throwing into doubt the future contribution of shale gas to the nation's energy supply. Also, on May 12, 2011, the French National Assembly (the powerful lower house of parliament) voted 287 to 146 to ban hydrofracking in France, becoming the first nation to do so.[47] Some analysts have also indicated that the prospects for shale gas have been oversold, saying that many wells come up dry and that others decline swiftly after initial exuberance.[48] It is too early to determine whether such assessments will prove accurate, but it is also premature to conclude that shale gas will provide a vast increase in the world's net energy supply, as suggested by its proponents.

Here, then, is the bottom line: Any expectations that ever-increasing supplies of energy from existing sources—oil, coal, natural gas, and nuclear power—will satisfy rising demand in the decades ahead are likely to be disappointed. Instead, recurring shortages, rising prices, and mounting discontent are likely to be the thematic drumbeat of the globe's energy future. By the middle of 2011, the world faced continuing unrest in the Middle East, a grim outlook for nuclear power, and a severe electricity shortage in China.

Girding for War

Given this picture, we can expect a constant struggle among consumers for access to adequate energy supplies and among the suppliers for production assists of one sort or another, be they subsidies, tax breaks, or exemption from environmental restrictions. For the major energy firms, especially those in private hands, this will result in a Darwinian struggle for survival, with stronger firms absorbing weaker ones or pushing them to the wayside. A vivid foretaste of this process was provided by the feeding frenzy that erupted over BP's hasty sale of selected oil and gas reserves in the summer of 2010, when it was forced to set aside $30 billion to pay for future claims arising from the *Deepwater Horizon* disaster and the resulting oil leak in the Gulf of Mexico.[49] Many companies, including ExxonMobil, considered an opportunistic takeover of BP at that time[50]; although no company went through with such a move, BP is still seen as being vulnerable to predatory acquisition. Given the likelihood of more energy shocks and disasters in the future, corporate takeovers and other mercenary behavior will become ever more common.

It is entirely possible, moreover, that the struggle among major consuming states for access to contested energy supplies could become more fierce and violent. This could occur, for example, in the East China Sea, where China and Japan both seek to develop an undersea natural gas field in an area claimed by both and where the two have each deployed air and naval forces in a provocative fashion; should tensions rise, muscle-flexing of this sort could lead to a clash at sea and armed escalation.[51] A similar danger could arise in the South China Sea, where China is contesting ownership of undersea oil and gas fields with Vietnam, Malaysia, and the Philippines. Conflict over contested offshore oil and gas supplies could also erupt in other areas, including the Caspian Sea, the Persian Gulf, and the Arctic Ocean. While eschewing any intention of using force to enforce their claims in these areas, many of the countries involved in such disputes are expanding their naval forces and increasing their military presence in the contested areas—thus increasing the risk of provocation and armed conflict.[52]

In addition to the struggle for survival among major energy producers and the competitive battles among major energy consumers, the next thirty years will see an intense struggle for market share among the various forms of energy (and the companies that stand beyond them). This will be a struggle unlike anything we have seen in recent times, as established sources of energy—oil, coal, natural gas, and nuclear power—fight to retain their dominant position in the face of growing competition from alternative forms of energy and new forms of energy fight to supplant them. Such a struggle is inevitable, as we have seen, because the existing forms are incapable of satisfying the world's surging demand for energy and because their climate and hazardous effects will provoke ever-increasing regulatory constraints. So other forms of energy will have an opening to secure ever-expanding market share. To secure a bigger share, however, they will have to fight both the older forms of energy and each other—for subsidies, tax breaks, regulatory advantages, and so on.

This struggle will be brutal, relentless, and profound in its ultimate effects. For the victors, the rewards are likely to prove immense: profits, wealth, and power beyond anything one could imagine. For the losers, the consequences are no less massive, involving bankruptcy, marginalization, and ridicule. And, for all the rest of us, the outcome will determine how our energy needs will be met in the future, by whom, and with what economic and environmental consequences. On this, hinges the future of the planet.[53]

The Leading Contenders

At this point, it is utterly impossible to determine which fuel (or group of fuels) will emerge as the victor of the new Thirty Years' War. At this point, many forms of energy are jockeying to replace oil and coal in the lead positions, but none stands out as the obvious winner. Here are some of the leading contenders.

Natural gas: Although natural gas is a fossil fuel, many energy experts and political leaders view it as a "transitional" fuel because it releases less carbon dioxide and other greenhouse gases (GHGs) when consumed than oil and coal. Also, global supplies of gas are much greater than previously believed because the use of new technologies—notably horizontal drilling and hydraulic fracturing—allows the exploitation of shale gas reserves once considered inaccessible.[54] For example, in 2011, the DoE predicted that gas would provide 24 percent of America's energy in 2030, while coal would supply 21 percent; just five years earlier, it predicted that coal would top gas in 2030, 26 to 21 percent.[55] (Oil is the number one source in both surveys.) Some now speak of a "natural gas revolution" that will see this fuel overtake oil as the world's number one fuel, at least for a time.[56] But, as previously noted, fracking poses a threat to the safety of drinking water and may thus arouse widespread opposition. The economics of shale gas may also prove less attractive than currently assumed, leading to diminished expectations for future output.

Nuclear power: Prior to the March 11 disaster at the Fukushima Daiichi nuclear power plant in Japan, many analysts were speaking of a nuclear "renaissance," entailing the construction of hundreds of new reactors. Although some of these plants, including those under construction in China, are likely to be completed, plans for others, for example those in Italy and Switzerland, have been scrapped. And despite repeated assurances that US reactors are completely safe, evidence is emerging nearly every day of safety lapses at many such facilities. Given public concerns over the risk of a catastrophic accident, it is unlikely that nuclear power—as now conceived—will emerge as a big winner in 2041. However, nuclear enthusiasts—including President Obama—are championing the manufacture of small, "modular" reactors that could be built for a far lesser cost than current reactor types and would produce fewer radioactive wastes that would have to be disposed of. Although the technology of such "assembly-line" reactors has yet to be demonstrated, advocates say they could provide an attractive alternative to both large conventional reactors and coal-fired plants.[57]

Wind and solar: Make no mistake about it, the world will rely on wind and solar power for a far greater proportion of its energy in thirty years' time than it does today. According to the International Energy Agency (IEA), their share will jump from approximately 1 percent of the total world energy consumption in 2008 to a projected 4 percent in 2035 (this assumes some degree of government effort to boost such usage).[58] But when all is said and done, this will still constitute but a small share of the total world energy. For wind and solar to claim a substantially bigger share—as desired by many climate activists—greater effort will have to be invested in making these forms of energy accessible on a very large scale. This will require improvements in the design of turbines and solar collectors, improved energy storage (to retain energy collected during sunny and windy periods for release at night and in calm weather), and an expanded electrical grid (to distribute energy from areas of greatest sun and wind to areas of greatest need).[59] China, Germany, and Spain have been making these sorts of investments—giving them a possible advantage in the new Thirty Years' War— but it is unclear if their efforts will prove adequate or decisive.[60]

Biofuels and algae: Many experts see a promising future for biofuels, especially as "first generation" ethanol, based largely on the fermentation of corn and sugar, is replaced by second- and third-generation fuels, derived from plant cellulose ("cellulosic ethanol") and bio-engineered algae. Aside from the fact that the fermentation process requires heat (and so *consumes* as well as releases energy), many policymakers object to the use of food crops to supply raw materials for a motor fuel at a time of rising food prices. However, several promising technologies to produce ethanol by chemical means from the cellulose in non-food crops are now being tested, and it is possible that one or more of these techniques will survive the transition to full-scale commercial production.[61] At the same time, a number of companies, including ExxonMobil, are exploring the development of new breeds of algae that reproduce swiftly and can be converted into biofuels[62]; the US Department of Defense is also investing in some of these experiments.[63] It is too early, however, to forecast which (if any) of these various endeavors will pan out.

Hydrogen: A decade ago, many experts were talking of hydrogen's immense promise as a carrier of energy. Hydrogen is widely abundant in many natural substances (including water and natural gas) and produces no carbon emissions when consumed. However, it does not exist by itself in the natural world and must thus be extracted from these other substances—a process that requires energy and is thus not particularly efficient. Methods for transporting, storing, and consuming hydrogen on a large scale have also proved hard to develop. Considerable research has been devoted to techniques for overcoming these obstacles, and it is possible that significant breakthroughs will occur over the next few decades; at present, however, it appears unlikely that hydrogen will prove a major source of the world's energy in 2041.[64]

Other potential sources: Many other sources of energy are being devised and tested by scientists and engineers at the world's universities and corporate laboratories; some are also being evaluated on a larger scale in pilot projects of various types. These include, among others, geothermal, wave energy, and tidal energy. All of these tap into the planet's immense natural forces and are infinitely exploitable, with little risk of producing GHGs. However, the technology to tap into these resources is still (with the exception of geothermal) at an early stage of development, and it may take decades to devise effective means to harvest them. Geothermal energy does show considerable promise but has run into problems, given the need to drill deep into the earth—in some cases, triggering seismic events.[65] From time to time, one also hears of other prospects even more unfamiliar, yet possessing some hint of promise. At present, none of these approaches appears likely to play a significant role in 2041, but a technological breakthrough of one sort or another should not be ruled out.

Energy efficiency: Given the lack of an obvious winner among competing energy sources, it may well be that the most effective approach to energy consumption in 2041 will in fact be *efficiency*—achieving maximum economic output for minimum energy input.[66] Rather than investing in costly (and problematic) new energy systems, the lead players in 2041 may be the countries and corporations that have mastered the art of producing the most with the least. This would include innovations in transportation, building and product design, heating and cooling, and production techniques. This approach would also downplay reliance on imported materials like oil, whose procurement and protection entails vast national expenditures.

Each of these approaches is likely to attract various combinations of corporate and governmental backers, along with advocacy groups of one sort or another. All will strive to maximize the advantages enjoyed by their favored fuel or approach, while denying such advantages to competing fuels. Because many of the options identified above will only achieve full-scale development if lavish government backing and a friendly regulatory environment are assured, the world's future energy order will largely be determined by the outcome of these battles. To some degree, the ultimate outcome of this process will reflect the relative value of the fuel itself—but to a considerable degree it will also reflect the relative ability of these various camps to secure the necessary support and approval. Much of the fighting among major energy contenders during the next thirty years will consist of these sorts of financing, taxation, and regulatory struggles.

When the War is Over

In thirty years' time, the world will be a very different place. It will be hotter, storms will be more frequent and severe, and the shorelines of many low-lying nations will have narrowed. By this point, strict limitations on carbon emissions will be enforced in most of the world and the consumption of fossil fuels—except under controlled circumstances—will be actively discouraged. Oil will still be available to those who need (and can afford) it, but will not be the world's paramount fuel, as it is today. Instead, some other form of energy (or a combination of energy types) will dominate the international economic order, providing wealth and power to those companies and governments that constitute its principal suppliers.

No one can say today which of the contending forms of energy will prove dominant in 2040 and beyond. If I were to wager a guess, I would say that future world developments will favor energy systems that are decentralized rather than centralized, are easy to make and install, and require relatively modest levels of up-front investment. For an analogy, think of laptop computers of 2011 as against giant mainframes of the 1960s and '70s. The closer a supplier gets to the laptop model in the provision of energy (or so I suspect), the more success will follow.

From this perspective, giant nuclear reactors and coal-fired plants are, in the long run, less likely to thrive, except in places like China where authoritarian governments still call the shots. Far more promising, once the necessary breakthroughs come, will be renewable sources of energy and advanced biofuels that can be produced on a smaller scale with less upfront investment and can thus be possibly incorporated into daily life even at a community or neighborhood level.

Whichever countries move most swiftly to embrace these or similar energy technologies are likely to emerge at the end of the Thirty Years' War with vibrant, productive economies; countries that resist such adaptation, I suspect, will find themselves on the unenviable path of decline.

Notes

1. The author first argued this point in "The Global Energy Crisis Deepens," TomDispatch.com, (accessed June 5, 2011).
2. US Department of Energy, Energy Information Administration (DoE/EIA), *International Energy Outlook 2011* (Washington, D.C: DoE/EIA, 2010), Table A1, 157.
3. Ibid. See also 2010 edition of this document, Table A1, 131.
4. BP, *Statistical Review of World Energy, June 2011* (London: BP, 2011), 6, 9.
5. DoE/EIA, *International Energy Outlook 2011*, Table E2, 230–31.
6. Eric Watkins, "IEA to Release Oil Stocks to Offset Libyan Disruption," *Oil & Gas Journal*, www.ogj.com (accessed June 23, 2011).
7. As quoted in Emma Farge, "Oil Could Hit $200–$300 on Saudi Unrest—Yamani," *Reuters* (April 5, 2011), uk.reuters.com (accessed July 14, 2011).
8. For discussion, see "The Price of Fear," *The Economist*, March 2011, 29–31.
9. See Neil MacFarquhar, "In Saudi Arabia, Royal Funds Buy Peace for Now," *New York Times*, June 2011.
10. As quoted in Farge, "Oil Could Hit $200–$300 on Saudi Unrest—Yamani."
11. DoE/EIA, *International Energy Outlook 2011*, Table E2, 230.
12. Ben Casselman, "Facing Up to the End of 'Easy Oil'," *Wall Street Journal*, May 2011.
13. For background and discussion, see Matthew R. Simmons, *Twilight in the Desert* (Hoboken, N.J: John Wiley, 2005).
14. Jim Landers, "Saudi Show off $10 Billion Khurais Mega-Project to Ease Doubts," *Dallas Morning News*, June 2008.
15. See Casselman, "Facing Up to the End of 'Easy Oil'."
16. Jareer Elass, and Amy Myers Jaffe, *Iraqi Oil Potential and Implications for Global Oil Markets and OPEC Politics* (Houston: James A. Baker III Institute for Public Policy, Rice University, 2011), 29–30.
17. UN Department of Economic and Social Affairs, Population Division, "World Population Prospects: The 2010 Revision," esa.un.org (accessed October 5, 2011).
18. For background, see Neil King Jr., "Saudi Industrial Drive Strains Oil-Export Role," *Wall Street Journal*, December 2007.
19. DoE/EIA, "Japan," Country Analysis Brief, March 2011, electronic document, www.eia.doe.gov (accessed July 14, 2011).

20. Ibid.
21. Ibid.
22. Martin Fackler, "Japan to Cancel Plan to Build More Nuclear Plants," *New York Times,* May 2011.
23. See Eric Watkins, "Japan Eyes Changes in Energy Policy after Nuclear Catastrophe," *Oil & Gas Journal* (May 4, 2011), www.ogj.com on (accessed May 9, 2011).
24. See Hiroko Tabuchi, "Quake in Japan Causes Costly Shift to Fossil Fuels," *New York Times*, August 2011. See also "Noda Pledges New Basic Energy Plan by Next Summer, *Asahi Shimbun*, (September 13, 2011), ajw.asahi.com (accessed October 6, 2011).
25. See Stefan Nicola, "German Government Unveils 'Road Map into the Age of Renewable Energy'," *European Energy Review*, www.europeanenergyreview.eu (accessed November 12, 2010).
26. Helen Pidd, and Suzanne Goldenberg, "Germany Suspends Power Station Extension Plans as Nuclear Jitters Spread," *The Guardian*, March 2011.
27. "Swiss Cabinet Agrees to Phase out Nuclear Power," *Reuters* (May 25, 2011), www.reuters.com (accessed July 14, 2011).
28. "China Halts Approval of Nuclear Projects in Wake of Japan Reactor Disaster," Bloomberg News, March 2011, www.bloomberg.com (accessed July 14, 2011).
29. For background and discussion, see Kevin Jianjun Tu, and David Livingston, "Wenzhou Crash Shows the Dangers of China's Nuclear Power Ambitions," *China Brief,* Jamestown Foundation, (July 29, 2011), www.jamestown.org (accessed August 21, 2011).
30. See, for example, Rebecca Smith, and Mark Maremont, "Earthquake Risks Probed at US Nuclear Plants," *Wall Street Journal*, July 2011.
31. See: Brian Vastag, "At North Anna Nuclear Plant, Reassurances but no Final Data on Quake Impact," *Washington Post*, September 2011; and Matthew L. Wald, "After Quake, Virginia Nuclear Plant Takes Stock," *New York Times*, September 2011.
32. DoE/EIA, *International Energy Outlook 2011*, Table A8, 165.
33. "Central China Drought Worst in More than 50 Years: Reports," *The China Post*, May 2011, www.chinapost.com.tw (accessed on July 14, 2011).
34. Leslie Hook, "High Coal Costs Force China to Ration Electricity," *Financial Times*, May 2011.
35. Ibid.
36. For background, see Intergovernmental Panel on Climate Change, *Climate Change 2007: Impacts, Adaptation and Vulnerability*, Contribution of Working Group II (Cambridge: Cambridge University Press, 2008).
37. Ibid.
38. For background, see National Commission on the BP Deepwater Horizon Oil Spill and Offshore Drilling (National Commission), *Deep Water: The Gulf Oil Disaster and the Future of Offshore Drilling* (Washington, D.C: National Commission, 2011), 50.
39. See Steve Hargreaves, "Flooded Nebraska Nuclear Plant Raises Broader Disaster Fears," CNN Money, June 2011, money.cnn.com (accessed August 21, 2011).
40. DoE/EIA, *International Energy Outlook 2011*, Table A2, 159.
41. For an analysis of the future world oil supply and the growing role of unconventional sources of supply, see International Energy Agency (IEA), *World Energy Outlook 2010* (Paris: IEA, 2010), 101–77.
42. For discussion, see Scott L. Montgomery, *The Powers that Be* (Chicago, IL: University of Chicago Press, 2010), 62–67.
43. See Mountain Justice, "What Is Mountain Top Removal Mining?" Electronic Document, www.mountainjustice.org (accessed July 15, 2011).

44. For discussion, see Montgomery, *The Powers that Be*, 94–110.
45. See John Deutch, "The Good News about Gas," *Foreign Affairs* 90, no. 1 (January/February, 2011): 82–93.
46. See: Jim Efstathiou Jr. and Kim Chipman, "Fracking: The Great Shale Gas Rush," *Business Week*, March 2011, www.businessweek.com (accessed July 15, 2011); Ian Urbina, "Regulation Lax as Gas Wells' Tainted Water Hits Rivers," *New York Times*, February 2011.
47. Sophie Pilgrim, "French Lawmakers Ban Controversial Shale Gas Drilling," *France 24*, May 2011, www.france24.com (accessed July 15, 2011).
48. See Ian Urbina, "Insiders Sound an Alarm Amid a Natural Gas Rush," *New York Times*, June 2011.
49. See Terry Macalister, "Vultures Circle BP over Fears Its Days Are Numbered in US," *The Guardian*, June 2010, www.guardian.co.uk (accessed January 6, 2011). See also: Guy Chazan, and Gina Chon, "BP Sells $7 Billion of Assets to Help Fund Cleanup," *Wall Street Journal*, July 2010; and Michael J. de la Merced, "BP to Sell Big Stake in Oil Asset," *New York Times*, November 2010.
50. See Danny Fortson, and Dominic O'Connell, "Exxon weighs £100bn bid for BP," *The Times* (London), July 2010, www.thetimes.co.uk (accessed January 6, 2011).
51. See "Oil and Gas in Troubled Waters," *The Economist*, October 2005, 52–53.
52. For discussion, see: Michael T. Klare, *Rising Powers, Shrinking Planet* (New York: Metropolitan Books, 2008), 220–27.
53. The author first developed this argument in "The New Thirty Years' War," TomDispatch.com, (accessed June 26, 2011).
54. For discussion, see Montgomery, *The Powers that Be*, 81–93.
55. DoE/EIA, *Annual Energy Outlook 2011* (Washington, D.C: DoE/EIA, 2011), Table A1, 115, and earlier editions.
56. See Deutch, "The Good News about Gas."
57. See Matthew L. Wald, "Administration to Push for Small 'Modular' Reactors," *New York Times*, February 2011.
58. IEA, *World Energy Outlook 2010*, 82.
59. For background and discussion, see ibid, 303–38. See also Montgomery, *The Powers That Be*, 165–72.
60. See Pew Charitable Trusts, *Who's Winning the Green Energy Race?* (Washington, D.C. and Philadelphia: Pew, 2010).
61. See Montgomery, *The Powers that Be*, 177–82.
62. Kambiz Foroohar, "Exxon $600 Million Algae Investment Makes Khosla See Pipe Dream," Bloomberg News, June 2010, www.bloomberg.com (accessed October 10, 2011).
63. William Matthews, "From Algae to JP-8," *Defense News*, January 2009, www.defensenews.com (accessed October 10, 2011).
64. For background and discussion, see Montgomery, *The Powers that Be*, 184–98.
65. See ibid, 172–77. See also Domenico Giardini, "Geothermal Quake Risks Must Be Faced," *Nature*, 462 (December 2009): 848–49, www.nature.com (accessed October 10, 2011).
66. For discussion and proposals, see Amory B. Lovins, "More Profit with Less Carbon," *Scientific American*, September 2005, 74–83.

3

Listening to the Planet and Building a Sustainable Energy Economy

Daniel M. Kammen

Part 1: The Challenge of Dealing with a Problem for Which We Lack a Suitable Language

Energy is in the news every day, discussed and described in many ways, yet we remarkably struggle for even a basic currency in which we can reflect the energy system we wish to build. This simple fact drives many of the apparent paradoxes of our interest in clean, sustainable energy and our inability to launch a new scientific and industrial revolution to build this new energy economy. We have seen decades of papers extolling the opportunities and needs for a new energy system. However, the sad fact is that we, as a global society, have wasted many good years—decades actually—during which we could have launched an economy based on job creation and investments in human capacity and creativity. Instead, we have ignored the clear signals that nature has been sending us about its status and stresses we are placing on our rivers, oceans, skies, mountains, and on the health of every ecosystem.

There are many metrics we might cite that consistently tell us that we are approaching, or that we are at, or that we are beyond the carrying capacity of the planet in terms of flows of pollutants and our need for resources.[1] That is not to say we do not have options; we do, and we need to listen to the planet and ourselves, as well as put our ideas into practice.

Energy is the largest (legal) piece of the global economy, underpinning nearly all economic activity. The size and complexity of the energy industry have profound consequences on our natural environment. The Intergovernmental Panel on Climate Change (IPCC)—which shares the 2007 Nobel Peace Prize—has been issuing assessment reports on the state of climate science since 1990. Several thousand climate and energy experts participate in the IPCC assessment and reporting process during preparation of any one assessment report (AR). I have been involved in the preparation of several assessment reports since my first "special report" on technology transfer in 1999.[2] Assessment reports aim to provide an understanding of the contemporary knowledge of climate change science and indicate the extent of the problem of climate change. How big is the problem? Not only is energy a dominant part of our economy but its impact on the planetary system can also be seen in a few historical trends that are summarized in Figure 1. This figure illustrates the consensus data on three broad climate indicators (global average surface temperature, global average sea level, and Northern Hemisphere snow cover) and interjects the timing of each of the first four assessment reports.

In 1990, the First Assessment Report—the first vertical line in Figure 1—concluded that it would take another decade at least to see clear signs of climate change in the natural record. The key language—vigorously debated and worked over most carefully—developed over time. While the first report articulated limited language to describe climate change, the next installment—the Second Assessment Report, the second vertical line in Figure 1, five years later in 1995—advanced the clarity with a consensus statement that "the balance of evidence suggests discernible human influence." This was clearer than the earlier IPCC installment, but still far from sealed. The Third (third vertical line) and Fourth (last vertical line in Figure 1) Assessment Reports clarified things considerably, assigning an analytic confidence (66 percent in the Third Assessment Report) to the likelihood that the environmental change in Figure 1 is due to human activities (e.g., energy generation, agriculture, and forest disruption). Finally, the Fourth Assessment Report in 2007 stated that most of the warming is (90 percent confidence) due to human activity. At this point, the climate community branched out more strongly into social issues and arrived at the conclusion that the consequences of global warming *will most strongly and most quickly affect the world's poor.* Unfortunately, what this also means is that because we generally ignore the poor, we will ignore climate change for a longer time than we should.

This record of environmental change is already significant and has only just begun. What is needed in response is an ambitious set of individuals and networks of people and institutions committed to changing this *status quo* of our energy economy. Basic scientific breakthroughs are needed, for example, in energy storage and solar cells. There are business opportunities in smart-grid energy services and electric vehicles, and there are tremendous political wins for elected officials who truly champion a clean energy economy.

Thus, finding ways to change trends such as those given in Figure 1 becomes our collective challenge. The challenge of putting good ideas into practice extents itself throughout society: from the academic community, to the entrepreneurial private sector, to civil society at large, and to the public servants community. We

Figure 1
**Changes in three environmental indicators (A: global average surface
temperature, B: global average sea level, C: Northern Hemisphere snow cover),
1850–2010, and Vertical Lines Indicating the Dates of the First through the
Fourth Intergovernmental Panel on Climate Change (IPCC) Assessment Reports**

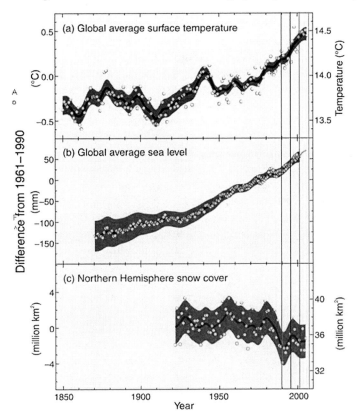

need to do better—not next year or next month, but right now. There are several
pathways that can be identified as ways forward.

First, we need a sustained and vibrant research base to understand our energy
options and their resulting climate impacts.[3] We also need something that may
sound very simple, in fact mundane,[4] for people in fields outside the energy
domain, such as agriculture and education, where "extension services" are
common. Farmers rely on information networks to plan their activities. Universities, community colleges, and night schools all focus on making continuing
education an available extension service. We need a mechanism to bring these
innovations from the laboratory to the market. In agriculture, every country
essentially has a network of agricultural extension services. Perhaps, because

much of the "modern" energy system has been managed by large, centralized energy utilities, the energy sector has failed to establish such extension networks. It is time to build one.

One of the most important tools that we need to develop is an economy-wide "appreciation" of the costs and benefits of our energy choices. That, in fact, is why I felt it was so critical for me to work at the World Bank at this time. Two elements of the World Bank stand out. First, the World Bank is significantly increasing its commitment and investment in energy efficiency and in clean and sustainable energy. Second, and of vital importance, banks are focused on the bottom line, pure and simple, and this is a virtue—the virtue of *clarity*. Our society, however, is so far only *casually* and *vaguely* interested in sustainability.

Beyond specific technical and policy innovations, there have been some important insights: such as the realization that distributed networks of energy suppliers and consumers (some of whom may be one and the same) could not only complement large, traditional energy systems—but that in many ways, distributed generation may deliver a superior performance. I liken this to a transition from conventional thinking. The old way was to view the energy grid as a one-way flow of energy *to* consumers. *Instead*, it could be similar to eBay, where anyone can buy and/or sell power, with the job of the utility—and the network regulator—being provision of fair and transparent rules for these transactions.

These (potentially critically important) innovations aside, our interest in sustainability (accessing the resources we need without degrading the opportunities of future generations) remains very casual. This is not because individuals are not worried about the world that their children will inherit. Poll after poll shows that when asked this question in isolation, people's response is that they are very concerned about the future. Nor are we afraid to make hard decisions for what we want—people and even governments (much to the surprise of some skeptics) do this all the time. People in a government generally work very hard for public good, and civil society and non-governmental groups put exceptional effort into innovating and giving voice to the watchdog role that every community, large or small, critically needs. It is true that immediate gratification versus long-term quality of life—and thus we must also address the complicated issue of discounting and of undervaluing the world in which we and then our children will live[5]—remains a problem for most people to keep squarely in mind. This issue, however, relates to the clarity that honest banks, and particularly the value that a clear and well-understood *currency*, can bring to our planning for the future.

As a society, we are only vaguely interested in sustainability because we collectively do not speak a language that permits us to value the world in which we live, except when we cut it down, spoil our waters with human and industrial effluent, or poison the skies with the waste of our energy generation. In contemplating this situation, my friend and colleague George Lakoff has made a remarkable and chilling observation.

It is clear to the environmental science community that nature is being degraded,[6] in fact destroyed by the current course of human action and neglect. Yes, it is true that the so-called industrialized nations emitted the majority of the greenhouse gas emissions, if we go back to the beginning of the industrial revolution. So, by one measure, they should "pay." Yet, the so-called developing nations will emit the bulk of greenhouse gases over the coming decades, so they should "pay it forward." What is more important than all this, however, is that we will all live in the world of the future. So, while we can argue about this issue all day, the solution must be a collective one.

We have established that anthropogenic climate change is the act of degrading our collective home, Planet Earth. The environment is a *complex system* that responds in ways that are not always predictable. Professor Lakoff's observation is that we actually lack a means to express—and thus fully understand—this situation. In other words, no language has a simple verb form that captures the effect of a system acting on an individual.[7] Certainly, there are ways to express this idea—notably, if we become anthropomorphic and refer to the planetary system as an entity: *Mother Nature*, for example. However, if we move beyond this view of the planet as a single, coordinated entity to the complex system that it is, we are not equipped to understand the process of this collective causality in terms of how it relates to us as individuals.

So, we are in a difficult place. First, we lack a *language*, and second, we lack something else that is vital to understand the planet. We lack a *currency* for determining the *values* of the planet, a clean energy economy, and our future.

Placing a price on greenhouse gas emissions to the atmosphere will not solve global climate change and environmental destruction by itself, but it gives us a language to express our values. Given that humans are social creatures that communicate constantly about every aspect of their individual and collective existence, lacking a means to communicate about the future not only indicates our shortsightedness but is also simply unacceptable.

Now, economists who are interested in valuing the planet will correct me here and say, "Actually, things are worse than you say." They, in fact, would be right. The true story is *not* that we do not value the environment in a positive way but that we reward damaging the environment, and hence ourselves. Rewarding waste is, in fact, *placing a negative value on the planet and saying financially that sustainability is a bad thing.* This is not to say that we are *intentionally* damaging our "nest," but that through our inactions, we are actually sending an economic and political signal that individual profit is more important than our collective well-being and that of the natural ecosystem. We can, for example, choose to invest in local job creation by supporting people and companies providing energy services without damaging nature. We know job creation benefits are real in terms of the higher numbers of jobs created in clean energy areas compared to polluting sectors.[8] This is not because "clean energy" is inherently superior; it is so because when ramping up a new field, greater investments in infrastructure and jobs

Figure 2
What Value Is There in Saving the Planet?

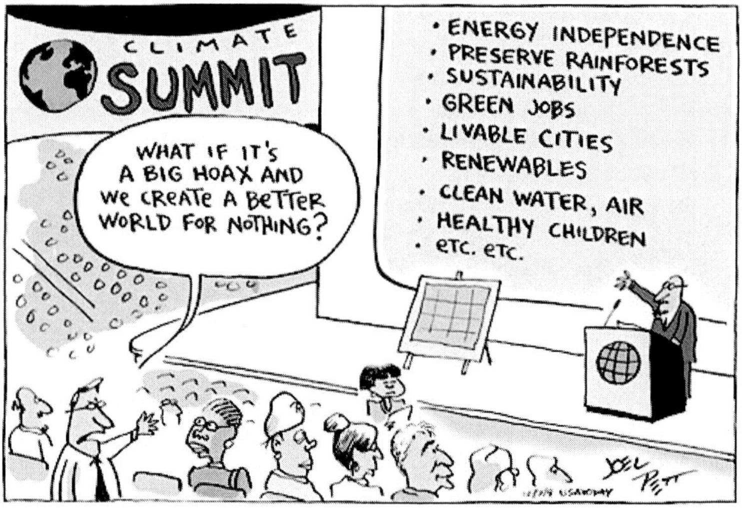

are needed. This means that when our energy dollars are put to productive use, and not simply used to increase our debt to the environment, we gain an added benefit. When adding up these environmental advantages of "going green," it is often difficult to determine why this transition is challenging. In this regard, I am reminded of a perplexing cartoon (Figure 2), particularly because we know, as we can see in Figure 1, that climate change is not a "big hoax."

Part 2: Putting Our New Language and Energy and Climate into Action

Let us hope that we are able to build and use this language of energy and environmental clarity. In this respect, the story begins to get brighter and brighter. Once we recognize that we lack (or lacked) language and financial metrics, the required action becomes not only clearer but also easier than many people think.

How can this be? First is the observation that sustaining the planet is good for us in the long-term (that should go without saying) and that wasting and polluting less yields many positive returns. One of the most important lessons of the rapidly expanding mix of energy efficiency, solar, wind, biofuels, and other low-carbon technologies is that the costs of deployment are lower than many forecasts, and at the same time, the benefits are larger than expected.

This seeming "win-win" claim deserves examination, and continued verification, of course.

Over the past decade, the solar and wind energy markets have been growing at rates over 30 percent per year, and in the last several years, growth rates of over 50 percent per year have taken place in the solar energy sector. This explosive and sustained growth has meant that costs have fallen steadily, and that an increasingly diverse set of innovative technologies and companies have been formed. Government policies in an increasing number of cities, states, and nations are finding creative and cost-effective ways to build these markets still further.

At the same time that a diverse set of low-carbon technologies is finding its way to the market, energy-efficient technologies (e.g., "smart" windows, energy-efficient lighting and heating/ventilation systems, weatherization products, and efficient appliances) and practices are all in increasingly widespread deployment. Many of these energy-efficient innovations demonstrate negative costs over time, meaning that when the full range of benefits (including improved quality of energy services, improved health, and worker productivity) are tabulated, some energy-efficiency investments are vehicles for net creation of social benefits over time.

Carbon abatement curves have become well-known since the Swedish power company Vattenfall collaborated with the McKinsey & Company to develop a set of estimates on the costs to deploy and operate a range of energy efficiency, land use, and energy-generation technologies. (They are actually just knock-offs of marginal pricing curves used in the electricity industry for decades, but context and timing is everything). These costs of conserved carbon curves depict the costs (or savings, in the case of a number of "negative cost" options such as building efficiency) as well as the magnitude (in gigatonnes) of abatement potential at a projected future time. In Figure 3, such an abatement curve for the entire world for 2030 is presented. The basic message is as follows: saving money often saves carbon emissions, if you are strategic about where to invest.

A World Bank-supported low-carbon development study shows that Mexico can reduce carbon emissions by 42 percent more than its target of 1,137 metric tons by 2030—477 million tons to be precise—by decisive action on multiple fronts. It can achieve this by moving in key areas such as improving bus systems, road and rail freight logistics, fuel economy standards, and vehicle inspection at the border, among others.

This is exciting news. It shows that significant—even dramatic—carbon reductions can be achieved by adjusting the use of existing technologies. Such adjustments can reduce costs too. These conclusions emerged from calculations based on a marginal abatement cost (MAC) curve, an analytical tool developed in 2008 by McKinsey & Company, and used by a team of experts studying Mexico's climate challenges headed by the World Bank.

The study in which this methodology was used, *Low-Carbon Development for Mexico* by Johnson, Alatorre, Romo, and Liu,[9] is one of a series of such studies financed by the Energy Sector Management Assistance Program (ESMAP), which includes Brazil and Nigeria (forthcoming).

This MAC tool has now been applied, with promising results, to two tiny communities of 1,100 people on Nicaragua's Atlantic coast. Results of a study

Figure 3
A 'Carbon Abatement Curve' Showing Financial Cost (+) or Savings (-) for a Range of Efforts, Projects, and Programs with the Unit Financial Impact per ton of Carbon Dioxide not Emitted to the Atmosphere. Examples Exist for a Range of Countries, Including Brazil, China, Mexico, the United States, and the United Kingdom, with the List Expanding Every Day

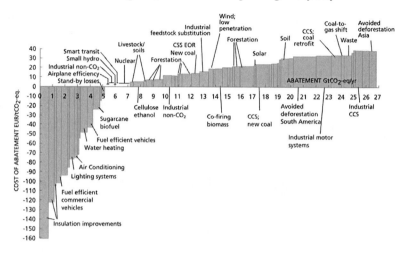

published November 26 in *Science* magazine demonstrate that low-carbon rural energy services can be delivered at cost savings in cases where communities utilize diesel-powered, isolated electricity grids.[10]

The study on which I was working with Christian Casillas before I joined the World Bank in September will (we hope) spur efforts elsewhere to build similar community-level carbon abatement and energy service tools. This could mean that communities that were often ignored or lumped together as "those billions without modern energy" can create their own locally appropriate development goals, and groups working with them can develop energy solutions at a price lower than the one they are paying now.

In 2009, the rural Nicaraguan communities of Orinoco and Marshall Point, which share a diesel microgrid, partnered with the national government and an NGO to implement energy efficiency measures including metering, which prompted residents to reduce wasteful use of electricity. Compact fluorescent light (CFL) bulbs were also introduced, biogas installations were put in place to partially replace the use of diesel.

After the government installed meters, energy use dropped by 28 percent, and people's electric bills dropped proportionately. Blue Energy, an NGO based in San Francisco, was able thereby to cut household energy use by another 17 percent with the provision of CFL bulbs.

Figure 4
Marginal Abatement Curve for Greenhouse Gas Emissions for a
Rural Community on the Atlantic Coast of Nicaragua

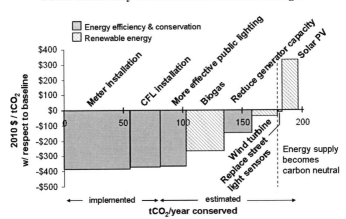

Source: Casillas and Kammen (2010)

The net result was reduced burning of diesel, even allowing for the fact that the community's reduced energy needs allowed the local energy supplier to run its generators two extra hours each day, providing longer service to customers. In the month after the conservation campaign, energy costs per household had dropped by 37 percent.

That the MAC curve can be used to analyze energy use in the community and pinpoint areas where investments would save the most energy and the most money for homeowners is something of a breakthrough. Until now, the model has been used mostly on a global or country-wide scale to target areas for carbon abatement. But now, it has gone local. This means that some of the world's poorest communities can reduce their energy costs by local action, which, when multiplied worldwide, could produce a global change and reduce carbon emissions.

The curves in Figure 4 illustrate the range of the existing low-carbon options, and that if we continue to build a menu of options that have been tested, vetted, and implemented, a new paradigm of clean energy development will have a solid economic footing in a wide range of national, city, and community environments.

Finally, let me conclude with a brief note on building a business model for clean energy.

This is a piece of the story that is left behind in many discussions: creating a new energy economy cannot be a battle between environmentalists saying we must "go green" and the business community saying we "cannot go green" today, or not that rapidly. In fact, there is such a great deal of emerging data—such as these marginal abatement curves—that if we manage the process of innovation and implementation well, we can find ways to both grow the economy and make it dramatically greener. Germany's experience in wind and solar, and sound urban and agricultural planning is a great example of doing both well. In fact, export earnings, job creation, and a stable economy in times of oil and gas shocks can be found in this emerging green economy as well.

Notes

1. Rokström, J. et al., "A Safe Operating Space for Humanity," *Nature* 461 (2009): 472–75.
2. Intergovernmental Panel on Climate Change Working Groups II and III, "Methodological and Technological Issues in Technology Transfer" (Cambridge UK and New York: Cambridge University Press, 2000).
3. Nemet, G. F., and D. M. Kammen, "U.S. Energy Research and Development: Declining Investment, Increasing Need, and the Feasibility of Expansion," *Energy Policy* 35, no. 1 (2007): 746–55.
4. Kammen, D. M., and Dove, M. R., "The Virtues of Mundane Science," *Environment* 39 no. 6 (1997): 10–15, 38–41.
5. Schelling, T. C., "The Cost of Combating Global Warming: Facing the Tradeoffs." *Foreign Affairs* 76 (1997): 8 –14.
6. Intergovernmental Panel on Climate Change (AR4), "Climate Change 2007: The Physical Science Basis," Contribution of Working Group I to the Fourth Assessment Report of the Intergovernmental Panel on Climate Change (Cambridge University Press, 2007) http://www.ipcc.ch.
7. Lakoff, G., "Women, Fire, and Dangerous Things: What Categories Reveal about the Mind," (Chicago, IL: University of Chicago Press, (1987)).
8. Wei, M., Patadia, S. and Kammen, D. M. "Putting Renewables and Energy Efficiency to Work: How Many Jobs Can the Clean Energy Industry Generate in the U. S.?," *Energy Policy* 38 (2010): 919–31.
9. Johnson, T., Alatorre, C., Romo, Z., and Liu, F., *México: Estudio Sobre Las Disminicución de Emisiones de Carbono* (Washington, DC: The World Bank, 2009).
10. Casillas, C. and Kammen, D. M., "The Energy-Poverty-Climate Nexus," *Science* 330 (2010): 1181–82.

4

Job Creation through a Green Energy Economy

Robert Wendling and Roger Bezdek

Although analysts have conducted renewable-energy (RE) and energy-efficiency (EE) industry studies and forecasts for more than the past three decades, no rigorous definitions currently exist for either of these industries or for their current size, structure, and composition. Researchers at Management Information Services, Inc. (MISI) have conducted the first comprehensive study of the size and breadth of the RE and EE ("RE&EE") industries and have created a standard definition for these industries, which provides comparability between data. Prior to this work, the basic information on these industries was not well-documented. For example, many studies have examined the potential of specific components of the RE industry (e.g., wind, photovoltaics, biomass, and so on), and experts have established long-term forecasts regarding the economic impacts of major proposed RE&EE policy initiatives and government spending programs. However, such analyses are of limited usefulness unless we have a better idea of the size and characteristics of the existing RE&EE industries. What does it mean to say that "experts predict that the number of jobs in the industries will increase threefold by 2020," when we do not know what the current employment base is?

At present, a rigorous, generally agreed-upon definition of what constitutes the RE&EE "industry" has yet to be developed. Obviously, the industry includes technologies such as wind energy, photovoltaics (PV), solar thermal energy, and biomass. But should all hydropower technologies be included, even large, environmentally threatening systems? What about geothermal energy? Daylighting? Climate-responsive buildings? Hydrogen?

This chapter serves to develop a rigorous definition of the RE&EE industries, estimate their current size and composition, and forecast their growth to

2030. It is anticipated that the findings reported here will become the standard for future economic analyses of the RE&EE industries in the United States and internationally.

Issues Involved in Defining and Specifying the Renewable-Energy and the Energy-Efficiency Industries

A rigorous definition of "renewable-energy and energy-efficiency industries" obviously includes a disparate collection of technologies such as solar, wind, PV, biomass/biofuels, etc. However, here we specify them to also include the following industries/technologies, among others:

- Hydroelectricity—small and large
- Geothermal
- Fuel cells
- Hydrogen
- Energy-conservation and energy-efficiency products
- Electric and hybrid vehicles
- Passive, solar/green, sustainable buildings, and energy-smart design[1]
- Daylighting

The shared concept underlying the renewable-energy and energy-efficiency industries as used here is appropriate for several reasons. First, many RE firms also offer energy-efficiency and conservation products and services. Distinguishing between RE and EE products, services, and sales would be virtually impossible. Second, RE and EE are closely related, share many goals, and are often offered as an integrated product. For example, solar buildings have to be extremely energy efficient. Similarly, energy-efficient structures often incorporate RE elements and features. Third, in some cases, there is no clear distinction between an "RE" product and an "EE" product. Examples include passive solar design, sustainable buildings, and daylighting. Fourth, RE and EE are large and growing industries that require accepted definitions for current and future economic researchers to follow. Finally, "RE&EE" is a much larger and more robust industry than the RE industry alone.

Thus, RE and EE industries include all aspects of the energy-efficiency industry, including energy-efficient buildings, firms offering energy audits and energy service contracts, and manufacturers, sellers, and installers of a wide array of energy-efficiency products and services. However, some difficult and complex decisions were necessary in developing rigorous definitions. For example:

- Windows and doors, gas and oil furnaces, home appliances, motors, etc. are offered at wide ranges of energy efficiencies—and prices. How do we evaluate and allocate such energy efficiencies? What constitutes an "energy-efficient" product? More "energy efficient" than what? The new generation of many products is more energy efficient than the previous generation. Where is the cutoff?

- Energy efficiency is currently a very powerful public relations and marketing strategy. Many products are advertised as being "energy efficient," and no one advertises a product as being "energy inefficient." Here, the authors had to exercise care to sort through these claims.[2]
- Many electric and gas utilities offer renewable and energy-efficiency products and services. Should these be identified, quantified, and included as part of the industry?
- Low-flow faucets, showerheads, and toilets conserve significant amounts of water. In doing so, they indirectly reduce energy requirements by reducing the amount of energy required to pump, transport, and process water. Should water conservation products thus be included in the definition of the RE&EE industry?
- Hybrid vehicles are a part of the RE&EE industry, but how are these to be disaggregated from the total operations of automobile manufacturers? More generally, in the definition outlined in this chapter, the authors distinguish between classes of vehicles on the basis of fuel efficiency. Obviously, a Hummer getting 11 miles per gallon (mpg) is not fuel-efficient, but a small vehicle getting 35 mpg is. Is the latter a part of the RE&EE industry? In the forecast scenarios devised for this chapter, the authors hypothesized a very substantial increase in US vehicular fuel-efficiency standards. In this scenario, vehicles at or above certain fuel-efficiency levels were included as part of the RE&EE industry.
- Are all recycling, reuse, and remanufacturing activities part of the RE&EE industry?

The desired definition needed to be general enough to apply to the entire RE&EE industry throughout the United States. The purpose of this requirement was to avoid the creation of a different industry definition for each state. Another major decision involved how to handle the characterization of RE and EE activities as engaged by federal, state, and local government, nonprofits, nongovernmental organizations (NGOs), foundations, etc. Should federal RE and EE R&D activities be included as part of the RE&EE industry? Is the National Renewable Energy Laboratory part of the industry? What about all numerous (and rapidly growing) state and local government RE and EE activities? What about all of the federal, state, and local government RE and EE trade, professional, and interest groups?

One may argue that it is favorable to restrict the industry definition to the activities of primarily private entities. However, this may not be desirable. First, doing so would exclude significant and important public, NGO, and nonprofit activities. Second, it could lead to contradictory results. For example, if PV panels are installed on a school by a private company paid with public funds, then under this definition, the installation activities would be included in the industry definition. However, if the PV panels are installed on a school by state or local government employees paid with public funds, then under this definition, the installation activities would not be included in the industry definition. Also imperative was the need to relate these industry and job categories to the North American Industrial Classification System (NAICS) standard.[3] The US Energy Information Administration (EIA) has made a start at this, but at present, the effort is incomplete.

Thus, for the definition offered in this chapter, the authors had to decide whether they wanted to measure the scale of activities and expenditures, or rather, only activities being conducted by private companies. These are difficult, complex, and critical questions, for which there is no single correct answer. Yet, with efforts described here to develop a rigorous definition of the RE&EE industry, this chapter seeks to flesh out what may become the standard in terms of any type of economic and job analysis of the industry as conducted by researchers in the future. In effect, the authors here are acting as the definer and "benchmarker" of the industry as it evolves.

The chapter utilized the following definitions to characterize jobs within the RE and EE industries:

- A job in the RE industry consists of an employee working on one of the major RE technologies included in this study. In addition, in this study, jobs in RE include persons involved in RE activities in the federal, state, and local governments, universities, trade and professional associations, NGOs, consultancies, investment company analysis departments, etc.
- A job in the EE industry consists of an employee working in a sector that is entirely part of the EE industry, such as an energy service company (ESCO) or the recycling, reuse, and remanufacturing sector. It also includes some employees in industries in which only a portion of the output is classified as within the EE sector, such as household appliances, HVAC systems, construction, etc.
- Finally, in this study, jobs in EE include persons involved in EE activities in the federal, state, and local governments, universities, trade and professional associations, NGOs, consultancies, investment company analysis departments, etc.

The US Renewable-Energy Industry

RE technologies are here defined to consist primarily of[4]:

- Hydroelectricity
- Biomass
- Geothermal
- Wind
- PV
- Solar Thermal
- Fuel Cells and Hydrogen

Except for hydro and industry biomass, RE's energy contribution in the United States is small but is growing rapidly. Some RE technologies, such as ethanol, biodiesel, and biomass-to-liquids, produce liquid fuels that directly replace imported oil. As shown in Figure 1 (see below), RE produced about 7 percent of the total energy in the United States in 2007.

Growth in the US Renewable-Energy Industry, 2006–2007

RE consumption in the United States declined 1 percent between 2006 and 2007 to 6,830 trillion Btu. However, this decrease resulted from a 14 percent decrease in hydroelectricity in 2007 due to reduced precipitation in several regions of the country. In contrast, both total and nonrenewable energy consumption increased 2 percent. Wide variation occurred in the consumption of individual RE sources. For instance, biomass-based energy increased 7 percent and wind-generated electricity increased 21 percent. Major increases in the consumption of biomass for biofuels (ethanol and biodiesel) were largely responsible for an increase in biomass in 2007.

From 2003 through 2007, the average annual growth rate of RE consumption was 3 percent, compared with just 1 percent for the total energy consumption. Biofuels, wind, and PV were largely responsible for the increase, with 5-year average annual growth rates falling in the range 25–30 percent.

Just over half of RE consumption occurred in the electric power sector in 2007, while the industrial sector was the second-leading consumer of RE, accounting for nearly 30 percent. The transportation, residential, and commercial sectors

Figure 1
Renewable Energy in the US, 2007

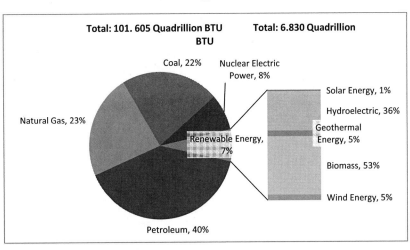

Source: US Energy Information Administration.

accounted for 9, 8, and 2 percent, respectively. While the electric power sector currently consumes the most renewable energy (at 51 percent), its use decreased 8 percent between 2006 and 2007. In 2003, the electricity sector accounted for 59 percent of the total renewable-energy consumption.

In contrast, RE consumption in the transportation sector increased 30 percent during 2007, and residential sector consumption grew 12 percent. Residential sector growth was due to significant increases in the following three energy sources: biomass, geothermal, and solar/photovoltaic. Commercial and industrial uses of RE changed modestly between 2006 and 2007 and have changed little as a fraction of the total renewable consumption since 2003. This scenario could change for the industrial sector if ethanol and biodiesel use continues to grow rapidly, resulting in increased feedstock consumption. This is especially significant in view of the fact that the largest biomass fuel consumed in the industrial sector, wood and derived fuels, has grown little since 1989 and appears to have peaked in 1997.

Within the electric power sector, wind energy consumption has grown each year since 1998. From 2003 to 2007, the share of wind energy of total renewable-energy consumption increased from 2 to 5 percent. For the first time, wind energy consumption in the electric power sector exceeded that of geothermal energy. Hydroelectricity accounted for 36 percent of total renewable consumption in 2007, down from 46 percent in 2003. However, hydro consumption is tied mostly to precipitation, which varies year to year, and few plants are being built or retired.

Electricity generation from all renewables decreased 9 percent in 2007 to 351 billion kWh, largely because of reduced precipitation. Excluding hydro-electricity, however, RE generation grew 7 percent. This gain was led by a 21 percent increase in electricity from wind and moderate increases in electricity from biomass waste. There has been little change in generation from the largest non-hydro renewable electricity sources, wood and derived fuels, since 2003.

With the exception of hydroelectricity, changes in renewable electricity capacity generally reflected generation changes in 2007. Total renewable electricity capacity increased 5 percent to 107 GW, led by a 38 percent (or 4,000 MW) increase in wind capacity. The total nonrenewable electric capacity rose just 1 percent to 892,000 MW.

The Renewable-Energy Industry, Revenues and Jobs

Table 1 below summarizes the status of the US RE industry in 2006 and 2007. In 2006, RE gross revenues totaled nearly $40 billion, the total number of jobs created by RE reached 450,000, and more than 90 percent of the jobs were in private industry. The majority of jobs were in the biomass sector (almost 70 percent), followed by the wind, hydroelectric, and geothermal sectors. In 2007, RE gross revenues totaled nearly $43 billion, the total number of jobs created by RE exceeded 500,000, and more than 95 percent of the jobs were in private industry.

Table 1
The Renewable-Energy Industry in the United States, 2006–2007

Industry segment	2006			2007		
	Revenues/ Budgets (billions)[a]	Industry jobs	Total jobs created	Revenues/ Budgets (billions)	Industry jobs	Total jobs created
Wind	$3.0	16,000	36,800	$3.3	17,300	39,600
Photovoltaics	1.0	6,800	15,700	1.3	8,700	19,800
Solar Thermal	0.1	800	1,900	0.14	1,300	3,100
Hydroelectric Power	4.0	8,000	19,000	3.5	7,500	18,000
Geothermal	2.0	9,000	21,000	2.1	10,100	23,200
Biomass						
Ethanol	6.3	67,000	154,000	8.4	83,800	195,700
Biodiesel	0.3	2,750	6,300	0.4	3,200	7,300
Biomass Power	17.0	66,000	152,000	17.4	67,100	154,500
Fuel Cells	0.9	4,800	11,100	1.1	5,600	12,800
Hydrogen	0.8	4,000	9,200	0.81	4,100	9,400
Total, Private Industry	35.4	185,150	427,000	38.45	208,700	483,400
Federal Government	0.5	800[*]	1,850	0.65	900[*]	2,100
DOE Laboratories	1.8	3,600[**]	8,300	1.9	3,800[**]	8,700
State and Local Government	0.9	2,500	5,750	0.95	2,600	5,800
Total Government	3.2	6,900	15,870	3.5	7,300	16,600
Trade and Professional Associations and NGOs	0.6	1,500	3,450	0.63	1,600	3,500
Total, All Sectors	$39.2	193,550	446,320	$42.58	217,600	503,500

[a] 2006 dollars
[*] Includes Federal employees and direct support contractors.
[**] Includes Federal employees, laboratory employees, and direct support contractors.
Source: Management Information Services, Inc., *Green Collar Jobs in the US and Colorado: Economic Drivers for the 21st Century*, prepared for the American Solar Energy Society, Boulder, Colorado, January 2009.

In terms of revenues in 2007, biomass power accounted for 41 percent of the total industry revenues and ethanol for 20 percent, followed by hydroelectric power (just over 8 percent), wind (just under 8 percent), and fuel cells and hydrogen combined (about 4.5 percent). Private industry accounted for just over 90 percent of the total RE revenues.

In terms of the total jobs created, the relative contributions of the RE sectors differ somewhat from the relative contributions on the basis of revenues. This is due to the fact that different RE technologies and industries have considerably different job creation effects. Three salient examples of this dynamic are that a) hydroelectric power generated 8.2 percent of RE revenues but 3.6 percent of the total RE jobs; b) biomass power generated 41 percent of RE revenues, and 31 percent of the total RE jobs; and c) ethanol generated 20 percent of RE revenues, and 39 percent of the total RE jobs. In all, private industry generated 90 percent of RE revenues, and 96 percent of the total RE jobs.

RE industry revenues in the US increased 8.7 percent from $39.2 billion in 2006 to $42.6 billion in 2007. As noted, hydroelectric production decreased in 2007. Excluding the hydroelectric sector, RE industry revenues increased 11.1 percent from $35.2 billion to $39.1 billion.

To eliminate the effects of inflation, growth rates must be compared in constant, real dollars. Converting the 2006 RE data to constant 2007 dollars indicates that in real terms, the total RE revenues increased 5.5 percent from $40.4 billion in 2006 to $42.6 billion in 2007. Excluding the hydroelectric sector, RE industry revenues increased 7.8 percent from $36.3 billion to $39.1 billion.

The real growth rate of US GDP between 2006 and 2007 was 2.19 percent. Thus, including hydro, the RE industry grew more than twice as rapidly as the overall US economy; excluding hydroelectricity, the RE industry grew more than three times as fast as the overall US economy. Furthermore, the biomass power sector is a significant part of the RE industry, but it grew little between 2006 and 2007. Excluding both hydro and biomass power, the US RE industry grew 15.4 percent between 2006 and 2007—more than seven times as fast as the overall US economy. Some sectors experienced very substantial growth. For instance, solar thermal grew more than 35 percent, biodiesel grew 30 percent, ethanol grew nearly 30 percent, and PV grew more than 25 percent (see Figure 2).

As shown in Figure 2, the percentage revenue increases in the different RE sectors varied widely. While the percentage growth figures are important, it should be noted that some of the most rapidly growing RE sectors, such as PV, solar thermal, and biodiesel, are very small and even relatively modest growth in total revenues will thus produce large percentage increases.

The dominant RE sectors sometimes differ from those that grow most rapidly. This is illustrated in Figure 3 below, which shows the total number of US jobs generated by the RE sectors in 2007. Despite the differential growth rates of the sectors between 2006 and 2007, job creation is dominated by ethanol and biomass power, followed far behind by wind, geothermal, PV, and hydroelectricity.

Figure 2
Increase in Real US RE Revenues, 2006–2007 (Constant 2007 dollars)

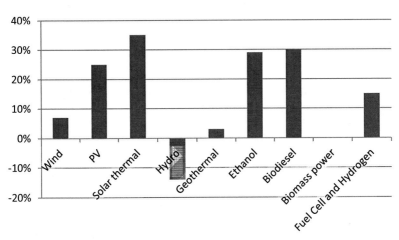

Source: Management Information Services, Inc. *Green Collar Jobs in the US and Colorado: Economic Drivers for the 21st Century*, prepared for the American Solar Energy Society, Boulder, Colorado, January 2009.

Figure 3
Total US Jobs Generated by Renewable-Energy Sectors in 2007

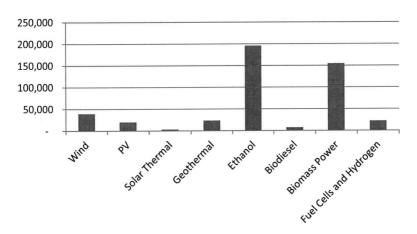

Source: Management Information Services, Inc. *Green Collar Jobs in the US and Colorado: Economic Drivers for the 21st Century*, prepared for the American Solar Energy Society, Boulder, Colorado, January 2009.

As noted, the vast majority of jobs created by RE are standard jobs for accountants, engineers, computer analysts, clerks, factory workers, etc., and most of the persons employed in these jobs may not even realize that they owe their livelihoods to renewable energy. This is illustrated in Table 2 below, which lists the jobs created by renewable energy in the United States in 2006 within selected occupations. For instance, this table shows that in 2006, RE generated in the United States:

- More jobs for shipping and receiving clerks (2,210) than for biochemists and biophysicists (1,580);
- More jobs for carpenters (780) than for environmental engineers (630);
- More jobs for truck drivers (9,500) than for forest and conservation workers (1,440); and
- More jobs for janitors (3,610) than for environmental science technicians (1,690)

Thus, many US workers are dependent on renewable energy for their employment, although they would often have no explicit way of recognizing that connection unless it is brought to their attention. This becomes especially clear when reviewing the job distribution for a typical renewable-energy company. Such an overview is presented in box 1 for a wind energy company.

Table 2
Renewable-Energy Jobs Generated in the United States in 2007 by Selected Occupations

Occupation	Jobs
Agricultural Equipment Operators	4,260
Biochemists and Biophysicists	1,580
Bookkeeping and Accounting Clerks	8,228
Business Operations Specialists	3,390
Carpenters	780
Chemical Technicians	1,880
Civil Engineers	3,080
Computer and IT Managers	1,210
Computer Programmers	2,660
Computer Software Engineers	3,260
Database Administrators	560
Electrical and Electronic Equipment Assemblers	840
Electricians	6,330

(Continued)

Table 2 (*Continued*)

Occupation	Jobs
Engineering Managers	1,350
Environmental Engineers	630
Environmental Science Technicians	1,690
Employment, Recruitment, and Placement Specialists	600
Forest and Conservation Workers	1,440
HVAC Mechanics and Installers	2,130
Industrial Engineers	1,340
Industrial Production Managers	760
Inspectors, Testers, and Sorters	2,400
Janitors and Cleaners	3,610
Machinists	1,820
Mechanical Engineers	1,950
Payroll and Timekeeping Clerks	1,160
Plumbers, Pipefitters, and Steamfitters	4,670
Purchasing Agents	1,280
Sales Representatives	4,140
Security Guards	1,310
Sheet Metal Workers	1,600
Shipping and Receiving Clerks	2,210
Surveyors	690
Tax Preparers	580
Tool and Die Makers	620
Training and Development Specialists	650
Truck Drivers	9,500

Source: Management Information Services, Inc. *Green Collar Jobs in the United States and Colorado: Economic Drivers for the 21st Century*, prepared for the American Solar Energy Society, Boulder, Colorado, January 2009.

Box 1
Example of a Wind Turbine Company

Thousands of RE&EE companies are located throughout the United States, and they generate jobs for more than nine million workers. Given the wide diversity in the size, function, and technologies of these companies, it is impossible to estimate the job profile of the "average" RE&EE firm. However, it is possible to identify the jobs and earnings profiles of typical types of firms involved in RE&EE-related areas of work. Table B-1 shows the 2007 occupational job distribution and employee earnings of a typical wind turbine manufacturing company. This table illustrates several important points.

First, firms working in RE&EE and related areas employ a wide range of workers at all educational and skills levels and at widely differing earnings levels. Second, in RE&EE companies, few of the employees are classified as RE or EE specialists. Most of the workers are in occupations labeled as machinists, engineers, laborers, clerks, book-keepers, accountants, maintenance workers, cost estimators, etc. All these employees owe their jobs and livelihoods to RE&EE, but they generally perform the same types of activities at work as employees in firms that have little or nothing to do with RE&EE.

This fact is illustrated in Table B-1. The occupational job distribution of a typical wind turbine manufacturing company differs relatively little from that of a company that manufactures other products. Thus, production of wind turbines and wind turbine components requires large numbers of engine assemblers, machinists, machine tool operators, mechanical and industrial engineers, welders, tool and die makers, mechanics, managers, purchasing agents, etc. They are "RE" workers only because the company they work for manufactures a renewable-energy product. Importantly, with the current national angst concerning the erosion of the US manufacturing sector and the loss of US manufacturing jobs, it is relevant to note that many RE technology sectors are growing rapidly.[5] These types of firms can help revitalize the manufacturing sector and provide the types of diversified, high-wage jobs that all US states seek to attract.

More generally, while traditional debate surrounding alternative energy has focused on applying new technology to offset traditional energy sources, RE&EE is more than a source of fuel or energy savings; rather, it is a source of jobs. As shown here, employment growth in the RE&EE industry varies for the different segments comprising the sector, but new breakthroughs in RE&EE technologies will come from the growing sectors of the industry, including architectural and engineering services, materials processing, and research and development. In addition, utilities are pioneering a number of alternative energy technologies, including superconducting power lines (which reduce the 20 percent loss of electricity due to transmission), as well as solar thermal, photovoltaic, and wind systems, and distributed power technologies. The latter will reduce losses from transmission and supply as more reliable localized generation enables power production all across the electrical system. Increasingly, however, RE&EE advances will come from all areas of the economy and may not be necessarily captured by traditional industry sources of energy technologies.

Table B-1
Typical Employee Profile of a 250-person Wind Turbine Manufacturing Company, 2007 (Selected Occupations)

Occupation	Employees	Earnings
Engine and Other Machine Assemblers	31	$36,900
Machinists	27	41,300
Team Assemblers	16	30,700

(Continued)

Table B-1 (*Continued*)

Occupation	Employees	Earnings
Computer-Controlled Machine Tool Operators	12	41,500
Mechanical Engineers	10	73,300
First-Line Supervisors/Managers of Production/ Operation?	10	60,800
Inspectors, Testers, Sorters, Samplers, and Weighers	8	41,100
Lathe and Turning Machine Tool Setters/Operators/ Tenders	6	40,800
Drilling and Boring Machine Tool Setters/Operators/ Tenders	4	40,500
Welders, Cutters, Solderers, and Brazers	4	40,600
Laborers and Freight, Stock, and Material Movers	4	30,300
Maintenance and Repair Workers	4	44,900
Tool and Die Makers	4	44,800
Grinding/Lapping/Polishing/Buffing Machine Tool Operators	4	35,500
Multiple Machine Tool Setters/Operators/Tenders	4	41,400
Industrial Engineers	3	71,900
Industrial Machinery Mechanics	3	46,900
Engineering Managers	3	110,600
Shipping, Receiving, and Traffic Clerks	3	32,700
General and Operations Managers	3	123,600
Industrial Production Managers	3	95,000
Industrial Truck and Tractor Operators	3	34,900
Purchasing Agents	3	57,100
Cutting/Punching/Press Machine Setters/Operators/ Tenders	3	32,000
Production, Planning, and Expediting Clerks	3	46,100
Milling and Planing Machine Setters/Operators/ Tenders	3	41,200
Mechanical Drafters	2	40,600
Customer Service Representatives	2	39,700
Bookkeeping, Accounting, and Auditing Clerks	2	36,300
Office Clerks, General	2	29,800
Sales Representatives, Wholesale and Manufacturing	2	56,300

(*Continued*)

Table B-1 (*Continued*)		
Occupation	**Employees**	**Earnings**
Janitors and Cleaners	2	30,200
Sales Engineers	2	73,900
Accountants and Auditors	2	61,000
Tool Grinders, Filers, and Sharpeners	2	44,800
Executive Secretaries and Administrative Assistants	2	44,000
Mechanical Engineering Technicians	2	51,900
Electricians	2	50,700
Other employees	48	50,600
Employee Total (126 occupations in the industry)	**250**	**$47,300**

Source: Management Information Services, Inc. *Green Collar Jobs in the United States and Colorado: Economic Drivers for the 21st Century*, prepared for the American Solar Energy Society, Boulder, Colorado, January 2009.

The US Energy-Efficiency Industry in 2006–2007

As discussed above, estimating the size of the EE industry is much more difficult than estimating the size of the RE industry. The RE industry is fairly well-defined and consists of distinct sectors such as wind, PV, biomass, hydro-electricity, geothermal, etc. The EE "industry," on the other hand, is much more nebulous and difficult to define, specify, and estimate. Certain elements are clearly part of the EE industry, such as federal and state EE programs, utility EE spending, the ESCO industry, and the recycling industry. However, most EE spending is included in partial segments of large industries such as vehicles, buildings, lighting, and appliances.

Table 3 below summarizes the status of the US EE industry in 2006 and 2007. In 2006, gross revenues of $933 billion and 8 million jobs characterized the US EE industry. These sales represent substantially more than the combined 2007 sales of the three largest US corporations—ExxonMobil, Wal-Mart, and GM ($905 billion).[6] Over 90 percent of the jobs were in private industry, and over 50 percent of the jobs were in the manufacturing sector. The second largest number of jobs were in recycling, reuse, and remanufacturing followed by the construction industry. Nearly 80 percent of the EE government jobs were in the state and local government.

The 2007 US EE industry is characterized by gross revenues of more than $1 trillion and 8.6 million jobs. Of this, more than 98 percent were in private industry and over 36 percent of the jobs were generated by the recycling, reuse, and remanufacturing sector. The second largest number of jobs was generated by the nondurable manufacturing sector, followed by the miscellaneous durables

Table 3
The Energy-Efficiency Industry in the United States, 2006–2007

Industry segment	2006			2007		
	Revenues/Budgets (billions 2006 dollars)	Industry jobs (thousands)	Total jobs created (thousands)	Revenues/Budgets (billions 2007 dollars)	Industry jobs (thousands)	Total jobs created (thousands)
ESCOs	3	19	44	3.8	23	53
Recycling, reuse, and remanufacturing	275	1,310	3,013	290	1,372	3,154
Vehicle manufacturing	73	165	380	86	193	443
Household appliances and lighting	22	86	198	35	134	308
Windows and doors	12	51	117	13	54	123
Computers, copiers, FAX machines, etc.	90	312	718	105	360	828
TV, Video, and Audio equipment	45	183	421	48	193	447
HVAC systems	12	45	104	13	47	108
Industrial and related machinery	19	76	175	21	82	187
Miscellaneous durable manufacturing	105	389	894	110	397	901
Nondurable manufacturing	220	528	1,214	218	518	1,183

(Continued)

Table 3 (*Continued*)

Industry segment	2006			2007		
	Revenues/Budgets (billions 2006 dollars)	Industry jobs (thousands)	Total jobs created (thousands)	Revenues/Budgets (billions 2007 dollars)	Industry jobs (thousands)	Total jobs created (thousands)
Utilities	2	14	32	2.2	14	32
Construction	41	253	582	48	288	660
Total, Private Industry	919	3,431	7,892	993	3,675	8,427
Federal government EE spending	3.3	15	35	3.8	16	37
State government EE spending	3	28	64	3.2	29	65
Local government EE spending	2.3	21	48	2.4	22	50
Total Government	8.6	64	147	9.4	67	152
EE Trade and Professional Associations and NGOs	0.5	3	7	0.52	3	7
Total, All Sectors	$932.6	3,498	8,046	$1,002.92	3,745	8,586

Source: Management Information Services, Inc. *Green Collar Jobs in the United States and Colorado: Economic Drivers for the 21st Century*, prepared for the American Solar Energy Society, Boulder, Colorado, January 2009.

manufacturing sector, and sectors manufacturing computers, printers, copiers, etc. Relatively few jobs were generated by the ESCO sector, the utilities sector, or the government sectors.

The total EE industry revenues increased 7.5 percent from $933 billion in 2006 to $1,003 billion in 2007. Converting the 2006 EE data to constant 2007 dollars indicates that in real terms, the total EE revenues increased 4.4 percent from $961 billion in 2006 to $1,003 billion in 2007. The total number of jobs created (directly and indirectly) by EE increased by more than 800,000.

The US Renewable-Energy and Energy-Efficiency Industries

Table 4 summarizes the US RE&EE industries in 2006 and 2007. Total revenues in the two industries increased from $972 billion in 2006 to $1,046 billion in 2007. Direct employment in the industries increased from 3.69 million in 2006 to 3.96 million in 2007. The total number of jobs created increased from 8.5 million in 2006 to 9.1 million in 2007. RE grew more rapidly than EE and slightly increased its share of total revenues of the two industries from 4.0 percent in 2006 to 4.1 percent in 2007. RE also slightly increased its share of the total jobs generated by the two industries from 5.3 percent in 2006 to 5.5 percent in 2007.

Industry Forecast Scenarios

As part of the study explored in this chapter, three forecast scenarios were developed: a Base Case, a Moderate Scenario, and an Advanced Scenario.

The Base Case essentially represents a "business as usual" (BAU) case scenario that assumes that there will be no change in RE&EE policies and that

Table 4
Summary of the US Renewable-Energy and Energy-Efficiency
Industries in 2006–2007

Industry	2006			2007		
	Revenues (billions)	Industry jobs (thousands)	Total jobs created (thousands)	Revenues (billions)	Industry jobs (thousands)	Total jobs created (thousands)
Renewable Energy	$39.2	196	452	$42.58	218	504
Energy Efficiency	932.6	3,498	8,046	1,002.92	3,745	8,586
Total	$971.8	3,694	8,498	$1,045.50	3,963	9,090

Source: Management Information Services, Inc. *Green Collar Jobs in the US and Colorado: Economic Drivers for the 21st Century*, prepared for the American Solar Energy Society, Boulder, Colorado, January 2009.

no major RE&EE initiatives will be implemented over the next two decades. Furthermore, the scenario assumes that the RE&EE industries continue to develop according to the general trends and rates of growth experienced over the past two decades. Under the Base Case, RE&EE development continues to be a small portion of the US economy. The Base Case is used as a comparison against the two alternative scenarios. The Base Case indicates that without any substantial change in policy, RE&EE is not expected to significantly increase its share in the US energy market.

The Moderate Scenario assumes that various moderate and incremental (above the Base Case) federal and state RE&EE initiatives are put in place over the next two decades. Additionally, the Moderate Scenario assumes the existence of policies such as R&D, tax incentives, RPS mandates, and externalities pricing and a continuation of the positive policies that are in place, plus market conditions favorable to renewables. The scenario is based on various "mid-range" estimates incorporating modest initiatives.

The Advanced Scenario "pushes the envelope" on the RE&EE industry with current or impending technologies and requires favorable market conditions and a sustained commitment of public policy to ensure that RE&EE achieves higher levels of contribution to the US energy market. The scenario assumes that RE&EE industries are available to take the United States in a new direction and that appropriate, aggressive public policies at federal and state levels are required and must be sustained over the next two decades. In the Advanced Scenario, the driving factor may be fossil fuel shortages and price increases, security concerns, recognition of global warming, etc. The outcomes of the scenario represent a dramatic indication of what would be possible under an aggressive renewable-energy scenario and includes what may be realistically feasible both economically and technologically in such a "crash" scenario.

Within the context of marketplace uncertainties, the major determinant of future market share for RE&EE is public policy.[7] The three scenarios indicate what a BAU scenario might look like if no policy changes are implemented and the potential for more aggressive renewable-energy strategies. However, it must be recognized that achieving any scenario is subject to significant uncertainties in key market drivers. In this, important factors are the volatility in oil and gas prices,[8] the pace and scale of action on climate change, the extent of technology breakthroughs, and federal, state, and local government RE&EE policies and incentives.

Despite this vulnerability to contingent developments in the market, the approach presented here has several advantages. For one, the approach has been used by MISI in previous studies and has been vetted in the peer-reviewed literature. This offers credibility to the approach.[9] Also, the MISI approach has been analyzed by the American Council for an Energy-Efficient Economy (ACEEE) and found to be valid and credible.[10] Second, the MISI analysis has been reviewed by Al Gore's staff and has been used to help formulate some of Mr. Gore's energy policy recommendations.[11] Third, the Base Case provides a baseline forecast of the RE&EE industry and employment in the United States over the next two decades. Fourth, the Moderate and Advanced Scenarios demonstrate the relative employment and jobs benefits to the United States of different levels of RE&EE initiatives and policies. Finally, differences between the estimates and forecasts for the three scenarios indicate the marginal differences from various levels of RE&EE development.

Development of these scenarios provides insight into major policy areas such as resource depletion, climate change, balance of payments, national security, and RE&EE employment. The economic implications revealed under different scenarios—and the impact of coming down the different experience curves at different annual rates—are especially useful.

US Renewable-Energy and Energy-Efficiency Forecasts to 2030

Some of the major results of the study are presented here in the following tables. While Table 5 shows a summary of the US RE&EE industries in 2030, Tables 6 and 7 provide a more detailed overview of the revenues and jobs created in both the renewable-energy sector and the energy-efficiency sector under the three scenarios.

The US Renewable Energy Industry in 2030

Table 6 illustrates that the growth of the US RE industry will be significantly affected by government policies and incentives. The size of the industry in 2030 under the Advanced Scenario is nearly six times as large as that under the Base Case. More importantly, some RE sectors under the Advanced Scenario grow much more than others. In fact, wind is sixteen times larger under the Advanced Scenario than under the Base Case, geothermal is fourteen times larger, the fuel cells sector is nine times larger, biodiesel is six times larger, biomass power is

Table 5
US Renewable-Energy and Energy-Efficiency Industries in 2030

	Revenues (Billions of 2007 Dollars)			Total jobs created (Thousands)		
	Base case	Moderate scenario	Advanced scenario	Base case	Moderate scenario	Advanced scenario
RE	$98	$212	$560	1,305	2,846	7,328
EE	1,868	$2,036	$3,734	14,953	16,658	29,878
Total	$1,966	$2,248	$4,294	16,258	19,504	37,206

Source: Management Information Services, Inc. *Green Collar Jobs in the United States and Colorado: Economic Drivers for the 21st Century*, prepared for the American Solar Energy Society, Boulder, Colorado, January 2009.

Table 6
The US Renewable-Energy Industry in 2030

Industry segment	Revenues (Billions of 2007 Dollars)			Total jobs created		
	Base case	Moderate scenario	Advanced scenario	Base case	Moderate scenario	Advanced scenario
Wind	$5.6	$22	$89	66,200	257,000	1,040,000
Photovoltaics	13.5	27	45	206,000	415,000	700,000
Solar Thermal	0.2	0.9	29	3,800	17,000	540,000
Hydroelectric Power	4.8	5.1	6.8	22,400	24,200	32,300
Geothermal	2.9	8.2	40	29,000	85,000	415,000
Biomass						
Ethanol	22.6	45	82	530,000	1,050,000	2,000,000
Biodiesel	1.3	2.7	7.6	25,100	56,900	160,000
Biomass Power	32.3	68	160	282,000	603,000	1,420,000
Fuel Cells	5.2	14.1	45	68,600	158,000	505,000
Hydrogen	4.1	12.2	36	47,200	143,000	420,000
Total, Private Industry	92.4	205.2	540.4	1,280,300	2,809,000	7,232,300
Federal Government	0.8	1	2.8	3,000	3,100	8,550
DOE Laboratories	2.3	2.6	7.8	11,000	12,300	36,100
State and Local Government	1.5	2.2	5.7	7,000	11,800	29,400
Total Government	4.6	5.8	16.3	21,000	27,200	74,050
Trade and Professional Associations and NGOs	0.8	1.5	3.6	4,700	9,400	21,300
Total, All Sectors	$97.8	$212.5	$560.3	1,305,400	2,845,700	7,327,650

Source: Management Information Services, Inc. *Green Collar Jobs in the United States and Colorado: Economic Drivers for the 21st Century*, prepared for the American Solar Energy Society, Boulder, Colorado, January 2009.

five times larger, and PV and ethanol are more than three times larger illustrating the projected effect of an aggressive support scenario. Similarly, Table 6 shows a wide variation in 2030 job creation between the Base Case and the Advanced Scenario. The largest differentials in terms of the numbers of jobs created are in the ethanol, biomass power, and wind sectors. The largest differentials in terms of the percentage increases in jobs created are in the solar thermal, geothermal, and wind sectors.

The US Energy-Efficiency Industry in 2030

Table 7 illustrates that the growth of the US EE industry will be significantly impacted by government policies and incentives. The size of the industry in 2030 under the Advanced Scenario is twice as large as that under the Base Case. Some EE sectors under the Advanced Scenario grow much more than others. For example, the EE construction sector is more than four times larger under the Advanced Scenario than that under the Base Case, the EE vehicle sector is nearly four times larger, the EE utility sector is nearly three times larger, and the EE windows and doors sector is more than 2.5 times larger.

Table 7
The US Energy-Efficiency Industry in 2030 (Billions of 2007 Dollars)

Industry segment	Revenues (Billions of 2007 Dollars)			Total jobs created (Thousands)		
	Base case	Moderate scenario	Advanced scenario	Base case	Moderate scenario	Advanced scenario
ESCOs	7.2	8	14	98	121	196
Recycling, reuse, and remanufacturing	546	580	618	5,220	5,178	5,732
Vehicle manufacturing	154	189	570	740	912	2,770
Household appliances and lighting	51	62	124	435	528	1,064
Windows and doors	27	33	71	250	298	645
Computers, copies, FAX machines, etc.	180	171	190	1,360	1,321	1,446

(Continued)

Table 7 (*Continued*)

Industry segment	Revenues (Billions of 2007 Dollars)			Total jobs created (Thousands)		
	Base case	Moderate scenario	Advanced scenario	Base case	Moderate scenario	Advanced scenario
TV, Video, and Audio equipment	98	104	200	870	922	1,775
HVAC systems	29	34	76	240	276	633
Industrial and related machinery	43	52	114	360	623	931
Miscellaneous durable manufacturing	205	233	522	1,640	1,840	4,230
Nondurable manufacturing	410	451	784	2,120	2,429	4,070
Utilities	5.2	6	14	75	98	204
Construction	93.3	130	373	1,240	2,964	5,186
Total, Private Industry	1,848.7	2,053	3,670	14,648	16,270	28,882
Federal Government EE Spending	7.2	8	24	70	83	222
State Government EE Spending	6.2	7	19	120	148	380
Local Government EE Spending	5.1	6	17	100	130	335
Total Government	18.5	21	60	290	361	937
EE Trade and Professional Associations and NGOs	1	2	4	15	27	59
Total, All Sectors	$1,868.2	$2,076	$3,734	14,953	16,658	29,878

Source: Management Information Services, Inc. *Green Collar Jobs in the United States and Colorado: Economic Drivers for the 21st Century,* prepared for the American Solar Energy Society, Boulder, Colorado, January 2009.

Implications for Skills, Training, and Educational Requirements

Renewable-Energy, Energy-Efficiency, and Related Occupations

Occupational data demonstrate that RE&EE industries create a variety of high-paying jobs, many of which take advantage of manufacturing skills currently going unused as manufacturing continues to undergo restructuring in the United States. Regions with traditional manufacturing economies can recruit RE&EE companies to take advantage of their highly skilled workforces, since as illustrated in Table 8, wind turbine manufacturing requires plant operators, machinists, mechanics, engineers, welders, etc.

As shown in Table 8, wages and salaries in many sectors of the RE&EE and related industries are higher than the average wages in the United States. Although many high-tech industries almost exclusively require highly educated workers with master's or doctoral degrees, as noted, the RE&EE industry requires a wide variety of occupations. Nevertheless, many occupations in the RE&EE industry include jobs which require associate's degrees, long-term on-the-job training, or trade certifications, including engineers, chemists, electrical grid repairers, power plant operators and power dispatchers, chemical technicians, mechanical engineering technicians, and RE&EE technicians, all of which pay higher than US average wages.

Table 8
Renewable-Energy, Energy-Efficiency, and Related Occupations: Wages,
Educational Requirements, and Growth Forecasts (Selected Occupations)

Occupation	10 year % Growth forecast	Median salary	% With bachelor's degree	Minimum education
Materials Scientists	8	$75,800	94	Bachelor's
Physicists	7	93,300	92	Doctoral
Microbiologists	17	64,600	96	Doctoral
Biological Technicians	17	37,200	60	Associate's
Conservation Scientists	6	54,800	88	Bachelor's
Chemists	7	64,800	94	Bachelor's
Chemical Technicians	4	40,900	27	Associate's
Geoscientists	6	74,700	94	Doctoral

(Continued)

Table 8 (*Continued*)

Occupation	10 year % Growth forecast	Median salary	% With bachelor's degree	Minimum education
Natural Science Managers	14	101,000	90	Bachelor's
Environmental Eng. Technicians	24	42,800	18	Associate's
Soil and Plant Scientists	20	59,100	64	Bachelor's
Mechanical Eng. Technicians	12	47,400	18	Associate's
Environmental Sci. Technicians	16	39,100	47	Associate's
Biomedical Engineers	31	76,900	60	Bachelor's
Chemical Engineers	11	80,800	92	Bachelor's
Mechanical Engineers	10	78,600	88	Bachelor's
Electrical Engineers	12	77,700	83	Bachelor's
Environmental Engineers	14	76,000	82	Bachelor's
Computer Scientists	26	95,900	67	Doctoral
Life and Physical Sci. Technicians	20	46,100	50	Associate's
Utility Plant Operatives	4	54,100	10	OJT
HVAC Technicians	12	38,300	14	OJT
Energy Audit Specialists	18	40,300	18	OJT
Forest and Conservation Workers	6	27,500	8	OJT
Refuse and Recycling Workers	5	26,400	2	OJT
Insulation Workers	6	$30,800	2	OJT

Source: Management Information Services, Inc. *Green Collar Jobs in the United States and Colorado: Economic Drivers for the 21st Century*, prepared for the American Solar Energy Society, Boulder, Colorado, January 2009.

Note: OJT refers to on the job training.

Emerging Renewable-Energy and Energy-Efficiency
Jobs, Occupations, and Skills

Growth in the RE&EE sector of the US economy will lead to vast new employment opportunities as businesses expand to meet the new, clean, and sustainable energy requirements. Jobs will be created across a new spectrum of work activities, skill levels, and responsibilities. Many of these jobs do not currently exist, and their occupational titles are not identified by the occupational classifications and standards of the federal and state government. In addition, many of these new jobs require a different set of skills from those required for current jobs, and training requirements must be assessed so that this rapidly growing sector has an adequate pool of trained and qualified job applicants. At some point in the future, many of these occupations will grow in the number of employees classified in the occupation, and the federal government will add them to the employment classification system. Until that time, economic and employment analysis and forecasting is usually conducted using the current set of US Labor Department occupational titles.[12]

Tables 9 through 17 identify, by occupational title, some of the new RE&EE jobs that will be created in the expanding RE&EE energy economy.[13] The listings of jobs spans a broad range of functions, starting with the mainstream RE technologies of solar thermal and PV, including traditional energy-efficient technologies related to buildings and transportation, as well as a range of jobs related to climate change, carbon capture and storage (CCS), and carbon markets. Many new occupations originating from consulting, research, government, and nonprofit institutions are listed separately and have not been specifically classified in a particular RE or EE sector or green technology area.

New occupational titles are listed in the first column of each of the tables. The average salary, listed in the second column, represents the average of the starting salary and highest salary for that occupation. Wages may be 15–20 percent lower at the beginning of employment and may rise to a level 15–20 percent higher as the person becomes an experienced employee. In addition, wages and salaries are usually much higher in urban areas than in rural areas.

The third and final column lists the minimum recommended educational attainment to gain entry into that occupation, and a recommended degree is listed for the advanced educational requirements. Obviously, employers will not hold fast to these recommendations, but this information can be useful to educational planners in providing an idea of the knowledge and skills that the employer is seeking in a candidate. Note that the education requirements listed include no requirement (N/A), high school degree or General Education Development (HSD/GED), and apprenticeship/trade school (TS) to a Master's degree. With the more advanced (Bachelor's degree and up) college requirements, some standard abbreviations were used to further define the recommended degree: CE, ME, EE—for chemical, mechanical, and electrical engineer degrees, respectively. Also note that many jobs can be filled by a candidate with one of several related science or engineering degrees and they are listed generically as such.

Emerging Renewable-Energy Occupations, Salaries, and Skills and Education Requirements

Some of the emerging job opportunities and corresponding salaries and education/training requirements in RE are presented in Tables 9 to 11.[14] The information presented in the tables clarifies that salaries vary widely from the range $20,000–$25,000 for solar and biomass technicians to nearly $140,000 for a director of wind development. Also, educational requirements span the gamut from apprenticeship/TS and HSD/GED to advanced college degrees. However, a wide variety of jobs and education training requirements exist, and many jobs do not require college degrees. Similar jobs in different RE industries can have different salaries and education/training requirements. For example, a solar lab technician may require an associate's degree and earn a salary of nearly $41,000, whereas a wind field technician may require only HSD/GED and earn a salary of less than $26,000. Similarly, a hydroelectric engineer with a bachelor's degree may earn more than $87,000, whereas a solar energy engineer with a bachelor's degree may earn only a little more than $71,000. Numerous career paths allow employees with apprenticeship/TS and HSD/GED credentials to earn relatively high salaries; examples include positions of solar operations engineer, solar thermoelectric plant manager, solar residential or commercial electrician foreman, wind field service technician, hydroelectric plant installation technician, and geothermal plant efficiency operator.

Table 9
Emerging Jobs, Salaries, and Educational Requirements
in the Solar and PV Industry

Occupational Title	Average salary	Minimum education
Solar fabrication technician	$23,092	HSD/GED
Solar lab technician	$40,664	Associate's
Solar hot water heater manufacturing technician	$45,264	Associate's
PV fabrication and testing technician	$45,264	Associate's
Solar energy system installer helper	$23,092	HSD/GED
Solar energy system installer	$31,372	HSD/GED
Solar and PV installation roofer	$35,144	HSD/GED
Solar residential installation electrician	$44,344	HSD/GED

(Continued)

Table 9 (*Continued*)

Occupational Title	Average salary	Minimum education
Solar commercial installation electrician	$44,344	HSD/GED
Instrumentation/Controls/Electrical systems technician	$35,144	HSD/GED
Solar commercial installation engineering technician	$47,104	Associate's
Solar residential installation electrician foreman	$58,236	HSD/GED
Solar commercial installation electrician foreman	$58,236	Apprenticeship/TS
Solar commercial installation engineer	$74,796	Bachelor's (EE)
Solar energy systems designer	$47,104	Apprenticeship/TS
Solar thermoelectric plant manager	$74,520	Apprenticeship/TS
Solar operations engineer	$87,400	HSD/GED
PV solar cell designer	$77,280	Master's (Science)
Solar energy engineer	$71,300	Bachelor's (Engineer)
PV power systems engineer	$75,440	Master's (EE)
Residential/Commercial solar sales consultant	$59,800	Bachelor's (Business)

Source: EDF and Management Information Services, Inc. *Green Collar Jobs in the United States and Colorado: Economic Drivers for the 21st Century*, prepared for the American Solar Energy Society, Boulder, Colorado, January 2009.

Table 10
Emerging Jobs, Salaries, and Educational Requirements in the Wind Industry

Occupational title	Average salary	Minimum education
Wind turbine machinist	$30,452	Apprenticeship/TS
Wind turbine sheet metal worker	$33,212	Apprenticeship/TS
Wind turbine engineering intern	$6,440	HSD/GED
Wind farm electrical system designer	$59,800	Bachelor's (Engineer)
Wind turbine electrical engineer	$87,400	Bachelor's (EE)
Wind turbine mechanical engineer	$87,400	Bachelor's (ME)
Wind field technician	$25,852	HSD/GED

(*Continued*)

Table 10 (*Continued*)

Occupational title	Average salary	Minimum education
Junior renewable-energy technician	$29,532	N/A
Wind generating installer	$31,372	HSD/GED
Electro-mechanical wind turbine technician	$35,144	Associate's
Wind field operations manager for commercial applications	$48,024	Bachelor's (Engineer)
Wind field service technician	$44,344	Apprenticeship/TS
Wind power plant project engineer	$69,000	Bachelor's (Engineer)
Director of wind development	$138,000	Bachelor's (Business)

Source: EDF and Management Information Services, Inc. *Green Collar Jobs in the United States and Colorado: Economic Drivers for the 21st Century*, prepared for the American Solar Energy Society, Boulder, Colorado, January 2009.

Table 11
Emerging Jobs, Salaries, and Educational Requirements
in the Biogas, Biomass, and Hydrogen Industries

Occupational title	Average salary	Minimum education
Landfill gas collection system operator	$40,664	Apprenticeship/TS
Landfill gas system technician	$23,092	HSD/GED
LGE plant installation, operations, engineering and management	$69,000	Bachelor's (Engineer)
Animal waste biomethane gas collection system technician	$40,664	Apprenticeship/TS
Biomass collection, separation and sorting	$21,252	HSD/GED
Biomass plant operations, engineering and maintenance	$69,000	Bachelor's (Engineer)
Hydrogen power plant installation, operations, engineering, and mgt.	$69,000	Bachelor's (CE)
Hydrogen plant operator and operations manager	$94,208	Bachelor's (Engineer)

Source: EDF and Management Information Services, Inc. *Green Collar Jobs in the United States and Colorado: Economic Drivers for the 21st Century*, prepared for the American Solar Energy Society, Boulder, Colorado, January 2009.

Emerging Energy-Efficiency Occupations, Salaries, and Skills and Education Requirements

A similar picture emerges from the evaluation of the energy-efficiency sector (see Tables 12 through 14). The tables present a selection of the emerging EE job opportunities and corresponding salaries and education/training requirements. Several lessons can be drawn from this information. First, the variation in salaries is similar to the renewable-energy sector. For instance, energy field auditors, agricultural workers, and recycling collections drivers receive less than $25,000, while environmental construction engineers receive about $115,000. Second, the educational requirements show a similar variety, spanning from apprenticeship/ TS and HSD/GED certification to advanced college degrees. Third, many jobs do not require college degrees. This is especially true in the green buildings sector. However, most of the jobs in the transportation and energy storage industries tend to require college degrees. Salaries tend to be higher in the transportation and energy storage industries than those in the green buildings or waste management industries. Finally, numerous career paths allow employees with apprenticeship/ TS and HSD/GED certifications to earn relatively high salaries; examples include the positions of field energy consultant, building maintenance engineer, roofing and skylight installer, and hydrogen pipeline construction worker.

Table 12
Emerging Jobs, Salaries, and Educational Requirements
in the Green Buildings Industry (Selected Occupations)

Occupational title	Average salary	Minimum education
Field energy consultant	$60,076	HSD/GED
Energy conservation representative	$48,024	HSD/GED
Energy manager and analyst	$82,800	Bachelor's (various)
Environmental compliance specialist	$46,184	Bachelor's (Science)
Engineering intern	$6,440	HSD/GED
Water systems designer and engineer	$36,984	Apprenticeship/TS
Site supervising technical director	$45,264	Bachelor's (Engineer)
Refrigeration engineer	$73,876	Bachelor's (Engineer)
Civil engineer	$72,220	Bachelor's (CE)
HVAC engineer	$77,556	Apprenticeship/TS
Electrical engineer	$71,760	Bachelor's (EE)
Residential green building and retrofit architect	$90,620	Bachelor's (Architect)

(Continued)

Table 12 (*Continued*)

Occupational title	Average salary	Minimum education
Indoor and outdoor landscape architect	$68,080	Bachelor's (Science)
Industrial green systems and retrofit designer	$90,620	Bachelor's (Architect)
Senior HVAC engineer	$89,700	Bachelor's (ME)
Environmental construction engineer	$115,000	Bachelor's (CE)
Energy engineer	$71,300	Bachelor's (Engineer)
Structural design engineer	$74,980	Master's (CE)
Home improvement retrofit trainee	$29,532	N/A
Insulation installer	$20,332	HSD/GED
Water purification systems service technician	$39,744	HSD/GED
Building maintenance engineer	$64,676	HSD/GED
Machinist	$30,452	Apprenticeship/TS
Welder	$29,532	Apprenticeship/TS
Carpenter	$35,144	Apprenticeship/TS
Electrical system installer	$44,344	Apprenticeship/TS
HVAC service technician	$48,944	Apprenticeship/TS
Roofing and skylight installer	$52,624	Apprenticeship/TS
Weatherization operations manager	$80,040	Bachelor's (various)
Residential energy field auditor	$24,012	Associate's
Industrial energy field auditor	$24,012	Bachelor's (Science)
Auditing services sales consultant	$59,800	Bachelor's (Business)
Renewable-energy consultant	$80,500	Master's (Science)

Source: EDF and Management Information Services, Inc. *Green Collar Jobs in the United States and Colorado: Economic Drivers for the 21st Century*, prepared for the American Solar Energy Society, Boulder, Colorado, January 2009.

Table 13
Emerging Jobs, Salaries, and Educational Requirements in the Transportation and Energy Storage Industries

Occupational title	Average salary	Minimum education
Automotive plant assembly	$48,024	HSD/GED
Diesel retrofit installer	$41,584	HSD/GED
Electric vehicle electrician	$44,344	HSD/GED

(*Continued*)

Table 13 (*Continued*)

Occupational title	Average salary	Minimum education
Hybrid powertrain development engineer	$69,000	Bachelor's (Engineer)
Air pollution specialist	$63,480	Bachelor's (Science)
Senior automotive power electronics engineer	$69,000	Bachelor's (Engineer)
Diesel retrofit designer	$75,440	Master's (Engineer)
Bus system operator	$25,852	N/A
Train system operator	$40,664	HSD/GED
Biofuel plant field and operations engineer	$69,000	Bachelor's (Engineer)
Biofuel plant field technician	$46,184	HSD/GED
Biodiesel/biofuel technology and product development manager	$59,800	Bachelor's (Science)
Alternative fuels policy analyst and business sales	$55,200	Bachelor's (Business)
Agricultural/farm worker	$21,252	N/A
Civil Engineer—agriculture/irrigation/water supply	$92,000	Bachelor's (CE)
Fueling station designer and project engineer	$73,600	Bachelor's (Engineer)
Fuel transporter—trucker	$36,984	HSD/GED
Alternative fueling station operations	$29,532	HSD/GED
Hydrogen pipeline construction	$46,184	HSD/GED
Program manager, environmental construction	$72,220	Bachelor's (Engineer)
Urban planner	$55,384	Bachelor's (Urban Planning)
Environmental engineering manager	$66,838	Bachelor's (Science)
Environmental planner	$61,916	Bachelor's (Science)
Civil engineer	$72,220	Bachelor's (CE)
Energy infrastructure engineer	$77,280	Bachelor's (ME)
Environmental engineer	$48,024	Bachelor's (Engineer)
Battery design engineer	$69,000	Bachelor's (CE)
Battery testing technician	$33,212	Associate's
Battery manufacturing technician	$24,932	HSD/GED

Source: EDF and Management Information Services, Inc., *Green Collar Jobs in the United States and Colorado: Economic Drivers for the 21st Century*, prepared for the American Solar Energy Society, Boulder, Colorado, January 2009.

Table 14
Emerging Jobs, Salaries, and Educational Requirements
in the Waste Management Industry

Occupational title	Average salary	Minimum education
Recycling collections driver	$19,412	N/A
Recycling center operator	$26,772	N/A
Hazardous materials removal worker	$41,584	HSD/GED
Hazardous waste management specialist	$55,384	Bachelor's (Science)
Solid waste (energy) engineer	$73,600	Bachelor's (Engineer)
Nuclear waste management engineer	$89,700	Bachelor's (NE)
Operations maintenance worker for water services	$36,064	HSD/GED
Associate engineer—wastewater treatment	$70,840	Bachelor's (ME)
Wastewater engineer in industrial facilities	$80,500	Bachelor's (Engineer)
Wastewater plant civil engineer	$71,760	Bachelor's (CE)

Source: EDF and Management Information Services, Inc., *Green Collar Jobs in the United States and Colorado: Economic Drivers for the 21st Century*, prepared for the American Solar Energy Society, Boulder, Colorado, January 2009.

Emerging Green Occupations, Salaries, and Skills and Education Requirements in Consulting, Research, Nonprofits, Government, and Carbon Management

When looking at the emerging occupations in the consulting, research, non-profit, government and carbon management sectors, the following becomes clear: a) the average salaries are considerably higher than those in the RE or EE sector with most jobs paying in excess of $40,000; and b) education requirements are relatively stringent and most of the jobs require at least a bachelor's degree (see Tables 15 through 17). With a few exceptions such as environmental technician, forestry worker, and power system operator, there is a lack of employment opportunities and career pathways for persons with apprenticeship/ TS and HSD/ GED qualifications.

Table 15
Emerging Green Jobs, Salaries, and Educational Requirements
in the Consulting, Research, and Nonprofit Industries

Occupational title	Average salary	Minimum education
Environmental technician	$42,780	Associate's
Air quality control engineer	$92,000	Bachelor's (CE)
Greenhouse gas emissions permitting consultant	$63,940	Bachelor's (Science)
Air pollution specialist	$63,480	Bachelor's (Science)
Air resource engineer	$72,220	Bachelor's (Engineer)
Emissions accounting and reporting consultant	$64,400	Bachelor's (various)
Greenhouse gas emissions report verifier	$55,200	Bachelor's (Science)
Water/wastewater quality consultant	$75,440	Bachelor's (Science)
Water resource consultant	$75,440	Bachelor's (Science)
Senior environmental consultant	$78,660	Bachelor's (Science)
Waste reduction, energy efficiency, and expert consultant	$64,400	Bachelor's (Science)
Renewable-energy consultant	$75,440	Master's (various)
Conservation policy analyst and advocate	$41,400	Bachelor's (Science)
Climate change and energy policy specialist and advocate	$41,400	Master's (Various)
Water resources policy specialist and advocate	$41,400	Bachelor's (Science)

Source: EDF and Management Information Services, Inc. *Green Collar Jobs in the United States and Colorado: Economic Drivers for the 21st Century*, prepared for the American Solar Energy Society, Boulder, Colorado, January 2009.

Table 16
Emerging Green Jobs, Salaries, and Educational Requirements in Government

Occupational title	Average salary	Minimum education
Energy commission specialist	$70,380	Bachelor's (various)
Smart grid engineer	$91,080	Master's (EE)
Power system operator	$50,784	HSD/GED

(Continued)

Table 16 (*Continued*)

Occupational title	Average salary	Minimum education
Marine/fisheries biologist	$52,624	Bachelor's (Science)
Water resource engineer	$63,940	Bachelor's (Science)
Environmental research manager	$73,876	Master's (Science)
GIS specialist	$47,380	Bachelor's (Geography)
Engineering geologist	$62,836	Bachelor's (Engineer)
Urban planner	$55,384	Bachelor's (Urban Planning)
Urban renewal planner	$74,704	Master's (Urban Planning)
Conservation of resources commissioner	$90,344	Master's (various)
Power systems operator and instructor	$50,784	HSD/GED
Air quality specialist and enforcement officer	$61,916	Bachelor's (Science)
Air emissions permitting engineer	$64,676	Bachelor's (Science)
Chemist	$63,940	Bachelor's (Chemistry)
Economist	$74,060	Bachelor's (Economics)

Source: EDF and Management Information Services, Inc. *Green Collar Jobs in the United States and Colorado: Economic Drivers for the 21st Century*, prepared for the American Solar Energy Society, Boulder, Colorado, January 2009.

Table 17
Emerging Jobs, Salaries, and Educational Requirements
in the Carbon Management Industry

Occupational title	Average salary	Minimum education
Carbon capture power plant installation, operations, eng. & mgt.	$69,000	Bachelor's (Engineer)
Carbon sequestration plant installation, operations, eng. & mgt.	$69,000	Bachelor's (Engineer)
Geologist & hydrogeologist	$66,010	Bachelor's (Science)
GIS specialist	$47,380	Bachelor's (Geography)

(*Continued*)

Table 17 (*Continued*)

Occupational title	Average salary	Minimum education
Environmental health & safety engineering manager	$76,360	Bachelor's (Science)
Environmental health & safety lead	$81,420	Master's (Science)
Plant technical specialist—safety instrument testing & repair	$64,400	Bachelor's (various)
Safety investigator—cause analyst	$88,320	Bachelor's (various)
Plant supervising technical operator	$52,624	Bachelor's (Engineer)
Plant safety engineer	$90,620	Bachelor's (various)
Soil conservation technician	$36,984	Bachelor's (Science)
Soil conservationist	$51,704	Bachelor's (Science)
Forestry conservation worker	$34,132	HSD/GED
Conservation forestry consultant	$74,796	Associate's
Forestry supervisor	$41,584	Bachelor's (Forestry)
Restoration planner	$72,220	Master's (Science)
Power marketing specialist	$63,480	Bachelor's (various)
Energy trading specialist	$63,480	Bachelor's (various)
Carbon emission specialist	$63,480	Bachelor's (various)
Market & rate analyst	$72,680	Bachelor's (Business)
Economist	$74,060	Bachelor's (Economics)
Emissions reduction credit marketer & market analyst	$72,680	Bachelor's (Business)
Emissions reduction credit portfolio manager	$46,460	Bachelor's (Business)
Emissions reduction project developer specialist	$63,480	Bachelor's (various)
Emissions reduction project manager	$78,200	Bachelor's (various)
Environmental sampling technician	$35,144	HSD/GED
Climatologist	$69,000	Bachelor's (Science)
Environmental scientist	$69,000	Bachelor's (Science)
Environmental engineer/scientist intern	$6,440	HSD/GED

Source: EDF and Management Information Services, Inc. *Green Collar Jobs in the United States and Colorado: Economic Drivers for the 21st Century*, prepared for the American Solar Energy Society, Boulder, Colorado, January 2009.

Conclusions

Findings and Implications

Several key findings and implications can be extracted from the presented information. In particular, the research summarized here is path-breaking in several respects. For one, this is the first time that RE&EE industries have been rigorously specified and actual, comparable sales and employment data derived for two years—2006 and 2007. Second, the RE&EE industries have been disaggregated in detail by technology, sector, sub-industry, and jobs—total jobs and jobs by occupation and skill.

Finally, these data have been forecast to 2030 on the basis of different scenarios relating to alternative government policies and incentives.

Major findings of the presented research are that RE&EE are *already* major US economic drivers, creating more than nine million jobs, and that RE&EE creates many jobs at all levels of skill, training, and salaries. Most jobs created are for occupations such as accountants, welders, factory workers, truck drivers, and computer analysts—not "solar energy specialists" or "energy-efficiency engineers." Per dollar of expenditure, RE&EE can create relatively large numbers of jobs compared to other initiatives. Many RE&EE programs are "shovel ready" and can be implemented expeditiously, for example, weatherization.

This work represents a major contribution in demonstrating how important RE&EE is to the US economy and labor market, and provides the industry specifications and benchmarks that can be used in all related studies conducted subsequently. The research disaggregates the RE&RE industries into their main components such as wind, PV, biofuels, fuel cells, recycling/ remanufacturing, construction, electronics, and vehicles. Revenues, jobs, and occupational and skill requirements are estimated for each component and forecast to 2030.

An important finding derived here is that the longer the United States delays the implementation of ambitious RE&EE programs and incentives, the more difficult it will be to achieve the goals outlined here for 2030. Time is of the essence: All RE&EE programs and initiatives take years to be implemented and ramped up, and in all RE&EE sectors, the largest gains are made in the years immediately preceding the target year of 2030, with the single largest gains in 2029 and 2030. It is the large gains in these last years that are lost by starting the advanced program in, for example, 2015 instead of 2014, and it will be very difficult to recoup these gains. The longer the United States delays this implementation, the more difficult it will be to achieve the RE&EE goals outlined here for 2030.

Research in this field aims to assist education professionals to create programs that will facilitate the emerging RE&EE industry. A major finding that emerges is that education and training programs will have to be developed and expanded in the near future to facilitate the anticipated growth of RE&EE. However, most of these programs should, in the immediate future at least, focus on the EE sector simply because there are nearly twenty times as many EE jobs in United

States as RE jobs. RE jobs will increase in the future in percentage terms, but the overwhelming number of jobs created over the next two decades in the RE&EE sector will be related to EE.

Of course, many of the RE&EE jobs, skills, and education and training requirements overlap. For example, the largest number of jobs that will be created in the RE&EE sector are related to energy-efficient construction and green buildings, but green buildings contain important elements of both RE&EE. Another example of RE&EE overlap is the rapidly growing market for "Green IT"—which involves the use of computer resources in an energy and environmentally efficient way. Green criteria include energy efficiency, using low-emission building materials, recycling, using RE technologies, and other green strategies. It is thus clear that the jobs, skills, and education and training requirements required for green buildings, green IT, and other sectors and technologies contain important elements of both RE and EE.

Challenges and Opportunities

The challenge is to identify the types of jobs, skills, and education and training requirements corresponding to the employment opportunities that will be created by RE&EE in the coming decades. We provide here information that can assist US labor market and education planners in identifying these opportunities. We have identified over 160 detailed RE&EE occupational specialties and corresponding salaries and education and training requirements. These illustrate that EE&RE currently, and will increasingly in the future, create numerous job opportunities for workers in many different sectors at all education and skill levels with a wide range of salaries.

RE&EE can create new jobs in the United States, and these industries generate skilled, well-paying jobs, many of which are not subject to foreign outsourcing. These jobs consist of both college-educated professional workers, many with advanced degrees, and highly skilled, technical workers, with advanced training and technical expertise, many of them in the manufacturing sector. The opportunity is that US education and training programs can be calibrated to address these emerging new energy economy jobs. This would prepare millions of prospective employees for new jobs and viable long-term career opportunities in rapidly expanding RE&EE fields. RE&EE thus generates jobs that are disproportionately for highly skilled, well-paid technical and professional workers, who provide the foundation for entrepreneurship and economic growth. These are the high-skilled, high-wage technical and professional jobs that all states and regions seek to attract.

Notes

1. Energy smart design facilitates the efficient use of energy resources through intelligent building design and the utilization of renewable-energy and energy-efficient building components and systems.

2. For example, several years ago MISI conducted an audit of the mandated RE & EE programs in New Jersey for the New Jersey Board of Public Utilities. We found that some utilities in the state were classifying natural gas fuel cells as "renewable."
3. www.census.gov/EPCD/www/NAICS.HTML.
4. Some RE applications contribute to both RE and EE. For example, in this study, daylighting is implicitly included in the energy-efficient construction sector, and plug-in electric vehicles are a component of the energy-efficient vehicles sector.
5. For example, wind power is among the most rapidly growing sources of electrical power in the world."
6. *Fortune*, April 30, 2007.
7. American Wind Energy Association, "Wind Energy Production Tax Credit," (September 2008); Navigant Consulting, "Economic Impacts of Tax Credit Expiration," (February 2008); University of Colorado, Denver, School of Public Affairs, "Presidential Climate Action Plan," (December 2007).
8. Bezdek, R., Hirsch,R., and Wendling, R., *The Impending World Energy Mess* (Toronto, Canada: Apogee Prime Press, 2010).
9. Bezdek, R., Wendling, R., and DiPerna, P., "Environmental Protection, the Economy, and Jobs: National and Regional Analyses," *Journal of Environmental Management* 86, no. 1 (January 2008): 63–79; Bezdek, R., and Wendling, R., "Jobs Creation and Environmental Protection," *Nature* 434, no. 7033 (March 2005): 678; Roger Bezdek, R., and Wendling, R., "Potential Long-Term Impacts of Changes in U.S. Vehicle Fuel Efficiency Standards," *Energy Policy* 33, no. 3 (February 2005): 407–19.
10. Erhardt-Martinez, K., and Laitner, J. A., *The Size of the US Energy Efficiency Market: Generating a More Complete Picture*, American Council for an Energy-Efficient Economy, Report E083, (May 2008).
11. Bezdek, R. H., "The US Renewable Energy and Energy Efficiency Industries: What Role Can They Play in the Climate Crisis? Will they be Part of a New American Industrial Revolution?" Al Gore Climate Summit (Nashville, Tennessee, 2008).
12. *2000 Standard Occupational Classification Code*, US Department of Labor, Bureau of Labor Statistics.
13. We derived these data primarily from the Environmental Defense Fund's September 2008 report, *Green Jobs Guidebook: Employment Opportunities in the New Clean Economy*. We revised and adjusted these estimates to account for gaps in the information to relate specifically to the industries, technologies, and sectors of interest in this chapter. A variety of other necessary changes were made to make this information useful and relevant here.
14. Additional tables on emerging jobs can be found in: EDF and Management Information Services, Inc. *Green Collar Jobs in the United States and Colorado: Economic Drivers for the 21st Century*, prepared for the American Solar Energy Society, Boulder, Colorado, January 2009.

5

The Link between Energy Efficiency, Useful Work, and a Robust Economy

John. A. "Skip" Laitner

In 1900, the United States was an emerging economic powerhouse with a population of just seventy-six million people—less than one-fourth of the population that resides here today. The size of the economy, measured by its gross domestic product (GDP), was just 3 percent of the size of today's number-one-ranked economy in the world. And in 1900, the United States used about 9.6 quadrillion British thermal units (BTUs) of purchased energy, including coal, natural gas, oil, and wood. If the many different forms of energy were converted to a corresponding amount of petroleum, the energy needed to power our economy in 1900 was the equivalent of about 4.5 million barrels of oil per day. That is about one-tenth of today's current energy use. Perhaps most surprisingly, in 1900, the United States wasted more than 97 percent of those various energy resources in maintaining its economy.

The good news is that the United States improved the overall level of energy efficiency by an average of 1.58 percent per year over the period 1900 through 2010—even as the larger productivity of the nation's economy grew by about 2 percent annually over the same time horizon. Indeed, as discussed later in this chapter, the data suggest that the growth in energy efficiency has enabled much of the country's larger economic productivity and growth in personal income. Also evident is that the United States will need to look for ways to maintain, if not accelerate, these efficiency trends, if it is to maintain a robust and prosperous economy.

For all of those many annual improvements since 1900, however, the overall energy efficiency of the economy remains, surprisingly, less than 14 percent.[1] In short, about 86 percent of the energy that is consumed producing the nation's goods and services continues to be wasted. In some ways, this is not significantly better than the conditions existing more than a century ago. Admittedly, not all of that energy can be recovered and put to more productive uses in a cost-effective manner. Yet a large number of studies suggest that we can do much better if we make better choices and smarter investments. Recent assessments by the American Council for an Energy-Efficient Economy[2] and the University of Toronto's L. Danny Harvey[3], among others, suggest that we have the technology and wherewithal to improve the nation's energy efficiency. On the basis of pure economics, we can perhaps quadruple our current level of efficiency by the year 2050—should those prospects be actively developed.[4] And the more complete set of energy-efficiency opportunities are likely to be enabled by what Jeremy Rifkin calls a transition to the Third Industrial Revolution.[5] On the other hand, if we continue along the business-as-usual path, the evidence suggests that the robustness of the economy may be substantially weakened.

This chapter advances three critical points. The first is that the level of energy efficiency has been a primary driver of economic productivity, especially since the advance of the Industrial Revolution that began in the mid-eighteenth century. Second, if a robust economy is to be maintained over the next decades, we are required to simultaneously decrease the costs of energy services while also enabling a paradoxical increase in "useful energy" consumption. The key to resolving this paradox is reducing the waste associated with our larger demand for raw energy. Finally, there remain significant opportunities to improve system energy efficiencies in ways that maintain the robustness of future economic activity.

Energy Efficiency and Economic Productivity

Economists and policy analysts formulate many of their insights based on data collected by the Energy Information Administration (EIA). As it turns out, the energy data collected by the EIA provide us with only part of the story regarding how energy moves the economy forward. The EIA routinely gathers annual data on the physical units as tons of coal, cords of wood, gallons of gasoline, therms of natural gas, or kilowatt-hours (kWh) of electricity. All these different energy forms have an equivalent heat value that allows us to compare a gallon of gasoline with, say, one kWh of electricity. In the majority of other countries of the world, these energy forms are converted using a standard energy unit called the joule. In the United States, the British thermal unit (BTU) is the heat equivalent of the joule. One BTU is roughly the amount of heat given off by the burning of a wooden kitchen match.[6] There are about 124,238 BTUs of heat equivalent in a gallon of gasoline—or the heat energy that might be provided by the burning of 124,238 wooden kitchen matches. Similarly, there are approximately 3,412 BTUs for every kWh of electricity delivered to a home or office building. And comparing electricity and gasoline, we might say that a gallon of gasoline is the heat equivalent of 36.4 kWh delivered to the end user. We will return to this discussion in more detail later in the chapter, but if we include the energy that is wasted in the generation, transmission, and distribution of that electricity, we will find that the production of electricity required an average of 10,697 BTUs per kWh. In other words, the 3,412 BTUs bundled in a single kWh available for use in the home is only 31.9 percent of the total energy needed to create and distribute that electricity. In this case, one might say that a gallon of gasoline is the heat equivalent of 11.6 kWh when compared to the energy that is necessary at the generation source.[7]

In the various methods of tracking energy consumption, energy is measured as heat. But the real question surrounds the ability of energy to do work. In simple words, work might be defined as lifting a weight against the force of gravity or overcoming sliding friction. From an economic perspective, work is defined as the energy that is actually used to transform matter into goods and services. For all practical purposes, useful work can be divided into three primary categories. The first category consists of what we might call "muscle work" or work that might be carried out by people and animals. In 1900, for example, there were an estimated forty-three million horses, mules, oxen, and other (non-milk) cattle on farms.[8] A large number of these were draft animals providing work, while another share of animals supported transportation. This did not include urban and work animals, which also provided substantial labor and transportation services that in turn bolstered economic value. Today, very few animals provide work or transportation services, but the number of working people has grown, and the latter now provide an estimated 250 billion hours of work each year, compared to less than one-third that amount in 1900 (see Chapter appendix).

The second category of useful work is mechanical and electric power. This work is done by a variety of steam and gas turbines, gasoline and diesel engines, and electric generators. Meanwhile, the last category is some combination of low- or high-temperature heat that is delivered to the point of actual use.[9]

As the EIA now tracks the data for many different amounts of purchased energy,[10,11,12] preliminary estimates for the year 2010 suggest the United States consumed about 98 quadrillion BTUs, or quads, of total energy in that year. One quad is roughly 8 billion gallons of gasoline, which is, at current levels of energy efficiency, sufficient fuel to run some 15.4 million cars for one full year of typical driving. It is also enough energy to provide the full energy needs for about 5.2 million households in a given year. Moreover, it is enough energy to power $135.4 billion of annual economic activity within the United States. But this is only part of the story.

Ayres and Warr[13] have documented a more detailed accounting of the total energy that is actually used for heat, mechanical power, electricity, light, and muscle power. While the EIA and other data (summarized in the appendix to this chapter), for example, suggest that economic activity in the United States

Figure 1
Useful Work (Energy) versus US Gross Domestic Product (GDP)

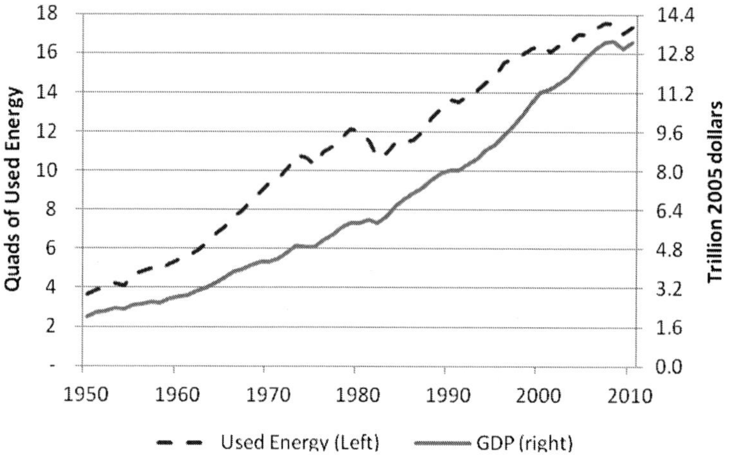

Source: Author calculations based on EIA data (2011a) and Ayres and Warr (2009), as updated from 2005 to 2010

required a total of 98 quads (rounded) in 2010, the Ayres–Warr data—with their more complete accounting of actual energy use needed to power the full economy—indicate that the United States required something more in the range of 124 quads (also rounded) of total energy use. More critically, only 17 quads of that total energy consumption were actually used and useful in the production of the typical basket of consumer goods and services. As Figure 1 above highlights, it is the raw energy converted to useful work that drives economic activity, as typically measured by the nation's GDP (again with relevant data shown in the Chapter appendix). And as Ayres and Warr further clarify, the reason the raw energy "inputs do not explain economic growth is that their inefficient conversion leads to a large fraction of waste heat (and other wastes, like ash) that does not contribute to the economy but actually creates health problems and costs of its disposal."[14]

As a complement to Figure 1, Figure 2 draws on the raw data contained in the Chapter appendix to highlight the role of useful energy over the period 1950 through 2010. This is done from two related perspectives: (a) useful energy per capita within the United States, and (b) useful energy per average hour of work activity. A quick observation shows a highly similar pattern for lines in Figure 2, namely the rather steep increase in useful energy per year during the 1950 to 1980 time period. By approximately 1980, however, the amount of useful energy flattens out on both per capita and per work hour basis. The immediate

Figure 2
Useful Work Energy Per Person and Work Hour (Index 1950 = 100)

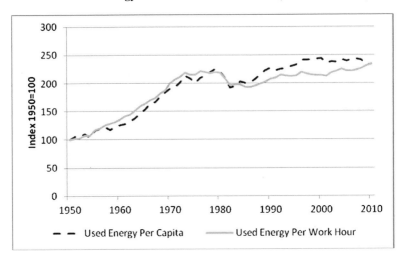

Source: Author calculations based on EIA data (2011a) and Ayres and Warr (2009), as updated from 2005 to 2010

suspicion is that the flattening demand for energy per capita must be affecting other aspects of the economy as well. And this is further borne out in Figure 3 and Table 1 below.

Figure 3 highlights the very close and significant impact of useful energy—that is, the actual work that is being done compared to the total energy being consumed. It is useful energy, not wasted energy, that drives productivity gains within the larger economy. This makes sense when we think of economic productivity as increasing the output per person, or output per hour of labor, by tapping into greater levels of cost-effective energy services. It actually does take energy to make the economy go and expand. Yet, increasing the efficiency with which we use that total energy will drive down overall costs so that we can more easily afford to integrate even more useful energy to continue economic momentum. Again, Figure 3 highlights that very powerful connection.

Table 1 further highlights the connection between useful energy and productivity, but in a slightly different way. In this case, we provide data that compare the growth rates for population (column A); improvements in the rate of conversion of total energy into useful energy or work (column B); growth in economy-wide productivity (column C); and finally, the overall growth in the nation's GDP (column D).[15] In this case, however, we are looking at two periods from 1950 to 1980 (data row 1), and from 1980 through preliminary estimates for 2010 (data row 2).

Figure 3
Useful Energy Per Person (Index 1950 = 100)

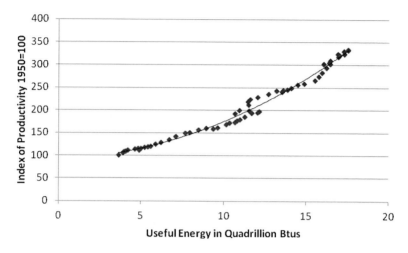

Source: Author calculations based on EIA data (2011a) and Ayres and Warr (2009), as updated from 2005 to 2010

Table 1
Key Historical Growth Rates in Energy and Economic Productivity

Compound average growth rate	(A) Increase in US population (%)	(B) Useful energy efficiency improvement (%)	(C) Annual rate of productivity improvement (%)	(D) Annual GDP measured in trillion 2005 $ (%)
(1) From 1950 to 1980	1.34	1.45	2.25	3.63
(2) From 1980 to 2010	1.03	0.42	1.72	2.77

Note: Derived from data in Appendix A with data from 1900 taken from both Ayres and Warr (2009) and EIA (2011a)

Of immediate interest is the link between the improvement in the conversion of energy into useful energy (work) reported in column B as it relates to economic productivity in column C, and finally, to the growth in the nation's economy (column D). The bad news is that the earlier period from 1950 to 1980 showed a larger improvement in columns B, C, and D. In the last thirty-years, covering the years 1980 to 2010, however, the entire economic process shows a weakening trend—especially as the economy-wide improvement in productivity fell from 2.25 to 1.72 percent. The reason for this weakening appears to be the result of a slowing rate of improvement in converting total "raw energy" into actual "useful energy" so that we can cost-effectively and simultaneously increase the magnitude of useful energy (or work expressed as quads of energy). The lesser rate of improvement, in turn, weakens the nation's larger economic productivity, and therefore, its total GDP.[16] Presumably, as the rate of efficiency conversion increases, the quantity of total energy required to power the economy will decline—even as the amount of useful energy continues to grow.

In effect, data in Figures 1 through 3, as well as in Table 1, all reveal two important aspects of US economic activity. First, when one properly measures the links between useful energy and economic output, it becomes clear that the overall level of energy efficiency in the United States has been stagnating at a rather anemic 13–14 percent for the past twenty years or so (notably over the period 1990 through 2005 with a preliminary update by the author through 2010). As suggested in the third data row of Table 1, the stagnating rate of converting total energy into useful energy or work does appear to constrain economic productivity. Second, if we merely double our current level of efficiency, we may continue to see a weaker economy over the next several decades. Indeed, the evidence points to a need to at least triple or quadruple the productive conversion of raw energy, so that a greater magnitude of cost-effective "useful energy" is available to power a more robust economy.

So, with this backdrop now in place, one might ask the question, how much of a difference can this make? For example, instead of an economy that expands at an annual rate of 2.6 percent as most recently suggested by data from the EIA[17] between now and 2040, what if the economy grew at a slightly less impressive rate? What if it grew at, say, 2.3 percent annually?[18] Following the data reported here, the reason for the lagging growth is that without a greater stimulus of energy efficiency to drive down the full cost of energy services, and in turn, drive up the larger productivity of the US economy, the nation's use of capital goods and labor will be insufficient to generate as many goods and services as might be possible with some other higher level of productivity.

At this point in the research, it is admittedly hard to know with confidence what the precise impact of lagging energy productivity might imply for the growth of the US economy. But one can gain further insights by looking at the data in Table 1 and examining the period 1980 through 2010. On the basis of this data, we can estimate the productivity elasticity as a function of improvement in the efficiency conversion captured in column B during the last thirty years. We can then compare this to the larger productivity growth of the economy found in column C in order to derive a working elasticity of 0.74.[19]

As it turns out, the most recent Department of Energy *Annual Energy Outlook* forecast for 2012[20] suggests a rate of improvement that would increase the nation's overall energy efficiency by a factor of about 1.8 times compared to the 2010 level of performance. On the basis of the historical evidence suggested in the appendix, however, a business-as-usual (BAU) energy-efficiency improvement of 1.8 times implies an overall economic productivity improvement of about 1.54 times that of today's economy.[21] With the population expected to increase about 32 percent over the next thirty years, the total GDP will expand at a rate of about 2.3 percent annually between now and 2040. As shown in the third data row of Table 2, this means a $2.1 trillion smaller economy than is now projected. The smaller economy might support 4.2 million fewer jobs. With a smaller economy, energy consumption would be just under 109 quads.

On the other hand, pushing the right mix of investments, infrastructure improvements and systems efficiencies, as suggested in the recent study by Laitner et al.,[22] could easily nudge the nation's energy efficiency to about 2.5 times of today's level.[23] In this case, the GDP (as shown in data row 4) would expand to as much as 3.2 percent per year, boosting economic activity to almost $5 trillion more than the standard reference case assumptions now projected for the year 2040. In this case, employment would expand by nearly 8.7 million more jobs, with a total difference of nearly thirteen million total jobs between the low- (row 3) and high-efficiency (row 4) scenarios. Even more interestingly, the total energy use could remain essentially flat over the next thirty years—even as the development of a more robust economy is ensured.

Table 2

Key Impacts of Energy-Efficiency and Economic-Productivity Scenarios

Scenario	(A) GDP Billion 2005 $	(B) Primary energy quads	(C) Jobs per $million of GDP	(D) Total jobs (Millions)	(E) Net change in jobs (Millions)
(1) Reference Case	13,221	97.77	10.52	139.1	n/a
(2) Year 2040 (EIA 2012b)	29,030	117.5	6.13	178.0	0
(3) Year 2040 (BAU Efficiency)	26,900	108.9	6.5	173.8	−4.2
(4) Year 2040 (High Efficiency)	33,851	100.4	5.5	186.7	8.7

How big is the Energy-Efficiency Resource?

Economist William Baumol and his colleagues (1989)[24] once wrote, "for real economic miracles one must look to productivity growth." As suggested by this chapter, by the large number of studies published by the American Council for an Energy-Efficient Economy,[25,26] Rachel Gold et al.,[27,28] and by many other recent studies,[29,30,31,32,33] if we prime "the productivity pump" with enhanced or expanded energy-efficiency provisions, we are likely to show a small but net positive impact for the economy. More critically, if we fail to accelerate the rate of efficiency gains compared to the last thirty years or so, we may risk a significantly less robust economy in the years ahead. Ensuring a more prosperous and economically sustainable future requires new thinking—in effect, a transition to what Rifkin called the Third Industrial Revolution.[34]

As illustrated throughout this chapter, past energy-efficiency investments have been a critical resource in promoting ongoing economic productivity throughout all levels of the economy. Further evidence from Ayres and Warr,[35] summarized in the data table found in Appendix A, suggests that a robust economy requires an even greater level of available and cost-effective improvements in energy efficiency on both the supply side (e.g., more energy-efficient combined heat and power technologies[36]) and the demand side (e.g., more energy-efficient lighting, consumer appliances, industrial processes, and transportation vehicles). Yet what is required is not just greater levels of energy efficiency, but the capacity for efficiency improvements to deliver useful energy services—in effect, the weighted share of both conventional energy resources and a mix of economy-wide energy-efficiency improvements—at a reduced cost compared to today. In short, energy efficiency is needed precisely because it remains the largest opportunity to reduce the total cost of energy services.[37] And it is the lower cost of total energy services that primes the productivity pump.

But some might ask the question, "Just how big is energy efficiency?" The surprising news, perhaps, is that the opportunity for gains in energy efficiency (energy productivity) is larger than one would imagine, albeit perhaps harder to achieve than initially expected. Among the more credible estimates is a study published by the National Renewable Energy Laboratory,[38] which suggested that if all commercial buildings were rebuilt by applying a comprehensive package of energy-efficiency technologies and practices, these structures could reduce their typical energy use by 60 percent. Adding the widespread installation of rooftop photovoltaic power systems could lead to an average 88 percent reduction in the use of conventional energy resources. Even more intriguingly, many buildings in the United States could actually produce more energy than they consume—in effect, transforming the nation's building stock into power plants.

The current electricity generation and transmission system in the United States now operates at an efficiency of about 32 percent, a level of performance that is essentially unchanged since 1960. What the United States wastes in the production of electricity today is more than what Japan uses to power its entire economy, according to calculations made by the author using various data from the EIA.[39,40] At the same time, a study published by the Lawrence Berkeley National Laboratory[41] suggests that a variety of waste-to-energy and recycled energy systems could pull enough waste heat from US industrial facilities and buildings to meet 20 percent of the nation's current electricity consumption. And that is only the beginning of potentially large efficiency gains in power generation. So combining even a 50 percent efficiency gain in US buildings with a minimum 25 percent productivity improvement in power production provides a total 60 percent efficiency gain (author calculations).

The good news is that larger efficiency gains are equally possible in the industrial and transportation sectors—if we are willing to thoroughly examine the possibilities. As but one example within the industry, the Massachusetts Institute of Technology (MIT) research scientist Daniel Cohn[42] suggests that new plasma gasification technologies could provide up to 40 billion gallons of liquid fuels from municipal and industrial wastes. That is about one-quarter of current US gasoline consumption. Meanwhile, in September 2009, Volkswagen introduced a sleek new two-passenger prototype car that achieved a phenomenal 240 miles per gallon (mpg).[43,44] But even if a typical car in 2040 achieves only a 50 mpg rating, then so long as new incentives exist to reduce driving by 20 percent while increasing the typical passenger load from 1.6 to 2 persons per car, fuel consumption stands to decrease by 72 percent (author calculations).

Moving beyond component- or device-efficiency improvements, there are significant system efficiencies that may contribute to future solutions as well. One recent study[45] completed for the Urban Land Institute identified a package of some fifty programs and policies that could reduce transportation-related greenhouse gas emissions 24 percent by 2050, through strategies to change travel behavior and land-use patterns. The emissions reduction hit 47 percent by adding road pricing techniques, ranging from pay-as-you-go insurance to charging Americans for every mile driven.[46] Adding improved fuel performance standards beyond what might occur through these behavioral and system efficiencies would further enhance the projected savings. Information-based transportation systems, meanwhile, might further add to that fuel savings. A 2007 study of the 272,000 traffic lights and signaling systems in the United States underscored significant benefits in the quality of life and environmental protection as a result of improved traffic patterns, which reduced unnecessary idling and traffic congestion. For typical households, improved traffic signal timing might save more than 100 hours per year in avoided car time. More to the point, however, reductions in fuel consumption of up to 10 percent might also be achieved.[47]

Why is there a need for greater efficiencies? The answer to this question bears repeating here. Larger gains in energy-efficiency improvements are required because they reduce the cost of energy services, which allows the productive use of all resources. So without greater efficiency, and the concomitant reduction in the cost of energy services (again, on both the end-use and supply sides), the economy may actually become much less robust. And, the economy may also become a bit smaller. The implication of this outcome is significant: If the economy contracts by even only tenths of a percent (which, as highlighted in footnote 5, might translate to $2.2 trillion or more), the United States will have fewer economic resources available to respond to phenomena such as climate change, or to engage needed improvements to the nation's infrastructure.

To explore this issue in greater detail, we can ask, what might it mean to have an economy that, yes, is significantly larger than the economy today, but one that grows less robustly than in the recent historical past? One may again turn to Table 2 for a useful example. If we look at Table 2 and compare the GDP column in data row 3 with comparable GDP data in row 4, we would find an economy that is $7 trillion smaller by the year 2040. A smaller economy, by definition, implies the availability of fewer financial resources to help us solve the many social problems that are likely to figure in the year 2040. By some estimates, about 18 percent of GDP in 2040 might be directed toward new investments in the US economy. And an estimated 12 percent of this GDP might be used to fund the many federal, state, and local government services on which the US population depends. Hence, an economy that is $7 trillion smaller means, on average, that there might be roughly $1.3 trillion less investment available to update and build new infrastructure and provide for new schools, hospitals, and factories. Similarly, an economy that is $7 trillion smaller also means that the United States might face approximately $800 billion less in government revenues, which could be used to retire the national debt, to maintain national security, to promote future research and development, and to maintain the health care system. So, this problem is not just an energy problem; in fact, its implications run deeper than many policymakers and business leaders may realize.

Mark Jacobson and Mark Delucchi[48] observe the possibility of efficiency, wind, water, and solar technologies providing 100 percent of the world's energy, eliminating all fossil fuels by 2030. They acknowledge that the scale of change is large, but they note that "society has achieved massive transformations before." They cite the World War II transition when "the US retooled automobile factories to produce 300,000 aircraft, and other countries produced 486,000 more." In other words, society has the technical capacity to move in this direction. Danny Harvey[49] notes that the opportunity may not be related to limitations of technology; rather, he suggests, it may be more about the lack of a trained, motivated, and properly equipped professional and construction workforce. John Laitner[50] observes that rather than practical limits as obstacles to further efficiency gains, the real limits of concern might be a lack of sufficient public policy to encourage further innovations. Although energy efficiency has been the workhorse of the US economy for many years, the economic potential of this resource has hardly been exhausted, and it stands poised to deliver an even greater contribution—if choices are made to develop this potential. And this is the huge task ahead.

Appendix A
Table of Key Population, Capital, and Energy Variables

Year	(A) Population Millions	(B) GDP Billion 2005 $	(C) Capital stock Billion 2005 $	(D) Labor hours Thousands	(E) Useful energy Quads	(F) EIA energy Quads	(G) Ayres-Warr Energy-quads	(H) Useful energy Efficiency
1900	76.4	431	2,009	41,624,200	0.4	9.6	14.8	2.5%
1950	152.3	2,006	7,187	122,010,000	3.6	34.6	45.6	8.0%
1951	154.9	2,161	7,480	128,936,000	3.9	37.0	48.3	8.1%
1952	157.6	2,244	7,774	130,225,000	4.0	36.7	48.2	8.3%
1953	160.2	2,347	8,106	131,695,000	4.2	37.7	49.3	8.6%
1954	163.0	2,332	8,403	127,837,000	4.1	36.6	48.3	8.4%
1955	165.9	2,500	8,759	132,056,000	4.6	40.2	52.3	8.9%
1956	168.9	2,550	9,077	134,253,000	4.8	41.8	54.2	8.9%
1957	172.0	2,601	9,387	133,038,000	5.0	41.8	54.4	9.2%
1958	174.9	2,578	9,632	128,388,000	4.9	41.6	54.4	9.0%
1959	177.8	2,763	9,970	133,073,000	5.2	43.5	56.5	9.3%
1960	180.7	2,831	10,288	134,030,000	5.4	45.1	58.4	9.3%
1961	183.7	2,897	10,600	133,119,000	5.6	45.7	59.3	9.5%
1962	186.5	3,072	10,974	136,464,000	5.9	47.8	61.7	9.5%
1963	189.2	3,207	11,383	137,552,000	6.2	49.6	63.9	9.8%
1964	191.9	3,392	11,846	141,126,000	6.7	51.8	66.4	10.1%
1965	194.3	3,610	12,373	145,977,000	7.1	54.0	68.9	10.3%
1966	196.6	3,845	12,933	151,085,000	7.7	57.0	72.3	10.6%
1967	198.7	3,943	13,452	152,752,000	7.9	58.9	74.4	10.6%
1968	200.7	4,133	14,004	155,517,000	8.5	62.4	78.3	10.8%

1969	202.7	4,262	14,552	159,301,000	8.9	65.6	81.9	10.9%
1970	205.1	4,270	15,016	156,510,000	9.4	67.8	84.4	11.1%
1971	207.7	4,413	15,514	155,821,000	9.6	69.3	86.1	11.2%
1972	209.9	4,648	16,095	160,121,000	10.1	72.7	90.0	11.3%
1973	211.9	4,917	16,745	165,239,000	10.8	75.7	93.4	11.6%
1974	213.9	4,890	17,268	165,842,000	10.7	74.0	91.7	11.6%
1975	216.0	4,880	17,675	161,091,000	10.3	72.0	89.8	11.5%
1976	218.0	5,141	18,167	165,755,000	11.0	76.0	94.3	11.6%
1977	220.2	5,378	18,762	171,589,000	11.3	78.0	96.7	11.7%
1978	222.6	5,678	19,452	179,664,000	11.7	80.0	99.0	11.8%
1979	225.1	5,855	20,151	184,524,000	12.1	80.9	100.2	12.1%
1980	227.2	5,839	20,691	184,008,000	12.0	78.1	97.7	12.3%
1981	229.5	5,987	21,215	184,394,000	11.5	76.2	95.9	12.0%
1982	231.7	5,871	21,619	181,648,000	10.7	73.2	92.8	11.5%
1983	233.8	6,136	22,135	184,908,000	10.9	73.0	92.9	11.8%
1984	235.8	6,577	22,862	194,236,000	11.5	76.7	97.0	11.8%
1985	237.9	6,849	23,651	198,678,000	11.5	76.5	97.0	11.8%
1986	240.1	7,087	24,452	201,005,000	11.6	76.8	97.6	11.9%
1987	242.3	7,313	25,218	206,453,000	12.1	79.2	100.3	12.0%
1988	244.5	7,614	25,974	212,615,000	12.7	82.8	104.5	12.2%
1989	246.8	7,886	26,703	218,492,000	13.2	84.9	107.0	12.3%
1990	249.6	8,034	27,369	218,868,000	13.6	84.7	107.0	12.7%
1991	253.0	8,015	27,853	215,771,000	13.5	84.6	107.2	12.6%
1992	256.5	8,287	28,370	215,945,000	13.9	86.0	109.0	12.7%

(Continued)

Appendix A (*Continued*)

Year	(A) Population Millions	(B) GDP Billion 2005 $	(C) Capital stock Billion 2005 $	(D) Labor hours Thousands	(E) Useful energy Quads	(F) EIA energy Quads	(G) Ayres-Warr Energy-quads	(H) Useful energy Efficiency
1993	259.9	8,523	28,993	221,000,000	14.1	87.6	111.1	12.7%
1994	263.1	8,871	29,686	227,916,000	14.5	89.3	113.2	12.8%
1995	266.3	9,094	30,448	233,531,000	14.9	91.2	115.5	12.9%
1996	269.4	9,434	31,320	236,446,000	15.5	94.2	119.0	13.1%
1997	272.6	9,854	32,189	243,372,000	15.8	94.8	120.1	13.1%
1998	275.9	10,284	33,271	248,610,000	16.0	95.2	120.9	13.2%
1999	279.0	10,780	34,478	253,474,000	16.2	96.8	123.0	13.2%
2000	282.2	11,226	35,752	256,852,000	16.5	99.0	125.6	13.1%
2001	285.0	11,347	36,875	253,714,000	16.1	96.3	123.4	13.0%
2002	287.7	11,553	37,907	250,412,000	16.4	97.9	125.2	13.1%
2003	290.2	11,841	38,952	249,114,000	16.5	98.1	126.1	13.1%
2004	292.9	12,264	40,051	251,911,000	16.9	100.3	128.8	13.2%
2005	295.6	12,638	41,139	255,818,000	16.9	100.4	128.5	13.2%
2006	298.4	12,976	42,310	260,502,000	17.3	99.8	126.9	13.6%
2007	301.3	13,254	43,365	262,272,000	17.5	101.5	129.0	13.6%
2008	304.1	13,312	44,151	259,843,000	17.5	99.4	126.2	13.9%
2009	306.7	12,987	44,515	246,990,376	17.0	94.7	120.2	14.2%
2010	309.7	13,240	44,882	249,147,239	17.3	97.9	124.2	13.9%

Notes: The data in columns A through E, G, and H are taken from Ayres and Warr (2009) with values transformed from 1990 (international development currency units called) Geary Khamis dollars and terajoules into 2005 U.S. dollars and quadrillion Btus for the economic and energy values, respectively. The original data are from 1900 to 2005 with preliminary updated values extended to 2010. For more details on this set of time series data, contact Skip Laitner at econskip@gmail.com.

Notes

1. Some may be surprised. The estimated efficiency in 1900 was only 2.48 percent. The calculation to estimate the current level of efficiency over the last 110 years at an average 1.58 percent rate of improvement is $2.48 \times 1.0158^{110} = 13.96$ percent efficiency in the year 2010.

2. Laitner, John A. "Skip", Steven Nadel, R. Neal Elliott, Harvey Sachs, and A. Siddiq Khan, *The Long-Term Energy Efficiency Potential: What the Evidence Suggests*, ACEEE Report E121 (Washington, DC: American Council for an Energy-Efficient Economy, 2012), http://www.aceee.org/research-report/e121.

3. Harvey, L.D. Danny, "Energy Demand Scenarios," In *Energy Efficiency and the Demand for Energy Services* (London: Earthscan Ltd., 2010).

4. Laitner et al., 2012.

5. Rifkin, Jeremy, *The Third Industrial Revolution: How Lateral Power Is Transforming Energy, the Economy, and the World* (New York: Palgrave MacMillan, 2011).

6. More formally, a British thermal unit is the amount of heat required to raise one pound of water by one degree Fahrenheit at a pressure of one atmosphere. There are approximately 1,055.056 Joules of heat in 1 BTU.

7. A quarter pound cheeseburger from McDonald's contains an estimated 510 food calories, which is equivalent to just over 2,000 BTUs. Expressed as an equivalent of electricity, the cheeseburger has an energy value of 0.593 kWh, about the same as the electricity required to light a 100-watt incandescent bulb for almost six hours.

8. Bureau of Statistics, *Statistical Abstract of the United States: 1900* (Washington, DC: US Department of the Treasury, 1901), 372

9. Ayres, Robert U., and Benjamin Warr, *The Economic Growth Engine: How Energy and Work Drive Material Prosperity* (Northampton, MA: Edward Elgar Publishing, Inc., 2009).

10. [EIA] Energy Information Administration, *Annual Energy Review 2010*. DOE/EIA-0384(2010) (Washington, DC: US Department of Energy, 2011a), http://www.eia.doe.gov/aer/.

11. [EIA] Energy Information Administration, *Short-Term Energy Outlook April 2012*. Washington, DC: US Department of Energy, 2012a), http://www.eia.doe.gov/steo/.

12. [EIA] Energy Information Administration, *Annual Energy Outlook 2012 Early Release*. DOE/EIA-0383er(2012) (Washington, DC: US Department of Energy, 2012b), http://www.eia.gov/forecasts/aeo/er/pdf/0383er(2012).pdf.

13. Ayres and Warr (2009).

14. Ayres and Warr (2009: 125).

15. Perhaps not immediately apparent, but the growth in GDP is a function of the product of the rate of population growth and the rate of growth in productivity; the latter can also be expressed as a growth in GDP per capita. So, for example, if the population growth in column A of Table 1 averages 1.03 percent in the period 1980 to 2010, and the productivity growth averages 1.71 percent, then $(1 + 0.0103)$ multiplied by $(1 + 0.0172)$ equals 1.0277, which results in a growth of 2.77 percent.

16. The US economy provided a GDP of $5,389 billion in 1980. It more than doubled in size, reaching a total of $13,248 billion by 2010 (with both values measured in constant 2005 dollars). Perhaps not immediately apparent, but had the US economy maintained a productivity improvement of 2.25 percent rather than 1.72 percent over the period 1980 to 2010, even with a smaller population grow rate, the GDP would have actually grown to an estimated $15,491 billion (also in constant dollars). In other words, a very small change in productivity would have meant an economy that was about $2.2 trillion larger than our actual record in 2010.

17. EIA (2012b).

18. This projection is drawn from the most recent forecast of Moody's Analytics (May 2012).
19. The compound average growth rate for efficiency improvement in column B in Table 1 over the thirty-year period 1980 through 2010 is 0.41 percent. The annual productivity improvement in the economy is 1.71 percent. The elasticity is found by taking the natural logs as follows: Ln(0.0171)/Ln(0.0041) = 0.74 (rounded).
20. EIA (2012b).
21. In rounded terms the calculation is $1.8^{0.74} = 1.54$. This implies an average annual productivity growth rate of about 1.46 percent compared to the historical productivity improvement rate of 1.72 percent in the preceding thirty years (see column C in row 2 of Table 1).
22. Laitner et al. (2012).
23. As noted in the data table for Appendix A, the current economy is about 13.9 percent energy-efficient. With appropriate investments in new technologies, systems, and infrastructure, the analysis here indicates that the level of efficiency might increase to about 35 percent by 2040. This increase is 2.5 times the current level.
24. Baumol, William J., Sue Anne Batey Blackman, and Edward N. Wolff, *Productivity and American Leadership: The Long View* (Cambridge, MA: The MIT Press, 1989).
25. Laitner, John A. "Skip," *The Positive Economics of Climate Change Policies: What the Historical Evidence Can Tell Us*, ACEEE Report E095 (Washington, DC: American Council for an Energy-Efficient Economy, 2009a).
26. Laitner, John A. "Skip," *Climate Change Policy as an Economic Redevelopment Opportunity: The Role of Productive Investments in Mitigating Greenhouse Gas Emissions*, ACEEE Report E098 (Washington, DC: American Council for an Energy-Efficient Economy, 2009b).
27. Gold, Rachel, Laura Furrey, Steven Nadel, John A. "Skip" Laitner, and R. Neal Elliot, *Energy Efficiency in the American Clean Energy and Security Act of 2009: Impacts of Current Provisions and Opportunities to Enhance the Legislation* (Washington, DC: American Council for an Energy-Efficient Economy, 2009).
28. Laitner et al. (2012)
29. [McKinsey], *Unlocking Energy Efficiency in the US Economy* (McKinsey & Company, 2009).
30. [AEF] Committee on America's Energy Future, *Real Prospects for Energy Efficiency in the United States* (Washington, DC: National Academies Press, 2009).
31. [APS] American Physical Society, *Energy Future: Think Efficiency* (Washington, DC: American Physical Society, 2008).
32. InterAcademy Council, *Lighting the Way: Toward a Sustainable Energy Future* (Amsterdam, The Netherlands: Royal Netherlands Academy of Arts and Sciences, 2007).
33. [Interlaboratory Working Group] Interlaboratory Working Group on Energy-Efficient and Low-Carbon Technologies, *Scenarios for a Clean Energy Future*, Prepared for the US Department of Energy, Office of Energy Efficiency and Renewable Energy (Oak Ridge, TN: Oak Ridge National Laboratory, 2000), http://www.ornl.gov/ORNL/Energy_Eff/CEF.htm.
34. Rifkin (2011).
35. Ayres and Warr (2009).
36. A building or an industrial plant typically will purchase electricity for its lighting and various motor systems, and it will also separately purchase natural gas for its various heating needs. As this chapter later illustrates, the electricity generation system is only 32 percent efficient, while many natural gas heating systems are 80 percent efficient. The weighted average of the two systems is about 55 percent.

On the other hand, onsite combined heat and power (CHP) systems use a single energy source to deliver both heat and power in ways that achieve system efficiencies of 70–90 percent or more.

37. In this regard, total costs include essentially three elements. The first is direct costs incurred when one purchases gasoline for cars or pays electricity bills. The second, indirect costs, may entail the need to install pollution prevention technologies and occupational safety systems in order to minimize health and environmental impacts. The third includes the actual costs of pollution because it affects our air and water and disrupts our personal health and the health of the larger environment. As but one example of how health costs can negatively impact the economy, a large number of studies have consistently shown that air pollution from the combustion of fossil fuels creates acute and chronic respiratory ailments (see, Muller, Mendelsohn, Nordhaus, 2011 and Greenstone and Looney, 2012). These, in turn, cause employees to lose time at work, which reduces the overall level of productivity within the economy.

    Muller, Nicholas Z., Robert Mendelsohn, and William Nordhaus, "Environmental Accounting for Pollution in the United States Economy," *American Economic Review* 101 (August 2011): 1649–75.

    Greenstone, Michael, and Adam Looney. "Paying Too Much for Energy? The True Costs of Our Energy Choices," *Dædalus* 141 (Spring 2012): 10–30.

38. Griffith, Brent, Nicholas Long, Paul Torcellini, Ron Judkoff, Drury Crawley, and John Ryan. *Assessment of the Technical Potential for Achieving Net Zero-Energy Buildings in the Commercial Sector* (Golden, CO: National Renewable Energy Laboratory, 2007).

39. EIA (2011a).

40. [EIA] Energy Information Administration, *International Energy Outlook*, DOE/EIA-0484(2011) (Washington, DC: US Department of Energy, 2011b), http://www.eia.gov/forecasts/ieo/.

41. Bailey, Owen, and Ernst Worrell, *Clean Energy Technologies: A Preliminary Inventory of the Potential for Electricity Generation* (Berkeley, CA: Lawrence Berkeley National Laboratory, 2005).

42. Cohn, Daniel R, Personal Communication and an ACEEE In-House Briefing, "Liquid Fuels from Municipal Waste Using Plasma Gasification." (Washington, DC, 2008).

43. Meiners, Jens, "Volkswagen L1 Concept: VW's one-liter car is finally on its way." *Car and Driver*, 2009, http://www.caranddriver.com/news/volkswagen-l1-concept-auto-shows.

44. As explained in the news article, essentially three ingredients are necessary to achieve this kind of improvement in fuel economy: a highly efficient powertrain, enhanced aerodynamics, and lightweight engineering.

45. Cambridge Systematics, *Moving Cooler: An Analysis of Transportation Strategies for Reducing Greenhouse Gas Emissions*, Prepared for the Moving Cooler Steering Committee (Washington, DC: Urban Land Institute, 2009)

46. Cambridge Systematics (2009)

47. [NTOC] National Transportation Operations Commission, *National Traffic Signal Report Card: Technical Report* (Washington, DC: Institute of Transportation Engineers, 2007).

48. Jacobson, Mark Z., and Mark Delucchi. "A Path to Sustainable Energy by 2030," *Scientific American* (November 2009): 58–65.

49. Harvey (2009).

50. Laitner, John A. "Skip", "How Far Energy Efficiency?" In *Proceedings of the 2004 ACEEE Summer Study on Energy Efficiency in Buildings* (Washington, DC: American Council for an Energy-Efficient Economy, 2004.

6

A Green Energy Manufacturing
Stimulus Strategy

Jon Rynn

Manufacturing and the Environment Need Each Other

At the most basic level, both manufacturing and environment exist in order to create. Manufacturing creates goods; the environment creates life. The long-term challenge is to create goods while maintaining creation of life. We need to create goods that are necessary to build a sustainable society. If we do not rebuild our man-made environment, the prospect for the future of our global civilization is not good, as several authors have documented.[1]

How would we rebuild the society? It is easier to look at the problem by dividing "society" into a set of sectors: transportation, energy, manufacturing, urban structure, buildings, and agriculture. Each sector of the economy relies on manufactured goods, and each sector requires a distinct set of manufactured goods to become environmentally sustainable. Let's look at each sector in turn.

Sustainable Sectors

First, dependence of transportation on oil is now 94 percent for the cars, trucks, trains, planes, and ships that are used as its main sources of machinery; even roads need oil in the form of asphalt.[2] Since oil will not last forever, is very polluting, and emits greenhouse gases, we need to switch to the use of renewable electricity for running this huge vehicle fleet, which will involve an increased production of electric trains, electric cars, smaller electric trucks, and ships, as well as planes kept aloft with sustainably produced biofuels.

Second, energy production is now dominated by fossil fuels, that is, petroleum for transportation, coal for electricity, and natural gas for electricity, heating, and cooking. Eventually, we will have to shift almost completely to renewable technologies such as wind, solar, and geothermal if we are to avoid disruptions caused by declining supply, climate change, and collapsing ecosystems. Instead of drilling rigs, mining equipment, refineries, and pipelines, we will need wind turbines, solar panels, and ground source heat pumps, along with a rebuilt national electric grid that carries electricity throughout the continent.

Third, manufacturing, service, and household economies use newly mined materials, directly or indirectly, as inputs and discard products after use and/or pollute in the process of production. Mining and trash disposal will have to be replaced by recycling systems and reusable design; polluting methods of manufacture will have to be replaced by clean methods.

Fourth, the bulk of any future building will need to be in the direction of denser town and city centers, and away from the low density of sprawl. The denser (i.e., the closer together and taller) the residential and commercial buildings, the easier it will be to adopt an electric transportation system, to economize on energy use (because large buildings are more energy efficient), and to restore natural ecosystems (because much less space will be needed).

Fifth, new buildings as well as most older structures will need to be retrofitted to be energy-efficient if not energy self-sufficient. This will require much of the equipment mentioned for energy, such as solar panels, but it will also require new materials and insulation.

Figure 1
**Creating a Sustainable World Will Require Transforming Many
Sectors of the Economy from Spatially, Energetically, and Materially
Wasteful Sectors into Efficient, Clean Ones.**

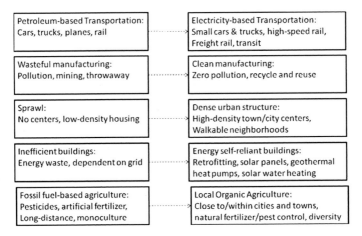

Sixth, provision of food will need to change from a long-distance, industrial model to a more local, organic, labor-intensive food system. Intensive agriculture that does not use pesticides or artificial fertilizers but requires large amounts of high-skill labor will require a different set of tools and infrastructure for a new set of farmers.

An economy that is producing machinery and goods for an electric transportation system, a renewable energy sector, a dense urban environment, a recycled-materials-based manufacturing sector, energy-efficient buildings, and sustainable agriculture—will provide the support for a thriving manufacturing sector if all of these systems are designed in a mutually self-reinforcing way.

Why Manufacturing Is Essential for a Wealthy Economy

We need manufactured goods to create an environmentally sustainable society, and we need an environmentally sustainable society in order to have a manufacturing sector. But does the economy really need a manufacturing sector? Aren't we a service economy now? Can't the United States just let everybody else manufacture all of those nice new green products and let us innovate and market?

The United States needs a strong manufacturing sector for a number of reasons. The case of the United States is a good one to examine because, for most of the twentieth century, the United States was the most dominant manufacturing economy in the world, and the decline in manufacturing has brought a host of problems in its wake.[3]

Buying More Than We Sell

The first problem associated with a declining manufacturing sector is that the United States has not been able to sell enough goods and services in exchange for goods and services from abroad, creating a huge trade deficit. The decline in manufacturing has contributed to this problem because international trade is mostly in goods, not services. That is, 80 percent of interregional trade is in goods, and only 20 percent is in services—as is the case with the United States.[4] The United States cannot possibly trade enough *services* for the volume of goods that it receives, and has instead been running up trade deficits, starting from soon after its manufacturing sector began to decline after 1968.[5] The United States has been making up for this shortfall ever since by exchanging dollars rather than goods. As its dependence on oil has expanded, it has provided dollars for oil as well.

So the trade deficit, which is the shortfall between what is bought from the rest of the world and what is sold, is a manifestation of both the economic unsustainability of the decline of manufacturing and the ecological unsustainability of the decline of petroleum output, which affected the United States first. Oil production peaked in the United States in late 1970,[6] even though it had been the "Saudi Arabia" of oil before then.

Figure 2
Trade Balance of the United States, 1968 to 2010, in Millions of Dollars

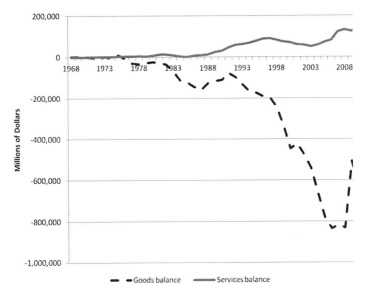

Source: U.S. Department of Commerce, Bureau of Economic Analysis, see note 5.

Figure 3
US Field Production of Crude Oil: Petroleum Production
in the United States Peaked in 1970

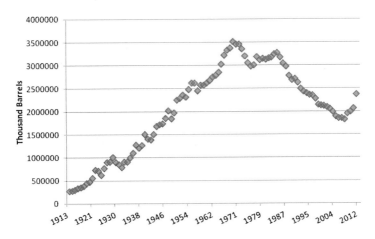

Source: U.S. Energy Information Administration (http://www.eia.gov/dnav/pet/hist/LeafHandler.ashx?n=pet&s=mcrfpus1&f=a)

Since the dollar is practically the only currency used to buy oil, the United States has been able to get away with paying for goods and oil with printed paper. If the dollar should ever stop being accepted in lieu of goods, then the United States will be in serious trouble economically because it does not produce the goods needed for the service sectors that have become the bulk of the economy.

Services Need Manufactured Goods

The second reason that manufacturing is important is that the service industries, which together constitute two-thirds (66.8 percent as of 2010) of the economy, are dependent on manufactured goods for their existence and technical progress. We can see why it is so dependent on a predictable supply of manufactured goods by looking at what is called the "value-added" percentage that a particular industry contributes to the economy. Often, writers discuss the total revenue of an industry, but by doing so, the writer includes inputs from other industries. For example, the manufacturing sector as a whole received a revenue of $4.5 trillion[7] in 2009, but only added $1.6 trillion to the economy[8] because the rest of the revenue, besides the value-added percentage, has been created by other sectors such as the service industries. By using value-added as a measure, we can get a better idea of how much a particular slice of the economy actually contributes to the economy as a whole.[9]

Retail and wholesale service sectors together contributed 11.3 percent to the value-added portion in the economy. They are clearly dependent on manufactured goods because they retail and wholesale these goods. If, say, the value of the dollar collapsed, Walmart and most other retailers would have much less to sell because imported manufactured goods would become too expensive. The transportation and warehousing sectors constitute 2.8 percent of the economy and use equipment which must be manufactured. Information industries, including publishing, software, TV, and phones, use an enormous amount of equipment, and constitute 4.5 percent of the gross domestic product (GDP). Healthcare services, at 6.8 percent of the value-added portion, also need equipment; a hospital without surgical instruments, diagnostic instruments, and pharmaceuticals would not be a hospital at all. Hotels and restaurants use buildings and food equipment, and constitute 2.7 percent of the value-added portion. Professional and business services, at 12 percent, either manage the use of equipment or use equipment for their operations. Other parts of the economy which are neither manufacturing nor service industries rely on considerable quantities of equipment and machinery: construction (3.8 percent), mining (1.7 percent), utilities (1.9 percent), and agriculture (0.9 percent).

Even real estate, at a whopping 13.2 percent of the value-added portion in 2009, basically involves buying and selling buildings which are assemblages of goods put together with construction machinery. Finance and insurance sectors (8.3 percent) take the surplus from all the other sectors of the economy. That is, they recycle the profits and rent all other industries generate. They use the wealth generated by the service industries which in turn depend on the manufacturing sector, as well as the wealth generated by manufacturing. The financial service industries do this by using computers and communications equipment.

The United States cannot assume that it can simply import all the equipment it needs. The United States may not have anything that the trading countries would want in return. Even if the United States were to sell all its assets, there are only so many to sell. Factories would be the best investment because a factory would create goods that could be sold abroad. But if the United States does not have many factories to buy and if it does not have the skilled workers and engineers needed to maintain world-class facilities, then foreigners would be less willing to hold dollars to buy assets. Thus, it is not only beneficial but also prudent to produce green technologies in the United States as a way to generate jobs.

Table 1
United States GDP by Industry, Value-Added, 2009. Services Constitute about Two-Thirds of the Economy, with Manufacturing and Other Production Comprising about One-Fifth, and Government the Rest

United States GDP by industry, value-added, 2009	
Agriculture & mining	3
Utilities	1.9
Construction	3.4
Manufacturing	11.2
Wholesale & retail	11.3
Transportation	2.8
Information	4.5
Finance and insurance	8.3
Real estate	13.2
Professional & business services	12
Health services	6.8
Hotel & food services	2.7
Entertainment, education, other services	4.4
Federal government	4.3
State & local government	9.3

Source: U.S. Department of Commerce, Bureau of Economic Analysis, see note 9.

Manufacturing Leads To Innovation

It is also becoming clear that by basing manufacturing and green manufacturing in the United States, the country retains its capacity for innovation as well. Put negatively, when American manufacturing industries are relocated abroad, US competence to create engineering and technological innovations weakens.[10] In addition, when one industry disappears, then others suffer because they lose the capacity to interact with "sister" industries. For instance, the automobile industry becomes less innovative partly because many of its support industries, such as domestic machine tools, have left.

The importance of keeping as many industries together as possible stems from the fact that an economy is an ecosystem of a kind, and manufacturing is an ecosystem within the wider economic ecosystem.[11] Like an ecosystem, an economy needs to have most or all of its main functioning parts in the same region in order to thrive. All the various parts of the economic system co-operate as much as they compete, and they need a certain closeness or proximity to other "co-evolved" industries in order to innovate and grow.

The United States was the first region to contain a full suite of modern manufacturing industries, and this power was the foundation of its rise as the most important economic, political, and military power.[12] The decline of this manufacturing base was the single most important reason for the decline of the middle class in recent years, and an aggressive program of green reindustrialization is now crucial rebuilding middle-class prospects.

Manufacturing Anchors Middle-Class Jobs

The "Great Recession" that started in 2008 is above all a problem of the lack of jobs. The employment picture has been transformed by the decline of manufacturing in the United States—the manufacturing sector is the main engine of job creation in a modern economy. While services have picked up much of the slack, the shortfall has been severe, and much of the service sector is composed of jobs that are lower paying than those typically associated with the manufacturing sector. Let us look at how different sectors have fared in the past several decades.[13]

From the 1950s to 1968, manufacturing as a percentage of employed persons barely declined from 28 to 25 percent. After 1968, however, the rate of decline roughly doubled, and manufacturing now constitutes only about 9 percent of the total US employment. By comparison, manufacturing in Germany still employs about 21 percent of German workers. In terms of GDP, the "value-added" percentage in the United States declined from about 25 percent in 1968 to 11.2 percent in 2009. So what sectors picked up the slack in terms of both GDP or overall national output, and employment?

The problem of the US economy is that the sectors that took manufacturing's share of GDP did not increase their share of employment, while the sectors that increased their share of employment did not increase their share of GDP. In other words, some sectors got richer, relatively, but did not increase their number of employees appreciably, and some sectors became relatively poorer per worker. Manufacturing, on the other hand, has for the most part pulled in as much income for employees as it has received from the economy—in other words, it is the quintessential middle-class sector.

Scholars of African development have used the term "structural disarticulation" to capture this concept of the skewing of an economy.[14] These scholars claim that there is an "unevenness in sectoral development," which in the African context means that most people are in the low-income-generating occupation of agriculture, while most of the wealth is generated by industry or mining. We can measure this with a "sectoral ratio", that is, we can find the ratio of the value-added percentage of the GDP, divided by the percentage of the total employment accounted for by a sector.

A sector with a sectoral ratio larger than one is richer, relative to the economy as a whole, while a sectoral ratio less than one indicates that a part of the economy contains a large amount of low-wage jobs. When a sector with a high ratio grabs more of the economy, then sectors that have lower ratios have less of the economy to provide to the same or more employees. So, the economy becomes more imbalanced as people are squeezed out of the middle class, particularly out of manufacturing.

Figure 4
Change in Employment for Selected Sectors, as a Percentage of the Work Force, from 1968 to 2009. Note the Shrinkage in Manufacturing and Rise in Low-Paying Services

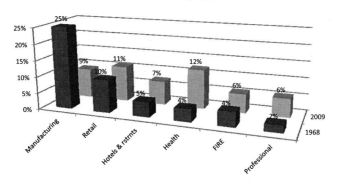

	Manufacturing	Retail	Hotels & rstrnts	Health	FIRE	Professional
1968	25%	10%	5%	4%	4%	2%
2009	9%	11%	7%	12%	6%	6%

Source: Department of Commerce, Bureau of Economic Analysis, see note 13.

Figure 5
Change in Gross Domestic Product (GDP) for Selected Sectors, as a Percentage of the GDP, from 1968 to 2009. Note the Decline in Manufacturing and the Increase in Low-Employment, High-Paying Services

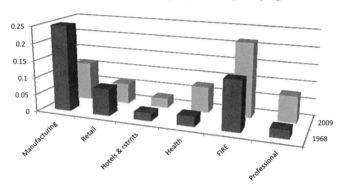

	Manufacturing	Retail	Hotels & rstrnts	Health	FIRE	Professional
1968	0.25	0.078	0.022	0.028	0.142	0.024
2009	0.112	0.059	0.027	0.075	0.215	0.076

Source: Department of Commerce, Bureau of Economic Analysis, see note 13.

Figure 6
Change in Ratio of Gross Domestic Product (GDP) to Employment Percentage for Selected Sectors, from 1968 to 2009. Because of their High Ratio, Since Finance, Insurance, and Real Estate (FIRE) Grew in GDP, They Didn't Add Much in Employment, Squeezing the Rest of the Workforce

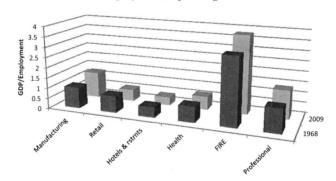

	Manufacturing	Retail	Hotels & rstrnts	Health	FIRE	Professional
1968	1	0.79	0.49	0.75	3.23	1.14
2009	1.24	0.55	0.38	0.63	3.77	1.36

Source: Department of Commerce, Bureau of Economic Analysis, see note 13.

Finance Went Up, Most Services Went Down

Finance, insurance, and real estate (FIRE) in the United States saw the biggest increase in its sectoral ratio; from 1968 to 2009, its share of employment rose from 4.4 to 5.7 percent, but its share of GDP rose from 14.2 to 21.5 percent. So, its ratio rose from about 3.2 to 3.8. Thus, FIRE grabbed 7.3 percent more of the economy, about half of the 14 percent of the GDP lost by manufacturing from 1968 to 2009.

Professional, technical, and scientific services moved from 2.1 to 5.6 percent of employment, while GDP rose from 2.4 to 7.6 percent for an increase in ratio from 1.1 to 1.4. Thus, about a quarter of the employment of the manufacturing sector loss went to this sector, which is composed of accounting, legal, advertising and management, and scientific and engineering consulting.[15] These are so-called "knowledge workers," who were supposed to take up the slack for manufacturing—which they only partially did.

On the other hand, several service sectors registered a *lower* sectoral ratio with increasing employment. The poster child for a lower standard of living is the hotel and restaurant sector. From 1968 to 2009, employment in this sector went from 4.5 to 7.2 percent, while the GDP rose from only 2.2 to 2.7 percent, with the ratio declining from 0.5 to 0.4. Health and social services saw a rise in employment from 3.8 to 11.9 percent, but only a rise in GDP from 2.8 to 7.5 percent, and thus, from 1968 to 2009, the ratio lowered from 0.7 to 0.6. Perhaps surprisingly, the percentage of retailing in the economy decreased, from 7.8 to 5.9 percent, but the percentage *employed* actually went up, from 9.9 to 10.8 percent. Thus, the sectoral ratio for retail jobs is at about one half the level of manufacturing jobs. If we add up the hotel, restaurant, health, and retail sectors, we see that from 1968 to 2009, the GDP rose from 12.8 to 20.5 percent for these sectors in the aggregate, while employment increased from 18.2 to 29.9 percent; the sectoral ratio for this group stayed about 0.70—and this included doctors; so, the relative income for most workers is much less. This sector gained almost 12 percent of the workforce, while manufacturing was losing about 15 percent. Thus, the rise in employment of a relatively low-paying set of industries accounts for most of the loss of the middle-class manufacturing sector (see Note 14 for sources).

Thus, a small part of the job force is doing much better, specifically in the "FIRE" sector, plus some technical occupations. However, manufacturing has declined and shed much of its workforce into sectors that are providing much less income per employee than manufacturing. These sectors include hotel and restaurant work, health care, and retailing.

Creating a Larger Middle Class

Owing to the dynamics described above, the economy is imbalanced. What it really needs, if it is to support a large middle class, is a large manufacturing sector contributing 20–25 percent to the GDP, with a small FIRE sector, and a reasonable services sector.

How does the United States grow manufacturing back to, say, the level of Germany, or 20 percent of the workforce? The United States would need to increase the share of employment in manufacturing by about 11 percent of the overall workforce, or some 14 million workers. Since the unemployment rate is 8.5 percent as of December 2011, another 5.2 percent of the employable population is employed part-time but wants to work full-time, and another 1.3 percent is too discouraged to look for work. For a total population of 22.2 million,[16] the United States could create 14 million jobs and not even exhaust the existing pool of unemployed or underemployed laborers.

It may also be possible for many millions of workers who have gone into restaurant, retail, or health jobs that pay less, to obtain jobs that pay more, and to increase their contribution to the long run vitality of the economy.

Green Manufacturing is Necessary for a Revival of Manufacturing

Manufacturing that is used to create a green society could generate about one-third of those 14 million jobs; a strategy to do so will be described in detail later. However, it is important to note that these green manufacturing jobs by themselves would not carry the entire load. This is to be expected because the manufacturing sector produces an enormous assortment of goods, from furniture to silverware, not all of which will be included in the transportation, energy, building, and food sectors which are the focus of green economics. Nevertheless, there are two important reasons why it would not be possible to revive manufacturing, and thus the economy, without spearheading reconstruction with green manufacturing.

Replace Resources with Machinery

First, green manufacturing will be required to replace the use of natural resources with machinery, and this need can lead to a large demand for manufactured products. In addition, manufacturing cannot continue if it depends on rapidly depleting resources; the yin of manufacturing requires the yang of a sustainable environment. A green energy economy will depend on the construction of machinery (equipment) such as wind turbines, solar panels, and even geothermal and tide/wave equipment in the U.S. Machines will create electricity by using, for all practical purposes, free fuel, that is, the wind, sun, and earth as energy sources. In a green economy employment will shift from the manning of drilling rigs and the maintenance of refineries and pipelines, to manufacturing wind turbines and solar panels, and then installing them on land or on buildings. Moreover, the processes for making manufactured goods will have to change in order to make recycling and reuse easier to carry out, thus giving engineers and machinery makers more work to create recyclable products.

In a sense, replacement of mining and fossil fuels with machinery will constitute the completion of the Industrial Revolution. While the Industrial Revolution began with the use of coal, this resource was favored because it was convenient and fitted well with the technological capabilities of the times. Prior to the use of coal an industrial revolution had occurred in many parts of the world, particularly in Europe, based mostly on the harnessing of wind and water power. Windmills ran grain mills and water wheels and provided the main power source for basic machining as well as textile and metal production before coal took over.[17]

But the essence of the Industrial Revolution was not coal, or later, petroleum, but machinery.[18] Industrial machinery is at the center of the manufacturing ecosystem because it is with machinery that we make all the products and the infrastructure that we actually use. Virtually no industrial machinery is used by consumers in everyday life; but without it, modern society would not exist. Within the industrial machinery "niche" there exists an even smaller and more critical central niche whose technologies can collectively reproduce themselves and produce the industrial machinery which indirectly powers industrial civilization.[19] These classes of machinery and equipment—such as machine tools, which create the metal parts of all other machines, including machine tools themselves—I call "reproduction machinery," as opposed to "production machinery," which is used to make the final consumer goods and infrastructure.

Machine tools are the master tools for shaping metal for embedding a form or structure on a material. An essential piece of reproduction machinery is the device that creates the energy that is used in the industrial process, and the first such device was the steam engine, invented by James Watt in 1776. We now use a particular kind of steam "engine," which is really a turbine, in electricity-generating plants which now use coal, hydropower, or other fuels. But what we are moving toward are renewable energy machines, such as wind turbines and solar panels, which for the first time are not tied to a fuel. They create energy, as far as we are concerned, from the machine alone—although the siting and actual production of the machine will depend on the characteristics of the wind and sun (and earth and water) in a particular location.

From Controlling Territory to Innovating

The effects of this shift on civilization will be enormous (Michael T. Klare, in this volume, explains the current position of fossil fuels and raw materials in the structure of international power). When economic power is based on a material that is specific to a particular territory on the surface of the planet, then economic power is determined by political control over that space—that is, political power determines economic wealth, not technological prowess, knowledge, or skill. This is an oversimplification because even the Saudis need sophisticated equipment to extract and refine their oil. But the fact remains that in the case of raw materials political power can be maintained simply by controlling a particular space.

When, on the other hand, economic power comes from knowledge and skill, then any country can learn to create that power, create wealth within their country, and can at least challenge their more powerful neighbors and retain some control over their own territory. Perhaps this is one of the reasons that the Chinese government has decided to encourage and create domestic renewable energy industries—they know that renewable technologies give them the independent capability to generate electricity, and this capability therefore limits the power that other countries have over the Chinese. The same considerations follow with regard to a shift from mining to recycling. Thus, a shift from fossil fuels to renewable energy technologies will have profound geopolitical repercussions and may well become a national security priority.

Creating Manufacturing Ecosystems

So producing energy from machines instead of fossil fuels will provide a market for more machinery, thus reinvigorating manufacturing in general. The second reason that green energy manufacturing can help resuscitate the wider manufacturing sector is that these final products, such as high-speed rail trains or wind turbines, will also serve as the center of their own industrial eco-subsystems, just as the automobile manufacturing ecosystem has served to anchor much of the American economy for a great portion of the post-World War II (WWII) period. For instance, Jonathan M. Feldman has shown how new transit manufacturing could help suppliers of components transition from brakes to motors, from using various kinds of steel to fabricating shells of subway cars, and from developing on-board information systems to manufacturing and installing electrical systems.[20]

Meanwhile, green manufacturing and recycling will be required to minimize the use of depleting and polluting resources. They will also be vital because large parts of the manufacturing landscape will change as the use of natural resources, particularly oil, declines, so that whole new industries will be needed to replace these. When fossil-fuel industries are replaced, the machinery subsystems that have been supporting the current machinery industries will again thrive.

A Green Economy Will Have a Different Mix of Industries

A shift to a green energy economy will lead to a shift in the distribution of employment, with less people needed in the transportation sector but more in the energy and construction sectors. In a green economy, significantly fewer cars may be needed, and if so, some other sectors must take on the employment role automobiles occupied in the post-WWII period. This could be accomplished by a combination of an expanded rail industry and a large wind turbine and solar panel industry, together with a scaled-back electric car industry and construction of dense town and city centers and intensive agriculture.

Cars support a vast ecosystem of parts makers, metal makers, electronics makers, tire makers, and so on. Makers of wind turbines, solar panels, rail, and electric cars together might be able to support a similar ecosystem. But why would the production of cars decline in a green economy, and why would this be a problem?

The Sunset of Petroleum

The main problem confronting the future of transportation is that oil will likely become more and more expensive and more and more difficult to extract. We have already seen the consequence of this situation in the Gulf of Mexico oil disaster,[21] and we are witnessing the unfolding of probably a bigger disaster in the development of tar sands in Canada.[22] Natural gas will probably never be able to replace oil as a source of transportation fuel, and we are also witnessing a potential slow-motion ecological disaster in the "fracking" of underground reservoirs of natural gas. Despite claims that fracking is a "game-changer," at best, fracking will make up for the loss of conventional natural gas drilling—and it is a more expensive process than before.[23] Ashley Dawson refers to the next phase of fossil fuel drilling as "extreme extraction."[24]

Another problem for the current transportation system is that it has never been shown that a good substitute exists for petroleum as utilized in the sector. Bio-fuels release more greenhouse gases than fossil fuels and have other devastating effects on ecosystems and agriculture.[25] Batteries, meanwhile, have never been shown to be able to store enough energy for the enormous demands of moving a multi-ton vehicle.[26] Thus, the future for large, fast-moving, long-distance personal vehicles looks grim.

The fact that transportation is based on oil is a cultural decision encouraged by economic self-interest. It is perfectly possible to have an all-electric transportation system[27] but this would involve profound changes to the spatial distribution of buildings in our society; in other words, the process of expanding sprawl would have to be reversed. It is beyond the bounds of this chapter to speculate on whether or how such a cultural shift will happen. However, besides the cultural changes such a shift would involve, an economic change likewise applies: we would not need nearly as many people or factories to make transportation machinery because a train-based society would be so much more efficient.

The Inefficiencies of the Automobile

The automobile is perhaps the most inefficient technology on the face of the earth to have achieved its stated purpose, specifically transportation of people and things (the same applies to trucks). First, vehicles are used only 4 percent of the time.[28] During the rest of the time, they use up parking space, thus wasting an enormous amount of space. Also, up to one-third of urban space is used for roads.[29]

A 4 percent rate of use means only one hour of use per day. Let's say that a good percentage of the population use their cars two hours per day, or even three.

This would lead to a usage rate of 8 or 12 percent—still very wasteful. Meanwhile, for a factory to be economical it generally must have a low "downtime"—ideally, machinery is used for at least sixteen hours per day, or even close to twenty-four hours. For the main transportation machinery of modern society to exemplify such low rates of usage is extremely wasteful.

The only thing more inefficient than a parked automobile is one that is being used. According to Amory Lovins, about 99 percent of the energy used to move an automobile is wasted; only 1 percent actually moves the person or things inside (about 70 to 75 percent is lost as heat by the engine while most of the rest is wasted in moving several tons of metal).[30] Moreover, most cars carry only one occupant, which is an inefficient method of transportation. But this is not the worst part: this form of transportation led to the deaths of 33,000 people in 2009, with more than 2 million people being injured that same year.

Jobs and the Modern Vehicle

So what would happen if all these inefficient vehicles were replaced by rail, plus some short-distance, small, slow electric cars? We would need consider-ably fewer factories to make transportation equipment. There are 877,000 jobs in the motor vehicle industry, 1.2 million in automobile dealerships, 800,000 in automotive repair, and 821,000 in gas stations.[31] So approximately 3 million people are directly engaged in manufacturing or servicing automobiles. Another 2 million jobs exist in the trucking and warehousing industries, alongside half a million in the aerospace industry and half a million in the airline industry.[32] Thus, a total of about 6 million jobs are associated with petroleum-based transport. Less vehicles would mean fewer jobs, although we cannot be sure exactly how many of these jobs will disappear until we see how technology for cars, trucks, and planes progresses.

Part of the explanation for the post-WWII boom was the enormous demand, and enormous workforce, created by this incredibly inefficient set of technolo-gies. The automobiles (and truck and plane) together were panaceas for a society where production was not a problem but demand and jobs were. We could call it the era of "automobile Keynesianism."

The economist John Maynard Keynes, in the 1930s, offered a theoretical justification for creating more demand than was currently in force, if by creating that demand, the idle capacity of factories and offices could be utilized to employ a considerable number of people. At least in the short term, for Keynes, it did not matter if this meant digging holes and filling them again, although doing something useful was clearly better.[33] So throughout the post-war period, for instance, some economists have accused the Federal government of practicing "military Keynesianism." This would mean using the military budget to create demand that would not be there otherwise, and thus, at least in the short-term, improving the performance of the economy.[34] In 1999, workers in defense-related industry numbered 2.2 million, with a total of 6 million in military-related occupations.[35]

After WWII, there were fears that the economy would slip back into Depression without the stimulus of military spending. Road building taken to an epochal level by the Interstate Highway System as well as a push for car ownership led to "automobile Keynesianism". This then led to "sprawl Keynesianism" as governments at all levels encouraged single-family home ownership. Sprawl encouraged more economic activity, including waste-energy usage. According to an Environmental Protection Agency (EPA) report, moving from a single-family home to an apartment building can easily save 50 percent of home-energy use per person.[36]

Greater distances traveled in most suburbs has led to an ongoing need for more roads, for consumption of more oil to drive long distances, and for more effective maintenance of automobiles owing to the extra mileage. In addition, because most single-family homes are far from commercial areas it takes more roads, miles, and maintenance to drive to malls instead of walking or taking transit to local stores. Additionally, freight rail is at least four times more energy-efficient per ton than trucking.[37]

Jobs and Denser Cities and Towns

So a large construction and transportation work force has been kept busy building and maintaining a system that uses many times more resources than an electric-rail-based system. However, a shift from an energy-inefficient urban structure to a denser, more energy-efficient form of urban development would require a construction boom. It has been estimated that while 30 percent of the population would like to live in a walkable neighborhood, only 5 percent are able to do so.[38] Let's assume that 25 percent of U.S. households, or about 25 million households, would like to live in a comfortable apartment building in a walkable neighborhood, but one that would have to be constructed where none exist now. This would require use of "infill" where there is no town center currently, or beefing up the existing town and city centers. Let's assume that we would perform standardization on a 250-unit apartment building—meaning that we would need 100,000 such buildings, sprinkled throughout metropolitan areas, to accommodate 25 million households. If this endeavor cost $50 million per building in construction expenditures (assuming $200 per square foot, for 1,000 square foot apartments[39]), we would need a total of $5 trillion spread out over ten years, or $500 billion per year. At $50,000 per job, including the jobs needed to make and ship the relevant materials, the result would be 10 million jobs per year for ten years. And that's just to house 25 percent of the public in a denser environment; looking ahead, we might expect the following decade to yield a similar boom if the next 25 percent of households faced overwhelming financial pressures from higher and higher gas prices in ten years' time.

Pursuing this exercise a bit further, we can imagine that these new buildings would be constructed with energy efficiency and recycling in mind. For example, each building might have a very deep geothermal heat pump providing both heating and cooling.[40] Geothermal heat pumps use the constant temperature of the ground several feet down to cool buildings in the summer and warm them in winter. They use about 50 percent less electricity than an electric heating and air-conditioning system, depending on the area, and with solar panels they could even provide heating and cooling without using the grid, at least during the day.

The use of such technologies is increasing. The Chinese are encouraging the installation of geothermal heat pumps[41] as well as solar hot water heaters on roofs; in the latter case, millions of these low-cost systems have already been placed on roofs in China, and these are becoming popular worldwide.[42] Also, each new building could be constructed as a huge "passivhaus," the German design that cuts energy use by as much as 90 percent, while providing healthy ventilation.[43] However, the inclusion of passivhaus or other efficiency methods adds at least 10 percent to the price of a house, and builders have been reluctant to make housing more energy-efficient because, in general, buyers do not seriously consider efficiency during purchase. Accordingly, a number of programs in the United States and Europe have been designed to overcome this problem.[44]

When it comes to improved residential structures, the use of large apartment buildings offers another advantage in that they simplify recycling. Although rarely done now, it should be possible to fit apartment buildings with waste-composting systems and even dry toilet composting. Meanwhile, recycling of paper, plastic, and metals is already fairly advanced in some apartment buildings, and it should be an easy step to recycle appliances, furniture, and other consumer goods from buildings in this manner. A United States that was recycling at least 75 percent of its materials could employ over 2 million people.[45]

Sustainable Agriculture

As the supply of petroleum becomes more and more unreliable, eventually up to 80 percent of the population might live in a dense community in order to avoid the need to rely on long-distance driving of a personal automobile. Much of the remaining 20 percent of the population might be involved in agriculture or some other more rural-based economic activity. Agriculture since the 1920s has become extremely productive in terms of labor but not in terms of land, and certainly not in terms of energy, water, and soil. In fact, water, soil, and biodiversity are catastrophically declining because of modern agriculture,[46] although all civilizations have been required to be careful. If water and soil are considered capital, then modern agriculture may be considered to have a negative net effect on global wealth. Currently, agriculture is dependent on petroleum for its soil and ecosystem-destroying pesticide production, on natural gas for its water-befouling artificial fertilizer, and on petroleum for the operation of its farm machinery and the movement of food throughout its average 3,000 mile journey to the end consumer.[47]

On average, it seems that 15 percent more labor is required for organic farming methods than is required for conventional approaches.[48] Organic food now constitutes about 4 percent of the food market.[49] If, ideally, all food was grown organically, then we would theoretically require 15 percent more workers than the number of individuals currently employed, or about 270,000 more farmers. Much of the transition of agriculture would entail minimizing the need to transport food. This would mean growing and processing food within easy reach of walkable neighborhoods, that is, within cities and towns themselves or in nearby farm belts, and additionally using very little land, as in Biointensive Agriculture.[50]

If anywhere close to 80 percent of the population could reside in dense, walkable neighborhoods, with much of the rest of the population living near gardens and farms in the nearby countryside, then vast stretches of American ecosystems, such as the prairie, might be able to reassert themselves. Perhaps a substantial part of the population would be involved in managing a revived wilderness, or in the resulting eco-tourism, particularly if cheap, fast, and comfortable rail made it easy to visit various parts of the country—sustainably, of course.

Sustainable Manufacturing

Recycling jobs would also include transporting disposed goods to be used as inputs for factories, which in turn should be located close to urban areas and freight rail networks, all connected to high-power renewable-energy grids. Thousands of factories could be built close to urban areas during the reconstruction period. These factories could be equipped to produce goods that are easy to recycle or reuse and emit very little or no pollution.[51] Also, a complete redesign of industrial processes, particularly in the chemical industry,[52] would employ the talents of thousands of engineers while the construction of new factories with machinery made in the United States could lead to employment for millions of people.

The need to replace resources with machinery and to replace old industries and infrastructure with new industries and infrastructure implies that green manufacturing could lead a wave of technological change and sustainable growth.

The Government Must Lead a Green Manufacturing Renaissance

David Leonhardt of the *New York Times* writes that the United States "has not developed any major new industries that employ large and growing numbers of workers. There is no contemporary version of the 1870s railroads, the 1920s auto industry, or even the 1990s Internet sector. Total economic output over the last decade, as measured by the gross domestic product, has grown more slowly than in any 10-year period during the 1950s, '60s, '70s, '80s, or '90s."[53]

A green energy and transportation industry boom would be a perfect successor to Leonhardt's list. In each case Leonhardt enumerates, the government was critical to industry breakout. Abraham Lincoln's main economic platform was to encourage the development of the railways, and the transcontinental railroad was completed during his administration. The U.S. government gave railroads land on either side of their rail lines, giving them an economic incentive and built-in profit for developing rail. Later, the rise of the automobile would have been impossible had the government not virtually given much of the public space over to roads, an act only topped by the government's eventual building of the roads themselves. This would include, by the 1950s, one of the biggest infrastructure projects in human history, the Interstate Highway System.

Today, zoning regulations in much of the country make it virtually impossible to reach centers of employment or access critical services without a car.[54] In the case of the internet, the government developed the system until it became commercially viable[55] and has received virtually no return on its investment, except the taxes, if it can collect them from the Microsofts, Googles, Apples, Ciscos, and other companies that have gone on to make billions from the public investment.

One may argue that it is disingenuous for critics to now talk about unfair subsidies to the green energy sector when one considers historical government assistance to many other industries. Even the oil industry could not have become the dominant force it is today without the aforementioned actions of the government in support of the road and highway system. In previous eras, as Leonhardt's list makes clear, the United States was on the cutting edge, doing much of the initial work to advance new technology, and its leadership in the modern context extended to high-speed rail and wind and solar technology. Yet now their development is being led by the Asians and Europeans. There is nothing wrong with other countries pursuing useful technologies, but the critical question is, why hasn't the United States kept up?

The Effect of the Military on Manufacturing

In the United States, the main governmental driver of manufacturing is the military. The internet was actually developed with funds from the research branch of the military, the Advanced Research Projects Agency (ARPA). Airplanes were heavily subsidized by the military, and the Interstate Highway System was partially justified based on its ability to transport tanks across the country. However, this emphasis on military production has over time warped the U.S. manufacturing sector in many ways.

As Seymour Melman sought to show in several books, the military–industrial complex, or "permanent war economy," as he called it,[56] has several unfortunate effects on manufacturing. First, a large percentage of scientists and engineers are soaked up by military production; generally, the pay is higher for military work, and thus, other firms are at a disadvantage in competing for the best engineering talent.[57] Second, the trillions spent to date on the military could have been better spent on rebuilding the infrastructure and manufacturing sector as a whole. But the third and probably the worst problem identified by Melman is the loss of competence in civilian manufacturing that accompanies a focus on military production.[58]

During the Cold War, we heard a steady drumbeat of criticism in the United States that the Soviet Union was less efficient because of central planning. The Soviet system output shoddy products and less of them, it was asserted, because the competitive discipline of the market was lacking. At the same time, many of these critics (not all) were advocating for larger and larger military budgets in the United States. But the military is, just like the Soviet system, itself a centrally planned economy, even if it is a smaller part. The same inefficiencies observed in the Soviet system also occur in the American military industrial complex—that is, military equipment is much too expensive, takes too long to make, and breaks down much more than it should.[59] This is partly because the military operates on a "cost-plus" system, that is, they can charge the government for any cost of building the equipment and then simply tack on a certain percentage as profit. The more expensive the output, the more profit there is to be made, and thus, there is an incentive to make equipment more and more costly.

The tragedy for the wider society is that once managers and engineers become used to a business culture in which cost is something to be maximized, not minimized, it becomes very difficult for a military equipment manufacturer or its managers and engineers to shift to a cost-minimizing culture. Thus, a significant percentage of the manufacturing economy, meaning those involved in military work, warp the industrial competence of much of the rest of the manufacturing economy. The U.S. Department of Defense procured $134 billion worth of manufactured goods in 2010[60] versus a manufacturing value-added of $1.6 trillion in 2009.[61]

The Military and a Green Economy

The dynamics explored above pose several consequences for a move to a green energy economy. First, the interest of the Department of Defense in becoming "green," for instance by using solar panels instead of incurring very steep costs from trucking in oil,[62] could prove to be a double-edged sword for the United States. On the one hand, money spent by the Pentagon will constitute an important market for nascent solar and wind manufacturers; on the other hand, these same manufacturers will get used to the "cost-plus" nature of Pentagon contracting. They will either decide to concentrate on the higher and less risky profits of military contracting, thus ceding the civilian market to other countries, or they will lose the competence to produce cost-minimized products in the civilian economy, or both.

Second, green manufacturers for the civilian economy face the disadvantage that much of the manufacturing ecosystem is devoted to cost-maximization, and thus, they will not be able to take advantage of the rich pool of subcontractors that the other developed countries maintain. In the United States, either many suppliers have gone out of business and are not available, or they have been pulled into the orbit of military production, thus making them effectively un-available as well.

Third, there is the wider problem that much of the military's claim to its hold over a significant amount of government resources is that the military protects our supplies of oil.[63] The less we need oil, the less we need the military. As Chuck Spinney has written, the military can be viewed as a vast network devoted to obtaining as much government revenue as possible.[64]

Positive Lessons for a Green Economy

Despite these problems, two major lessons can be gleaned for green energy advocates when considering the success of the military–industrial complex. First, perhaps the greatest weapon in the Pentagon's arsenal is not its nuclear weapons or aircraft carriers, but its carefully orchestrated placement of military factories and bases throughout the United States which engenders support for the military from the Representatives and Senators who are endowed with these job-creating assets. Politicians are perfectly willing to accept "socialism," that is, government control of economic activity, if it brings a predictable supply of high-paying jobs. The lesson for a green economy is that a similar network could be created for wind, solar, high-speed rail, and other green technologies. That is, factories for producing solar panels, wind turbines, rail equipment, and even materials to be used in energy self-sufficient apartment buildings could be distributed throughout the country with an eye to creating a self-sustaining political consensus within the Congress. In other words, the United States could build an "infrastructure–industrial complex."

But how would this strategy of institutionalizing the "political will" for a green infrastructure be implemented? First, a program of economic reconstruc-tion would require that all equipment and products bought with government financing would have to be made in the United States. Second, an overall plan of action would have to be designed for a time horizon of at least five years, and preferably, ten to twenty years. Then, the location of the factories used to create the wind, solar, or rail equipment could be proposed.

The second lesson of the military economy for green energy advocates is that it might be possible to convert military factories into green energy equipment factories (what Seymour Melman called "economic conversion"). That is, we can cut the military budget in order to serve much more pressing civilian needs but at the same time ensure that individuals dependent on military production for their livelihoods would retain good jobs. Through such efforts the economy's conversion—from one that is militarized, to one that is civilianized—can potentially gain significant and widespread support.[65] This concept of conversion could be extended to the fossil fuel and automobile industries as well.

Avoiding the Pitfalls of Government Planning

But if a green economy was at least partially planned, if there was an infra-structure–industrial complex, or a green economy–industrial complex, what would prevent the appearance of the same inefficiencies as demonstrated by the military or Soviet systems?

First, a continental system of reconstruction would need to be planned in a decentralized manner, that is, local and state governments would have to be inti-mately involved in the planning. Since local governments are more familiar with the needs of their communities and constituents than the Federal government, it would make sense to provide them with significant input into the process, includ-ing the ability to make the case for siting factories. With more eyes and hands involved in the planning process more transparency could be achieved. Ideally, the Federal government would present the broadest design plans possible, with the local governments filling in as much as possible. Second, an infrastructure–industrial system would not be cost-plus; rather, the contracts would be for a specific amount with no room to increase the cost. Third, there would be no need to "sole-source" the equipment, so that more than one company would provide the trains, wind turbines, and solar panels. In this way, if one company went out of business, or did not come through with its order, or otherwise violated its contract, other contractors could be engaged. Fourth, since the general population would be using the equipment or the output of the equipment, reliability stands to become a much larger factor than that in the military situation. Fifth, while some inefficiencies will remain—inefficiency and even some corruption are part of any human enterprise—that does not mean, however, that inefficiency and corruption should not be minimized. And transparency, more broadly in the current context, would be a very important part of a program of economic reconstruction.

Why Planning is Necessary

While several approaches may be pursued to decrease the inefficiencies which are a part of national planning, there are also ways in which national planning is much more efficient than the ad hoc development provided by the market. The national government can plan holistically and in the long term which the market cannot. For example, the Interstate Highway System was designed as a whole, not in pieces—and with a great deal of local input.[66] Similarly, design of a high-speed rail system or revived medium-speed rail could be planned by the Federal government, along with input from the state and local governments, implemented locally. In fact, a high-speed rail network could run alongside much of the Interstate Highway System and part of the interstate could be used for slower trains, as J. H. Crawford suggests.[67]

A national wind system would probably benefit the most from a national perspective, although that is not how the wind system is currently developing, since there is no Federal grand plan. Wind power becomes more reliable, as more wind turbines are available and are distributed in different environments. To put it most simply, since wind is always blowing somewhere, if placement of wind turbines is appropriate, wind will always be available for generating wind power.[68]

By designing and financing a long-term, continental plan for wind, the Federal government could guarantee a market for firms through means that would be much more effective than the current preferred method, a tax credit. If a company received a ten-year contract to build a certain number of wind turbines, then the company and its suppliers would be guaranteed a stable market. Also, by contracting for a large amount of wind turbines at the same time, economies of scale could be achieved.

Another advantage to a national wind plan would be to site the wind farms, not only to ensure that wind power is being continuously generated but also to avoid siting problems that occur in more localized situations. In a number of situations, local communities have expressed an interest in wind power but the placement of wind farms close to populated areas, or to the flight paths of birds and bats, or to forest stands, would cause environmental damage.[69] In contrast, if turbines are concentrated in the windiest parts of the Great Plains and Midwest, for instance, away from populated areas, many of these problems are resolved and local areas can have clean wind energy without nearby siting issues. Meanwhile, some companies are using more efficient wind machinery that requires rare earth metals; a national plan could require that these be used sparingly.[70] However, materials should not be a constraint in building wind turbines.[71]

National planning for a green energy system would ideally take place at an even higher level than a national wind system. The size of the wind system would very much depend on the characteristics of the rest of the system, for example, how much solar photovoltaic energy is used, while the design of an energy infrastructure would also depend on what kind of rail and electric car systems are envisioned. Local photovoltaic, ground source heat pump, and other decentralized electricity sources would lead to a smaller wind system but a national rail system, including local transit plus electric cars and trucks, would lead to additional wind energy needs.

A national energy strategy would have to include the reconstruction of the national electric grid.[72] Currently, electricity is moved from power plants to home via an electricity grid, that is, a network of wires that passes across the country. However, this network grew in a very ad hoc way without planning at the national level. In addition, the grid has not been properly maintained, partly because private utilities do not see much profit in doing so. It is even possible that private utilities would not mind the government taking over the construction and maintenance of the grid.[73] The advantage here is that the government could create a national design for an updated, much more efficient grid, and run and maintain it, thus providing a basic service for the entire country. To accomplish this goal perhaps the government could simply buy the grid from utilities.

We can look at the economy from an even higher vantage point and consider the benefits of planning the energy, transportation, and urban infrastructures simultaneously. If we plan to achieve a much smaller (or even zero) use of petroleum, then we need to reconfigure the transportation system to be electricity-based, with enough density in town and city centers to realize a train-centered transportation system. We would have to plan to build a large number of structures in these towns and cities. We would also need to move manufacturing and food production closer to the denser population centers. The Federal government could provide broad guidelines for a particular region for transportation, energy, urban layout, and even agriculture, as well as financing. Localities would actually design these town and city centers and the transportation, energy, and production configurations specific to these centers.

Planning the transformation of a significant part of the national infrastructure in no way implies central planning on the order of the Soviet model. The latter was in fact designed as a means to funnel most of the country's output into the military sector, and thereby encompassed the bulk of the economy.[74]

The larger the green economy, the greater the manufacturing capacity needed for the various green economy machinery. Moreover, as this green machinery market expands the system of suppliers that form the base of any manufacturing system will also expand. Currently, efforts are so scattered that few domestic manufacturers are convinced that there will be a long-term market for them.

Jonathan M. Feldman has written on the reasons for the anemic state of the domestic rail industry in the United States, and possible methods to revive it. A key finding is that the weakest element in the industry is the lack of a stable, long-term market which must be accompanied by a long-term relationship with local and national governments and trade unions.[75]

Currently, corporate America is sitting on almost $2 trillion of funds because firms do not perceive any profitable ways to invest their money.[76] And there may not be, at least without some larger national push—we are in a situation akin to standing at the bank of a river knowing that we will enjoy a better life if we can only get across the river; but we have no boat. A national program of economic reconstruction, financed by the Federal government, could be that boat.

The Market Can't Create a Green Economy on its Own

John Maynard Keynes and many others have pointed out that the private market can remain stuck at a suboptimal level unless the government gives the economy a "kick." Thus, not only was government spending necessary to pull the United States out of the Great Depression in the 1930s, but government financing, which began in Franklin D. Roosevelt's administration, was also necessary to kick-start the entire real estate financing system which led to home ownership for tens of millions of people. Before the government intervened in this market, most financing was for five years for a home, and repossession was common.[77]

In the same way, the government will be required to simultaneously create various pieces of the green economy in the early twenty-first century. Manufacturers need to know that a long-term market exists for their offerings, but currently small rail, wind, and solar markets offer little incentive. For instance, since the construction of wind farms is piecemeal at present, it cannot be as effective as a national system of coordinated development, so wind does not look as promising as it should. The same applies for a reconstruction of the national grid, high-speed rail, other electric rail systems, and solar manufacturing.

So far, however, only the Chinese have been willing to put forward the required investment.[78] The financial sector is not capable, or at least willing, to make a multi-trillion dollar investment in something new. They demonstrated a willingness to enter the real estate market only after the government intervened, first by creating the market, then by deregulating it. Only the government boasts the capability for the long-term planning and financing that is necessary to get us over the river and to a green economy. Once this new economy is constructed the private market will feel very comfortable when making shorter-term investments.

A Program for Creating a Green Energy Economy

Substantial employment and stimulative effects could ensue from a stimulus strategy for green manufacturing. For instance, let's start by looking more closely at a national wind system.

Interstate Wind System

Let's first assume that the entire supply of electricity in the United States was generated from wind, and just to make things easy, let's assume that the demand for electricity does not go up in the next two decades. The conventional estimate for the age at which a wind turbine needs replacement is twenty years. Let's assume that it would take twenty years to increase the percentage of electricity generated from wind, that is, from 1.9 percent in 2009[79] to 100 percent, say in 2034. Then we would need to construct 5 percent of a 100 percent national wind system every year, until the entire system was built by, say, 2032. After 2032, we would still indefinitely need to replace 5 percent of wind turbines each year.

The United States uses about 4,000 billion kilowatt hours (kWh) of electricity.[80] This huge number can be converted in many different ways: 4 million gigawatt hours (GWh), or 4,000 terawatt hours (TWh), or 4 petawatts. Here, we will use TWh because that is a concise way to keep track of how much electricity is used and generated.

Electricity sources are usually rated according to their hourly capacity to generate electricity. To make matters complicated, each hour for each source of electricity may generate a different amount of electricity, depending, in the case of wind, on how much wind is blowing and the size of the turbine. In 2009, wind generated 73.886 TWh[81] with a capacity of 34.296 gigawatts (GW),[82] which is a capacity factor of 24.5, that is, about one-quarter of the maximum capacity of wind is being used. The National Renewable Energy Laboratory (NREL) estimates that the average capacity factor for new wind turbines is about 39 percent.[83] Another advantage of Federal planning would be that turbine farms would be located in the best areas in terms of capacity usage, and the larger the turbine, the better the capacity factor.[84] But let's assume a middle ground between the current overall capacity factor and the technological best-case, and for convenience, let's say that we can figure on a 33 percent capacity factor.

So, if we take the total hours generated, 4,000 TWh, we know that if we wanted it all to come from wind, we would need a system that has the capacity to generate 4,000 × 3 = 12,000 TWh, so we would need an economic system that would have the capacity of generating 12,000 TWh. Divide this by the number of hours in a year, and we find that we need about 1,370 GW of wind-turbine capacity. Thus, while using *terawatt hours* as the main measure for actual output, it may be easier for the reader to use *gigawatts* when considering capacity.

Now, adding wind capacity is not the same as adding coal or natural gas plants because fossil fuel plants are running about 90 percent of the time, but wind may die down in one area completely. People do not want to deal with "intermittent" sources of power; they quite understandably want power all the time. According to engineering researchers[85] the intermittency problem is overcome by adding wind turbines over a larger area. Eventually if wind turbines are properly spaced all across the continent it should be possible to create a wind-based electrical system in which enough wind is blowing in most locations all the time so that no one is deprived of electricity.

Most studies of national wind power systems do not assume a 100 percent wind-based system. Gar Lipow[86] argues that with the addition of enough battery storage capacity, and perhaps adding more to the wind system than is strictly necessary, it should be fairly straightforward to construct an all-renewable system. Lipow also points out that some electricity is lost in the process of transmission, so—returning to our earlier calculations—we may add about 10 percent for transmission loss which gives us a round number of 1,500 GW for a built-up national wind system. For estimating the effect of a wind-based economy on the structure of employment we may use this figure even though further research is needed.

NREL estimates that some 4,300 full-time jobs are created per gigawatt of capacity of wind power.[87] In addition, according to a Renewable Energy Policy Project (REPP) report, "70 percent of the potential job creation is in manufacturing the components, 17 percent in the installation, and 13 percent in operations and maintenance."[88] The REPP study finds that there are 3,000 manufacturing jobs for every $1 billion in investment which the study's authors translate into 1 GW of capacity. Another way to confirm wind power employment levels is to look at the operations of Vestas, the largest wind turbine manufacturer in the world, based in Denmark. They have 21,000 employees, and in 2011 it was estimated that they will produce 6 GW of capacity,[89] or about 3,500 employees per gigawatt, close to REPP's estimate. So if the United States builds 1,500 GW of capacity, spread out over twenty years, then it would have to build 75 GW per year, translating to approximately 225,000 manufacturing jobs per year. This level of employment would be indefinite, as after twenty years the wind turbines would need to be replaced. The extra 1,300 full-time jobs per gigawatt would be in installation and operations and maintenance. Assuming for every gigawatt some 700 jobs in installation, which would stay constant, we would have 52,000 jobs in installation; however, the operations and maintenance would increase until the full 1,500 GW was installed, when we would need 900,000 permanent workers in these service occupations at the end of twenty years.

These estimates do not include the steel production needed to create wind turbines, which is a relatively low percentage of the total steel output. The United States alone produced about 80 million tons of raw steel in 2010, and the world total was 1,413 million tonnes.[90] NREL estimates a need for 114,000 tons of steel per GW capacity for wind;[91] if we need to create 75 GW per year, we need more than 8 million tons of steel per year which is only 10 percent of the US total. The steel industry claimed approximately 159,000 workers in 2008, so we would add only about 16,000 workers. Meanwhile, a considerable amount of fiberglass and concrete is used in wind turbines, but these also do not create any great pressure on resources, labor, or land.[92]

Estimates of the cost of a kilowatt (kW) of capacity for wind vary from $1,500 to $2,000.[93] Assuming $2,000 per kW, and 75,000,000 kW built per year, the required budget would be $150 billion per year, hardly a huge amount by national standards. If we assume 3-megawatt (MW) turbines, which are large in scale but not particularly cutting edge technologically speaking, we would need to construct 25,000 wind turbines per year for a total of 500,000 by the end of the twenty-year period. The current US average turbine size is 1.79 MW[94]; if we assume the use of 1.5 MW turbines, we would need to construct about 1 million in twenty years.

Advantages of Federal Ownership

It would be best if the Federal government financed and designed the system; ideally, the Federal government should also own the system for a few reasons.

First, as argued previously, it is much better if wind is sited according to a master plan from a national perspective in order to minimize the problem of inter-mittence. Only a national authority such as the Federal government can serve this role. Second, the Federal government can finance the construction and operation of a national wind system either with very low interest rates, or even out of general funds. If this system was funded without loans, the resulting electricity would be so cheap that the government could offer a set annual amount of electricity to each person for a very low price, or even for free, offering American citizens and businesses a higher standard of living. In order to avoid the problem of Jevons paradox,[95] that is, greater use of electricity owing to its lower cost, all electricity above a base amount could be charged at the full price of added capacity. Third, the Federal government would not need the expense of providing a return to investors which in the utility industry is generally around 10 percent.[96] A national wind system could thus be built at a cheaper rate by the government (albeit perhaps through the use of private contractors). Fourth, a wind system requires a rebuilt national grid. According to the Electric Power Research Institute, an upgraded smart grid would require between $338 billion and $476 billion.[97] Let's round that up to $500 billion. If we assume a twenty-year construction period, and $25 billion per year invested, and an average of $50,000 per job, then we could have 500,000 jobs per year upgrading the grid. In addition, a national set of large battery systems could be integrated into the network, providing another layer of protection against intermittency problems. Lipow estimates $1,000 per kW capacity,[98] or $1.2 trillion over, say, twenty years, or an extra $60 billion per year to add in a battery storage system. If this yielded 10,000 jobs per $1 billion invested (high capital manufacturing generally yields fewer jobs), we would have another 600,000 jobs per year making and installing batteries.

Jobs and a Green Economy

Construction of a national wind and grid system supports many other sectors. For instance, rail, transit, solar energy, heating and cooling, electric cars, and electricity for transportation all benefit from their integration into a national wind and grid system. Similar calculations can be performed for the other sectors, and these calculations are presented in Table 2 to provide an understanding of the potential and scope of a green manufacturing stimulus strategy. Other chapters in this volume also cover job creation (see Wendling and Bezdek, Chapter 4).

The calculations assume a timeframe of twenty years per project, where the total jobs per year indicate the number of jobs required after the completion of the twenty-year buildup program. Thus, the United States could achieve approximately 24 million middle-class jobs by the time the country has finished the construction of a green economy. More than five million could be employed in manufacturing, moving the nation almost half the way back to a full manufacturing economy.

This renewable energy part of this program would replace more than 1 million jobs lost in the fossil fuel industry. The construction component would also replace about half of 7.2 million jobs in the construction industry employed in expanding sprawl.[99] The recycling program would make up for more than 700,000 jobs lost in mining.[100] The various rail programs and electric car manufacturing would

Table 2
Summary of Program for Creating a Green Energy Economy

Industry	Total jobs per year	Manufacturing jobs per year	Cost per year/ billions
Wind 4,000 TWh	1,130,000	225,000	150
New electric grid with battery storage	1,100,000	300,00	85
100,000 250-unit apartment buildings	10,000,000	2,500,000	500
100% Organic agriculture	270,000	0	N/A
Recycling	2,000,000	200,000	100
17,000 mile high-speed rail system	600,000	90,000	301
High speed rail operations	1,000,000	0	0
Electric freight train system	500,000	125,000	25
Transit capital	300,000	300,000	60
Transit operating	1,300,000	0	200
Geothermal heat pumps	1,000,000	250,000	50
Solar 1,000 TWh	2,500,000	600,000	150
Weatherizing	1,000,000	250,000	25
Electric car	1,000,000	500,000	0
Total	**23,700,000**	**5,040,000**	**1,375**

replace most of the 6 million jobs that were counted previously in the vehicle manufacturing and services industries, most of which would probably disappear in a green economy. In other words, while approximately 10 million jobs would be lost, some 14 million more jobs would be created in a green economy, with a stronger manufacturing base than in the current US economy. These new green jobs will have indirect employment effects that are equivalent to or greater than the benefits of direct jobs; in other words, about 20 million extra jobs could be created as well through a green economy (because of the multiplier effect[101]).

Conclusion

In order to create a thriving national economy and a strong middle class, the United States needs to re-establish its manufacturing base. By engaging in a program of economic reconstruction as laid out in this chapter—transforming the transportation, energy, building, urban, manufacturing, and agricultural systems— the country creates a golden opportunity to fulfill these long-term goals. However, the Federal government, in concert with local and state governments, will have to engage in a minimum twenty-year construction program. The investment in this program should be close to $1 trillion (or more) in order to capture the benefits of mutually beneficial programs. Creation of an "infrastructure–industrial complex" to at least partially replace the military–industrial complex of the past decades would go far in bringing about the political will for such a transformation.

At the same time, by virtually eliminating the use of fossil fuels, by conserving our water, soil, and ecosystems, and by re-using our resources instead of throwing them away, we can prevent the ecological catastrophes of global warming, resource depletion, and ecosystem destruction.

We face a difficult set of challenges, both economic and ecological. Fortunately, we have the technologies, the resources, and the human talent to meet those challenges. Ultimately, ecological sustainability is the same as economic sustainability. The earth, the machine, and our species can co-exist peacefully, if we so choose.

Notes

1. Brown, L., *World on the Edge: How to Prevent Environmental and Economic Collapse* (New York, NY: W. W. Norton & Company, 2011); McKibben, B., *Eaarth: Makin: Making a Life on a Tough New Planet* (New York, NY: St. Martin's Griffin, 2011); Speth, J. G., *The Bridge at the Edge of the World: Capitalism, the Environment, and Crossing from Crisis to Sustainability* (Hartford, CT: Yale University Press, 2009).

2. U.S. Department of Energy, Energy Information Administration, *Annual Energy Review 2010*, Section 2: Energy Consumption by Sector (Washington, DC: Government Printing Office, 2011), Retrieved from http://www.eia.gov/totalenergy/data/annual/pdf/aer.pdf.

3. Rynn, J., *Manufacturing Green Prosperity: The Power to Rebuild the American Middle Class* (Santa Barbara, CA: Praeger Press, 2010), Part 2.

4. World Trade Organization, *International Trade Statistics 2011* (Geneva, Switzerland: World Trade Organization Publications, 2011). Retrieved from http://www.wto.org/english/res_e/statis_e/its2011_e/its11_toc_e.htm. See the book cover for aggregate view, and tables I.16 and I.4 for more detail.

5. U.S. Department of Commerce, Bureau of Economic Analysis, *International Economic Accounts website*, http://www.bea.gov/international/index.htm, Trade in Goods and Services. Retrieved from http://www.bea.gov/newsreleases/international/trade/trad_time_series.xls.

6. U.S. Department of Energy, Energy Information Administration, *Petroleum and other liquids website* at http://www.eia.gov/petroleum/, data tab, Crude Reserves and Production, Crude Oil Production website, View History link for 1920–2011, from U.S. production row, 2011.

7. Unless otherwise indicated, all "$" amounts signify U.S. dollars.

8. The figures will use 2009 as the base year. This data is available from the website of the Bureau of Economic Analysis, U.S. Department of Commerce. First go to this page, http://bea.gov/iTable/index_industry.cfm, then click on the "Begin using the data" button, then on the next web page click "Next Step," then click on "Value-added by Industry," and when that section opens up, click on "Value-added by industry" again. You will see several years of data. You can now click on the "GDP-by-Industry tab, and choose "Gross Output by Industry," and then the "Gross Output by Industry" link when that section opens up. "Gross Output" is total revenues, including all other

sectors, while "Value Added" just reports the value added only by the industry in question.

9. U. S. Department of Commerce, Bureau of Economic Analysis, "Annual Industry Accounts," *Survey of Current Business* (May 2011) 17, table 2. Retrieved from http:// bea.gov/scb/pdf/2011/05%20May/0511_indy_accts.pdf.

10. Pisano G., and Shih, W., "Restoring American Competitiveness," *Harvard Business Review* (July 2009).

11. Rynn, J., *Manufacturing Green Prosperity: The Power to Rebuild the American Middle Class* (Santa Barbara, CA: Praeger Press, 2010) Chapter 3.

12. Ibid, Chapter 7.

13. Data for this section can be found at the website of the Bureau of Economic Analysis, U.S. Department of Commerce. You can find this information on the following spreadsheets:

    http://bea.gov/industry/xls/GDPbyInd_VA_NAICS_1998-2010.xls
    http://bea.gov/industry/xls/GDPbyInd_VA_NAICS_1947-1997.xls
    http://bea.gov/industry/xls/GDPbyInd_FTPT_1948-1997.xls

14. Huang, J., "Structural Disarticulation and Third World Human Development," *International Journal of Comparative Sociology* 36 (1996): 164–83.

15. U.S. Census Bureau, *The 2011 Statistical Abstract of the United States*, "Accommodation, Food Services, and Other Services," table 1277. Retrieved from http://www.census.gov/prod/2011pubs/12statab/services.pdf.

16. U.S. Department of Labor, Bureau of Labor Statistics, "Employment Situation Summary for December 2011," http://bls.gov/news.release/archives/empsit_01062012.pdf (accessed January 6, 2012).

17. Pacey, A., *Technology in World Civilization: A Thousand-Year History* (Cambridge, MA: MIT Press, 1991).

18. Pursell, A., *The Machine in America: A Social History of Technology* (Baltimore, MD: Johns Hopkins Press, 2007).

19. Rosenberg, N., *Perspectives on Technology* (Cambridge, UK: Cambridge University Press, 1976). See in particular Chapter 1: "Technological Change in the Machine Tool Industry, 1840–1910."

20. Feldman, J., "From Mass Transit to New Manufacturing," *American prospect*, March 2009.

21. For a discussion of the inherent dangers of offshore drilling, see J. McQuaid, "The Gulf of Mexico Oil Spill: An Accident Waiting to Happen," *Environment* 360, http://e360.yale.edu/feature/the_gulf_of_mexico_oil_spill_an_accident_waiting_to_happen/2272/(accessed May 10, 2010); For a discussion of the Federal report on the disaster, see J. M. Broder, "BP Shortcuts Led to Gulf Oil Spill, Report Says," *New York Times*, September 2011.

22. Natural Resources Defense Council, *Tar Sands Invasion: How Dirty and Expensive Oil from Canada Threatens America's New Energy Economy* (New York, NY: National Resources Defense Council, 2010). Retrieved from http://www.nrdc.org/energy/files/TarSandsInvasion.pdf.

23. J. D. Hughes, *Will Natural Gas Fuel America in the 21st Century?* (Santa Rosa, CA: Post-Carbon Institute, 2011). Retrieved from http://www.postcarbon.org/reports/PCI-report-nat-gas-future-plain.pdf.

24. Dawson, A, "Extreme Extraction," *Counterpunch* September 2011. Retrieved from http://www.counterpunch.org/2011/09/09/extreme-extraction).

25. Howarth *et al.*, Rapid Assessment on Biofuels and the Environment: Overview and Key Findings, from Biofuels: Environmental Consequences and Interactions with Changing Land Use Proceedings of the Scientific Committee on Problems of the Environment (SCOPE) International Biofuels Project Rapid Assessment, September

22–25, 2008,Gummersbach, Germany, Howarth R. W. & Bringezu, S. eds., 2009. Retrieved from http://cip.cornell.edu/scope/1245782000).

26. De Decker, K., "The Status Quo of Electric Cars: Better Batteries, Same Range." *Low Tech Magazine*, May 2010. Retrieved from http://www.lowtechmagazine. com/2010/05/the-status-quo-of-electric-cars-better-batteries-same-range.html).

27. Rynn, J., "Singing the Transportation Electric: What would an all-electric transportation system look like?," presented at the conference "Sustainable Transit," held at the City College of New York, May 2010, by the University Transportation Research Center. Retrieved from http://economicreconstruction.org/sites/economicreconstruction.com/static/JonRynn/Jon_Rynn_Singing_Transportation_Electric.ppt.

28. Gore, A., *Our Choice: A Plan to Solve the Climate Crisis* (Emmaus, PA: Rodale, 2009), 286.

29. Manville, M., and Shoup, D., "Parking, People and Cities," *Journal of Urban Planning and Development* 131, no. 4 (2005).

30. Lovins, A., "Amory Lovins talks solutions," Harvard Press Office, March 2009. Retrieved from http://green.harvard.edu/amory-lovins-talks-energy-solutions.

31. U.S. Department of Labor, Bureau of Labor Statistics, *Occupational Employment Statistics*. Statistics are available on-line as follows: Vehicle manufacturing, http://www.bls.gov/oco/cg/cgs012.htm; for dealerships, http://www.bls.gov/oco/cg/cgs025.htm; for automotive repair and gas stations, first go to http://bls.gov/data/#employment, under the "Employment" heading, then the "Annual and other" heading, then click on "multi-screen" button, then click on the "multiple occupations for one industry" radio button, then click "continue" button, then pick an "industry sector," for example, "retail," then click "continue," then click "industry," for example, "gas stations," then click on the choice that picks all for the next two screens.

32. U.S. Department of Labor, Bureau of Labor Statistics, Occupational Employment Statistics. For trucking and warehousing, http://www.bls.gov/oco/cg/cgs021.htm; for aerospace, http://www.bls.gov/oco/cg/cgs006.htm; for the airline industry, http://www.bls.gov/oco/cg/cgs016.htm.

33. Keynes, J. M., *The General Theory Of Employment Interest And Money*, 1936.

34. Johnson, C., "Republic or empire: A National Intelligence Estimate on the United States," *Harper's Magazine*, January 2007, Retrieved from http://www.harpers.org/archive/2007/01/0081346); Engler, M., "War: The Wrong Jobs Program," *Foreign Policy in Focus*, November 2011.

35. Melman, S., *After Capitalism: From Managerialism to Workplace Democracy* (New York, NY: Alfred A. Knopf, 2001), 101 and 123. It has been very difficult to obtain hard numbers on military employment. Seymour Melman was a respected authority in the field, and these are his most recent numbers.

36. U.S. Environmental Protection Agency, Office of Policy, Office of Sustainable Communities, Smart Growth. "Location efficiency and Housing Type: Boiling it down to BTUs," 2011. Retrieved from http://www.epa.gov/smartgrowth/pdf/location_efficiency_BTU.pdf.

37. McCulloch, R., Pollack, E., and Walsh, J., "Full Speed Ahead: Creating Green Jobs Through Freight Rail Expansion," *Blue Green Alliance and the Economic Policy Institute*, May 2010, Retrieved from http://www.bluegreenalliance.org/admin/publications/files/BGA-Freight-Rail-Report-FINAL.pdf.

38. U.S. Environmental Protection Agency, Office of Policy, Office of Sustainable Communities, Smart Growth, G. Logan, S. Siejka, and S. Kannan, "The Market for Smart Growth," 2009. Retrieved from http://www.epa.gov/dced/pdf/logan.pdf; C. Leinberger, *Footloose and Fancy Free: A Field Survey of Walkable Urban Places in the Top 30 U.S. Metropolitan Areas*. Metropolitan Policy Program at the Brookings Institute (Washington, DC: Brookings Institute, 2007). Retrieved from http://www.

brookings.edu/~/media/Files/rc/papers/2007/1128_walkableurbanism_leinberg/
1128_walkableurbanism_leinberger.pdf.

39. EVStudio, "Construction Cost per Square Foot for Hotels, Motels and Apartments,"
August 2011. Retrieved from http://evstudio.info/construction-cost-per-square-foot-
for-hotels-motels-and-apartments/.

40. U.S. Department of Energy, Energy Efficiency and Renewable Energy, "Geothermal Heat Pumps," 2011. Retrieved from http://www.energysavers.gov/your_home/
space_heating_cooling/index.cfm/mytopic=12640.

41. U.S. Department of Energy, National Renewable Energy Laboratory, *Renewable
Energy in China: Development of the Geothermal Heat Pump Market in China.*
Document NREL/FS-710-39443. (Washington, DC: Government Printing Office,
2006) Retrieved from http://www.nrel.gov/docs/fy06osti/39443.pdf.

42. Brown, L., *Plan B 4.0: Mobilizing to Save Civilization* (New York: W.W. Norton &
Company, 2009), Chapter 5.

43. Gregor, A., "'Zero-Energy' Construction Crosses the Ocean," *New York Times*,
December 2011. Retrieved from http://www.nytimes.com/2011/12/04/realestate/
zero-energy-construction-crosses-the-ocean.html).

44. Baden et al. "Hurdling Financial Barriers to Low Energy Buildings: Experiences
Form the USA and Europe on Financial Incentives and Monetizing Building Energy
Savings in Private Investment Decisions," Proceedings of 2006 ACEEE Summer
Study on Energy Efficiency in Buildings, 2006. Retrieved from http://www.fsec.ucf.
edu/en/publications/pdf/FSEC-PF-396-06.pdf).

45. Goldstein, J. "More Jobs, Less Pollution: Growing the Recycling Economy in
the U.S.," *Blue Green Alliance*, November 2011, Retrieved from http://www.
bluegreenalliance.org/press_room/publications?id=0086).

46. Brown, L., *World on the Edge*, Chapters 2 and 3.

47. Pfeiffer, D., *Eating Fossil Fuels: Oil, Food and the Coming Crisis in Agriculture*
(Gabriola Island, BC: New Society Publishers, 2006).

48. Pimentel, D., Hepperly, P., Hanson, J., Seidel, R., and Douds, D., "Organic and
Conventional Farming Systems: Environmental and Economic Issues," Report 05-1
(Ithaca, NY: Cornell University College of Agriculture and Life Sciences, 2005),
Retrieved from http://ecommons.cornell.edu/bitstream/1813/2101/1/pimentel_
report_05-1.pdf).

49. Organic Trade Association, "Industry Statistics and Projected Growth," 2011.
Retrieved from http://www.ota.com/organic/mt/business.html.

50. Stewart, A., "The Man Who Would Feed the World: John Jeavons' farming
methods contain lessons for backyard gardeners too," *San Francisco Chronicle*,
April 2002.

51. Palmer, P., *Getting to Zero Waste* (Portland, OR: Purple Sky Press, 2005); McDonough
W., and Braungart, M., *Cradle to Cradle: Remaking the Way We Make Thing* (New
York, NY: North Point Press, 2002).

52. B. Gardiner, "Upcycling Evolves From Recycling," *New York Times*, November
2010; Heintz, J., and Pollin, R., "The Economic Benefits of a Green Chemical Industry in the United States Renewing Manufacturing Jobs While Protecting Health
and the Environment," *Blue Green Alliance and Political Economy Research Institute*, May 2011, Retrieved from http://www.bluegreenalliance.org/press_room/
publications?id=0070).

53. Leonhardt, D., "The Depression: If Only Things Were That Good" *New York Times*,
October 2011.

54. Jackson, K., *Crabgrass Frontier: The suburbanization of the United States* (Oxford,
UK: Oxford University Press, 1985) Chapter 13.

55. Castells, M., *The Rise of the Network Society: The Information Age: Economy, Society, and Culture Volume I* (Hoboken, NJ: Wiley-Blackwell, 2009).

56. Melman, S., *Permanent War Economy: American Capitalism in Decline* (New York, NY: Touchstone, 1985).

57. Melman, S. *Profits Without Production* (New York, NY: Alfred A. Knopf, 1988), Chapter 5; Dumas, L. "Finding the Future: The Role of Economic Conversion in Shaping the Twenty-first Century" In *The socio-economics of converting from war to piece* (Armonk, New York: M.E. Sharpe, 1995).

58. Melman, S., *After Capitalism*, Chapter 5, and S. Melman, *Profits without Production*, Chapters 9 and 10.

59. Melman, S. *After Capitalism*, Chapter 6.

60. U.S. Department of Defense, Procurement Programs (P-1), Office of the Under Secretary of Defense (Comptroller), 2011. Retrieved from http://comptroller.defense. gov/defbudget/fy2012/fy2012_p1.pdf.

61. Please see end note 8.

62. Rosenthal, E., "U.S. Military Orders Less Dependence on Fossil Fuels," *New York Times* (October 4, 2010).

63. Klare, M. "The Pentagon v. Peak Oil: How Wars of the Future May Be Fought Just to Run the Machines That Fight Them," *TomDispatch.com*, June 2007, Retrieved from http://www.tomdispatch.com/post/174810.

64. Spinney, F., "Why the War Machine Keeps on Running" *Counterpunch.org*, July 2011. Retrieved from http://www.counterpunch.org/spinney07052011.html.

65. Melman, S., "Beating Swords into Subways" *New York Times Magazine*, November 1978); J. Feldman, "The Conversion of Defense Engineers' Skills: Explaining Success and Failure Through Customer-Based Learning, Teaming and Managerial Integration," In *The Defense Industry in the Post-Cold War Era: Corporate Strategy and Public Policy Perspectives*, ed. Gerald I. Susman, and Sean O'Keefe (Oxford: Elsevier Science, 1998), 281–318.

66. U.S. Department of Transportation, Federal Highway Administration, "History of the Interstate Highway System," 2011. Retrieved from http://www.fhwa.dot.gov/ interstate/history.htm.

67. Crawford, J., "Interstate Rail: Adapting the Interstate Highway System to Rail Use," 2001 Retrieved from http://www.jhcrawford.com/energy/interstaterail.html.

68. Milligan, M et al., "Power Myths Debunked," *IEEE Power and Energy Magazine* November/December, 2009. Wind.

69. Colby, D., "Turbines and Health," *Magazine of the Association of Power Producers of Ontario* 25, no. 5 (November 2011), Retrieved from http://awea.org/blog/index. cfm?customel_dataPageID_1699=12329; R. Cohen, "Britain goes Nimby," *New York Times*, August 2011; E. Rosenthal, "Tweety Was Right: Cats Are a Bird's No. 1 Enemy," *New York Times*, March 2011; S. Wright, "The Not-So-Green Mountains," *New York Times*, September 2011.

70. Galbraith, K., "Wind Power Gains as Gear Improves," *New York Times*, August 2011.

71. U.S. Department of Energy, Energy Efficiency and Renewable Energy, *20% Wind Energy by 2030: Increasing wind energy's contribution to U.S. electricity supply*, Document DOE/GO-102008-2567.

72. Gail the Actuary, "The US Electric Grid: Will it be Our Undoing?—Revisited," *TheOilDrum.com*, August 2010. Retrieved from http://www.theoildrum.com/ node/6817; American Society of Civil Engineers, "Report Card for America's Infrastructure: Energy," 2011. Retrieved from http://www.infrastructurereportcard. org/fact-sheet/energy#conditions.

73. Makansi, J., *Lights Out: The Electricity Crisis, the Global Economy, and What It Means To You* (Hoboken, NJ: Wiley, 2007).
74. Melman, S. *After Capitalism*, Chapter 6.
75. Feldman, J., "The Foundations for Extending Green Jobs: The Case of the Rail-Based Mass Transit Sector in North America," *International Journal of Labour Research* 2, no. 2 (2010): 269–92.
76. Yang, J., "Companies Pile Up Cash but Remain Hesitant to Add Jobs," *Washington Post*, July 2010.
77. Jackson, *Crabgrass Frontier*, Chapter 11.
78. U.S. Department of Energy, Energy Information Administration, "Renewables and Alternative Fuels, Wind," January 2011. Retrieved from http://38.96.246.204/cneaf/ solar.renewables/page/wind/wind.html.
79. U.S. Department of Energy, Energy Information Administration, "Annual Energy Review 2010, Table 8.1. Electricity Overview, Selected Years, 1949–2010," (Washington, DC: Government Printing Office, 2011), 235. Retrieved from http://38.96.246.204/totalenergy/data/annual/pdf/sec8_5.pdf.
80. U.S. Department of Energy, Energy Information Administration, "Electric Power Annual 2009: Table ES1. Summary Statistics for the United States, 1998 through 2009," 2011, 9–11. Retrieved from //38.96.246.204/cneaf/electricity/epa/epaxlfilees1. pdf, also at http://38.96.246.204/cneaf/electricity/epa/epates.html.
81. See Note 78.
82. U.S. Department of Energy, National Renewable Energy Laboratory. "Utility-Scale Energy Technology Capacity Factors Energy Analysis," 2011. Retrieved from http:// www.nrel.gov/analysis/tech_cap_factor.html.
83. *Wind Energy by 2030*, Chapter 3, 61.
84. Archer, C., and Jacobson, M., "Supplying Baseload Power and Reducing Transmission Requirements by Interconnecting Wind Farms," *Journal of Applied Metereology and Climatology* 467 (November 2007): 1701–17. Retrieved from www.stanford.edu/group/efmh/winds/aj07_jamc.pdf; Milligan et al., "Power Myths Debunked."
85. Lipow, G., *Solving the Climate Crisis through Social Change: Public Investment in Social Prosperity to Cool a Fevered Planet* (Santa Barbara, CA: Praeger Books, 2012), Chapter 15.
86. Laxson, A., Hand, M. M, and Blair, N., "High Wind Penetration Impact on U.S. Wind Manufacturing Capacity and Critical Resources," Technical Report NREL/ TP-500-40482, 2006, Retrieved from http://www.nrel.gov/docs/fy07osti/40482.pdf.
87. Sterzinger, G, and Svrcek, M., "Wind Turbine Development: Location of Manufac-turing Activity," Renewable Energy Policy Project, 46, September 2004. Retrieved from http://www.repp.org/articles/static/1/binaries/WindLocator.pdf.
88. Vestas.com, "Employees," 2011. Retrieved from http://www.vestas.com/en/ about-vestas/profile/employees.aspx); Vestas.com, "Interim Financial Report for the Third Quarter of 2011," 2011. Retrieved from http://www.vestas.com/en/media/ news/news-display.aspx?action=3&NewsID=2879.
89. World Steel Association, "World Crude Steel Output Increases by 6.8% in 2011," *World Steel Association press release*, January 2012. Retrieved from http://www. worldsteel.org/media-centre/press-releases/2012/2011-world-crude-steel-production. html.
90. Laxson et al.*High Wind Penetration*, 18.
91. U.S. Department of Labor, Bureau of Labor Statistics, "Career Guide to Industries, 2010–11 Edition: Steel Manufacturing," 2011. Retrieved from http://www.bls.gov/ oco/cg/cgs014.htm; Laxson et al. *High Wind Penetration,* Chapter 5.

92. Windustry.org, "How much do wind turbines cost?" 2011. Retrieved from http://www.windustry.org/how-much-do-wind-turbines-cost; *20% Wind Energy by 2030*, 61; U.S. Department of Energy, National Renewable Energy Laboratory, Energy Analysis, "Utility-Scale Energy Technology Capital Costs," 2011, Retrieved from http://www.nrel.gov/analysis/tech_costs.html.

93. Vestas.com, "Procurement Turbines," Retrieved from http://www.vestas.com/en/wind-power-plants/procurement/turbine-overview.aspx#/vestas-univers.

94. Gelman, R., "*2010 Renewable Energy Data Book*," (U.S. Department of Energy. Energy Efficiency and Renewable Energy, Document DOE/GO-102011-3310, 62, 2011, Retrieved from http://www.nrel.gov/analysis/pdfs/51680.pdf.

95. National Renewable Energy Lab, Science and Technology, Dynamic Maps, GIS Data, and Analysis Tools. Wind Maps. "Wind Resource Map," 2009. Retrieved from http://www.nrel.gov/gis/pdfs/windsmodel4pub1-1-9base200904enh.pdf, via http://www.nrel.gov/gis/wind.html.

96. Owen, D., "The Efficiency Dilemma: If Our Machines Use Less Energy, Will We Just Use Them More?" *The New Yorker*, December 2010. For a rebuttal, see J. Barrett, "Debunking the Jevons Paradox: Nobody Goes There Anymore, It's Too Crowded," *ClimateProgress.org*, February 2011. Retrieved from http://thinkprogress.org/romm/2011/02/16/207532/debunking-jevons-paradox-jim-barrett/?mobile=nc.

97. Berndt, R. E., and Doane, M., "System Average Rates of U.S. Investor-Owned Electric Utilities: A Statistical Benchmark Study," MIT document MIT-CEEPR 95-005WP, June 1995. Retrieved from http://dspace.mit.edu/bitstream/handle/1721.1/50183/35719406.pdf, 9.

98. Behr, P., "Smart Grid Costs Are Massive, but Benefits Will Be Larger, Industry Study Says," *New York Times*, May 2011.

99. U.S. Department of Labor, Bureau of Labor Statistics, "Career Guide to Industries, 2010–11 Edition: Construction," 2011. Retrieved from http://www.bls.gov/oco/cg/cgs003.htm.

100. U.S. Department of Labor, Bureau of Labor Statistics, "Career Guide to Industries, 2010–11 Edition: Mining," 2011. Retrieved from http://www.bls.gov/oco/cg/cgs004.htm.

101. Bivens, J., "Updated Employment Multipliers for the U.S. Economy," Working paper 268, Economic Policy Institute, 2003.

7

Transitioning to Eco-Cities: Reducing Carbon Emissions while Improving Urban Welfare

Susan Roaf

Introduction

We live in a world where rapid change and unmanaged growth is driving the collapse of established systems, be they top-down macroeconomic failures of the economies such as those of Greece and Spain in 2012, the bottom-up collapse of the housing markets in the United States in 2007–2010, or the escalating failure of infrastructures and buildings to provide adequate shelter during extreme weather events (see Figure 1). While seeking to avoid such collapses, we must understand where economic weaknesses and related breaking points exist and how to measure them. In rapidly evolving systems, we must efficiently draw on experience to grow in successful directions. This chapter tackles the challenges of creating such measures and explores the form they might take, using case studies from Arizona and the city of Dundee in Scotland.

The chapter is divided into four parts. The first deals with concepts of resilience and adaptive capacity of individuals and populations within that built environment. Part two presents a case study of the failure of the Arizonan housing market in 2007–2010. This details the problem of asymmetric information, the ignorance by the majority of the potentially catastrophic impacts of rising energy, transport and water prices on their own household budgets, and raises the question of how this was allowed to happen in a so-called "responsible" society. Part three introduces the second case study of the city of Dundee in Scotland and a consideration of the role that solar technologies might play in alleviating

Figure 1
The Warming of the Climate Will Exacerbate the Nature and Rate of Collapse of the Whole Gamut of Our Social and Physical Systems within the Built Environment. Strengthening of These Systems Is Essential to Improve Their Adaptive Capacities and Resilience to Collapse

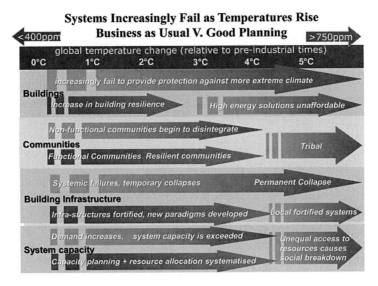

Source: Roaf et al. 2009.

fuel poverty in that city. The fourth and final part discusses the development of standardized metrics and indicators for describing the adaptive economic capacity of a population and testing the sensitivities of a group to a range of hazards that may threaten to cause the socioeconomic systems within which they operate to collapse. Such metrics and indicators would provide decision makers with the critical ability to test policies and strategies against the capacity of the system to absorb stress and its breaking points under a range of different conditions. While so doing, such metrics can be used to inform successful adaptation strategies for different populations.

The conclusions clearly point to the social and economic benefits of moving toward facilitating the adoption of wide-scale use of solar energy for domestic populations to reduce economic stress upon them and prevent the collapse of regional housing markets and with them their attendant communities.

Resilience

Systems all around us are beginning to very visibly fail, some in response to extreme weather events, some because of aging and poor infrastructures, and some because the capacity of those systems are being pushed beyond their limits by physical pressures and/or, by mismanagement and a lack of planning. The growing human and economic costs of system failures, be they local power outages or the tripping out of a grid feeding 700 million people in northern India, is driving the political imperative to build more "resilience" into the lives, landscapes, infrastructures, communities, and businesses of citizens around the world. Resilience is now a headline issue that sells popular books at airports[1] and is stimulating a new generation of resilience research and thinking. If so many clever people are putting their minds into the subject, why does the rate and scale of the collapse of systems around us continue to increase so alarmingly? This is a key question addressed in this paper in relation to housing and the energy used to run it.

Resilience is described differently by engineers, ecologists, or system scientists. It is, however, an attribute of social and physical systems that is increasingly sought by planners, architects, engineers, and politicians alike. The technical definition of resilience for material scientists *is the property of a material to absorb energy when it is deformed elastically and then, upon unloading, to have this energy recovered*—the ability to bounce back to normal after a testing event.

The amount of resilience is defined by material scientists as the maximum energy (stress) that can be absorbed per unit volume without creating a permanent distortion as can be calculated for metals, for instance, by integrating the stress–strain curve from zero to the elastic limit.[2] One premise of this paper is that we need to be able to define the amount of stress that can be absorbed in the housing sector before it begins to yield to that pressure. Another challenge is to develop methods for defining the bandwidth of the adaptive opportunities available for households to take the extra strain before the system collapses and does not "bounce back" and therefore suffers permanent distortion. Without being able to define those three properties of the system: yield behaviors of the system, adaptive capacity bandwidths and fracture points, planners, designers, and politicians have insufficient information to be able to act accordingly to maximize the adaptive capacity of the system and avert fracture and system collapse.

Figure 2 shows the characteristics that are involved in this definition. Stress is placed on the material that reaches a yield point; then, the bandwidth of strain is a range over which the material can absorb stress before it reaches its maximum strength and then fractures. In a metal, as with socioeconomic systems, the fracture points also change with the temperature of the material.

The rate at which stress is applied to an inanimate, non-learning metal does not affect the strain at which it fractures because the metal simply mechanically performs to its limits. In "learning" ecological systems, the rate of change, a product of the rise (scale of the stress) and run (scale of the strain) of the process over time, can be significant. Berkes et al.[3] point out that in a rapidly changing world, socioecological systems either evolve or do not survive. The function of this evolution is to increase the strain the system can absorb before collapse occurs.

Figure 2
The Metallurgists Definition: Resilience is the Area under the
Linear Portion of a Stress-Strain Curve

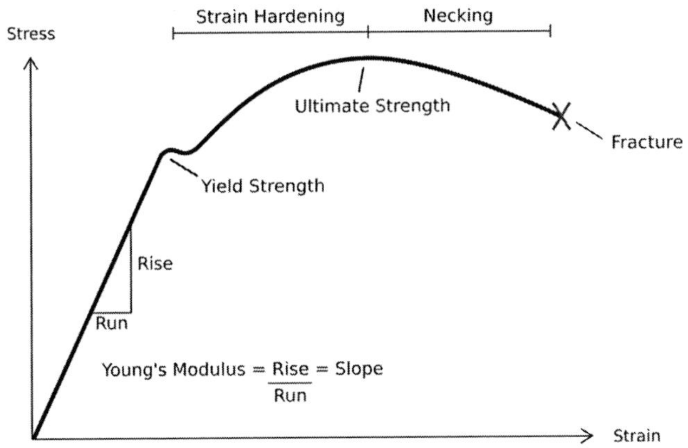

Source: Campbell, 2008.

The capacity, or maximum strength or ability to do work, of the whole system is obviously key in relation to the demands placed upon it. Authors have been warning of the risks posed by growing populations within the limited ecosystems of the planet for four decades.[4,5] However, the rate of change in the system is obviously also critical in "learning" systems, and Donella Meadows[6] argued that the buffeting of our socioeconomic systems by multiple stresses today makes it difficult to react "in time" to threats because the complex behaviors of systems that are dependent on the relative strengths of the multiple feedback loops in the system causing a chain of loops to dominate behaviors. The critical issue here is how confounding this complexity is.

Clarity is sought by some who believe that answers can be reached through disaggregation of the problem. "If resilience is to be measured operationally it must be in relation to a potential and specific change in the system."[7] The risk of a system breaking in relation to climate change in the built environment has been disaggregated and characterized by David Crichton as having three vectors:[8,9]

- *Hazard*: how bad is it going to get? How large is the stress placed on the system?
- *Exposure*: where are people situated in relation to that Hazard? How exposed is the system to the hazard and how likely is it to happen, how often and how hard?
- *Vulnerability*: how likely is the combination of the above to prove lethal? What strain can the system take? What are the adaptive opportunities and potentials of the system?

The impacts of resilience building actions must also be taken into account to ensure that the reinforcing of one part of the ecosystem in the short term does not threaten the survival of the whole ecosystem over time. "Resilience is the potential to sustain development by responding to, and shaping change in a manner that does not lead to the loss of future options. Resilient systems also provide capacity for renewal and innovation in the face of rapid transformation and crisis."[10] Here, the idea of innovation being a key tool for those trying to build resilience into the system is important, but where perhaps innovation is most needed is not in the development of physical innovations and "product" based approaches but in the resetting of the way in which we view the problem, and importantly the language in which we describe it.

Geoff Wilson[11] defined resilience as follows: "The capacity of a system to absorb disturbance and reorganize while undergoing change to still retain essentially the same function, structure and identity, and feedbacks . . . resilience is measured by the size of the displacement the system can tolerate and yet return to a state where a given function can be maintained." Perhaps, the presumption that the same functions, structures, and identities are to be maintained threatens the survivability of the structure. It does appear to presume that the way things were done before is how they should be done in the future, which is a counter-intuitive assumption in a system that has already failed in a rapidly changing world. C. S. Holling[12] noted that "Placing a system in a straitjacket of constancy can cause fragility to evolve," and all too often in recent years this fragility, or brittleness, of the fabric of the built environment has proved lethal in terms of the thousands of lives and buildings lost to flood, heat, cold, or winds as well as proving catastrophic for the livelihoods and communities involved.

Ecosystems today do have to evolve rapidly to survive but perhaps we should be more radical in the way we re-imagine our futures. Business-as-usual interests and thinking are, according to some, providing a damaging straitjacket to the evolution of systems. Ironically, market liberalization and programs of deregulation led by the Chicago School of Economics and their advocates, such as Milton Friedman and his adherents Margaret Thatcher and Ronald Regan, weakened the built environment by allowing the "lowest common denominator" solutions to prevail. For instance, in the building industry, this resulted in the prevalence of the cheapest and most brittle building types despite the apparent move in the United Kingdom toward more stringent regulations. An example is the Shard building in London, just completed, that is a tall, all glass, and steel tower with appalling environmental building performance and embodied energy costs that are many times those of a modest office block that will be refused planning permission because it is naturally ventilated. There is one rule for the rich here and one for the poor. Joseph Stiglitz[13] rails against the chronic damage done by those who have undermined the public sector, plundered financial markets, and been "seduced and neutered by lobbyists and their own avarice."

Garrett Hardin[14,15] developed the idea of the Tragedy of the Commons to describe the dilemma that arises from the situation in which multiple individuals acting independently and rationally consulting their own self-interest will ultimately deplete a shared limited resource, even when it is clear that it is not in anyone's long-term interest for this to happen. His ideas were taken up by many attempting to rationalize why, in the face of our growing knowledge base, decisions were being made that appeared to push progress into socially, environmentally damaging directions. All of us wanting and having more would inevitably push local ecosystems, and in turn, economies, beyond their capacities. Stiglitz[16] highlighted the profound dangers of "asymmetric information" where some individuals have access to privileged information that others don't causing unfair advantage that can result in system collapse in itself. Increasing numbers of authors suggest that the internal mechanisms of neoliberal and turbo-capitalism, which are devoid, in such ways, of values of fairness, trust, and civil responsibility, are to blame and will inevitably take us to the cliff edges highlighted by Hardin and Stiglitz.[17,18]

Many authors point to the confounding nature of the high level of complexity of the systems in play and their feedback loops, and the problems associated with our reliance on trend extrapolating models in forecasting and predicting the performance of systems. Robert Ayres[19] specifically points out that to forecast "turning points," it is necessary to get away from trend-based models as used in extrapolation but goes on to point out the weakness of trying to characterize too many complex nonlinear interactions with limited differential equations such as those used since the early years of Ecological modeling and in the original Limits to Growth model by Meadows et al.[20]. Ayres claimed that simple quantifiable models will not be adequate to identify timings and other attributes of Turning Points but that *"naive intelligence and intuition may be the best tool for coping with a very complex and non-deterministic future."*

Adaptive Capacity

In the background documents for the third report of the International Panel on Climate Change,[21] the concept of adaptive capacity was developed in relation to the vulnerability of populations. At a national level, vulnerability is determined by factors such as economic wealth, technological opportunities and infrastructural resilience, information and knowledge, and equity and social capital. Both determinants of, and indicators of, adaptive capacity are typically given at a national level, rather than in relation to the risk level of an individual sector, person, or household,[22] and definitions of key adaptation terms are not often associated with methods for their quantification.[23,24,25] Key data collection and analysis methods for adaptation typically deal, on one hand, with high-level measures such as gross domestic product (GDP), or are specific to local community-level factors.[26]

While adaptive capacity can cushion the impacts of climate hazards, Lea Berrang-Ford et al.[27] and others have noted that adaptive capacity does not necessarily translate into adaptation not least because of the "wicked" complexity of the issues involved. The role of a wild card such as background temperatures at which a particular hazard manifests itself for a set exposure and vulnerability level may prove to be a determining factor in system collapse, an idea that is developed in the Arizonan case study described below. The speed at which catastrophic events occur can inhibit effective action being taken in time and the routes between data collection, analysis to strategy development, and action may not exist or be difficult and slow to functionalize, and the inappropriate scale of available data in relation to the available policy tools for implementation is an issue. So, the actual scale, or grain, of available data in relation to the scale of the data requirements for policymaking and implementation are critical factors and need to be compatible.

The following case study sets out to look at the operational bandwidths for adaptive economic capacity of typical households in Phoenix, Arizona, which were crossed over the course of the catastrophic collapse of the Arizonan housing market in 2007–2008. The idea that Ayres' "naive intelligence and intuition," or perhaps "common sense" might have been applied to inform a more resilient development model for Arizona Valley is proposed.

Arizona: A Case Study of Housing Market Collapse

This case study arose from a period the author spent as visiting Research Fellow in the Herberger Institute Research Centre with Professor Harvey Bryan and Janet Holston at Arizona State University (ASU) in Tempe in the spring and summer of 2007. Having recently completed a book on climate change,[28] it appeared that Arizonans were potentially a) exposed and b) vulnerable to both oil shocks and spikes in oil prices and extreme weather events in the hot, dry desert climate for which Arizona is famous. At ASU, it became clear that the individuals and their families were often sitting on a financial fence where pressures in each direction could unbalance their monthly budgets for better or worse. Unstable oil prices at the time were particularly problematic.

The Peak Oil issue has been widely discussed for decades. Interest in the phenomenon of Peak Oil started with the work of M. King Hubbert in the 1950s.[29] By the 1990s, the related movement was a growing force.[30] The potential scale of the related impacts for keeping buildings and cities[31] comfortable during extreme weather events were beginning to surface, and by 2005/6, credible studies began to predict that the point of global maximum oil production was approaching.[32] Colin Campbell of the Association for Peak Oil (ASPO) had predicted a 2007/8 peak for global oil production, which was subsequently acknowledged as credible by a number of authorities.[33] By the turn of the millennium, this oil peak was being associated with the potential for "oil shocks" within the global economy.[34]

Teaching a course in the Global Institute of Sustainability (GIOS) at ASU in its inaugural year offered an opportunity to explore different aspects of the "sustainability" of communities across the valley with student groups. One such community was that of the new, idealized, out of town "sustainable" community of Verrado, which is part of the larger town of Buckeye.[35]

Verrado is a new master-planned community on 8,800 acres of land owned by Caterpillar, which in the mid 1990s went into partnership with the developer DMB to create the new town. The area was zoned for up to 14,080 homes. Since the 2008 housing crash, the planned numbers were scaled back to eleven thousand home sites. In reality, only around two thousand homes have been completed to date. The 2004 opening of the first homes was at the peak of the Arizona housing boom and at the top of the market; DMB claimed to be selling a hundred homes a week. DMB constructed parks, roads, a village center, a golf course, health club, and schools and wrote its own development and zoning codes that permit mixed use. There was a high ambition for this to be a truly "sustainable" community, a concept with many different definitions.[36] Designers applied the philosophy of New Urbanism to the town and included medium urban densities, walkable destinations, mixed-use buildings, and a traditional village center with pedestrian ways through connective parks, schools, and job centers. Students identified the following properties of the development:

Strengths

- Diverse land use within community
- Hundred year water rights secured
- Well-engineered wastewater reclamation system

Weaknesses

- Poor initial plant selection in common areas
- Overall local increasing demand on Colorado River water

Recommendations

- Replace common area plants with more suitable species.
- Realize reduced water and energy consumption.

In hindsight, these recommendations appeared toothless and reflected the prevailing mindset of "everything is alright" despite huge cracks already being visible in the developer model that was reshaping Arizona Valley.

The GIOS study did identify the problem of covering large distances to travel for work, amenities, or recreation, and in fact, the location of Verrado may well prove the most challenging obstacle it has to overcome in its future. It is located in an area with very few local jobs, 40 miles from Phoenix and 45 miles from Tempe and Scottsdale, the three largest local areas of employment. Some websites claim Verrado is only 25 miles from Phoenix making commuting appear more attractive and affordable. More credible sites claim the higher figure. The difference between the two mileage estimates for someone who, for instance, worked at ASU in Tempe, lives in Verrado, and drives a 20 mile per gallon SUV would be around 200 miles a week. At $4 a gallon of gas, this equates to an extra cost of around $40 a week and $150–$200 a month difference in commuting costs. A desire to manage information on this issue by developers wanting to sell homes is understandable. Here is a good example of "asymmetric information."

Buckeye is also in an area with extremely limited freshwater rights despite house purchasers being told that it has hundred years of water rights. Again—"asymmetric information." The high cost of domestic water supplies has proved an additional, often unforeseen, cost in moving to the town. Monthly water rates were always high, but in 2012, the Arizona American Water Company sought an immediate 83 percent increase in water rates, translating to roughly $25.00 per month more for the average water customer in Verrado and other communities. After much negotiation, a settlement was reached to phase-in annual increases that totaled 58 percent by the third year. Average water users could expect the following on their monthly water from April to July 2012: no increase; a $12.95 increase on monthly bills in July 2012 followed by a $3.09 increase in July 2013; and a further $3.09 increase in July 2014 (www.verrado.net).

The subject of water provision and rights in Arizona is an enormously complex one and well-covered by experts such as Robert Glennon[37,38] and James Powell,[39] who have written eloquently of the problems in books on America's water crisis and the excessive use of the finite ground water to spur on development growth. Water literacy will not only increasingly become a part of local, state, and national election cycles, but it is also clear from blogs for buyers wanting to buy homes in Verrado that the cost of water is creeping up their lists of priorities. People are now requesting information on the local water rates from estate agents as they consider property purchases. In 2007, researchers at ASU knew that Buckeye did not have the claimed adequate hundred years of fresh water rights, but they were not allowed to go public on such facts because of the power of the developers in the state. How the Arizona government, which had all the facts on hand, let this happen remains to be explored, although Andrew Ross[40] clearly describes the conspiracy of the influential in Arizona to promote unsustainable growth to the detriment of the masses and the enrichment of those with power. In addition, new legislation was introduced in 2007 to liberalize rules on the allocation of water to promote the interests of housing developers with prices for water per gallon set at half that of rainy Seattle where water use per person was half that of Arizona.

The energy–water nexus is the most hidden actor in the affordable housing crisis. Water, Glennon argued[41], is more important than oil in our society, since nearly every economic activity and production cycle requires water. Energy companies cannot produce electricity without water. The energy industry depends on water to process and transport petroleum, coal, natural gas, and other fuels like ethanol. Municipalities need this energy to pump water through distribution systems, treat drinking water, and manage storm water. As such, Glennon argues that sustainable water, energy, and economic policy should not be separated but are currently negotiated independently. But what in fact became the nemesis of such settlements as Verrado in 2007/8 was not the predicted and catastrophic collapse of water supplies[42] but an unparalleled spike in the price of oil.

The pressures faced by ordinary Arizonan families in paying monthly bills seemed clear to my common-sense view of the world, and on April 27, 2007, I presented the following tables at the Livability Summit organized by Valley Forward, a group of five hundred of the leading companies in Arizona Valley. With the rather imperfect tables for single-person outgoings in Verrado and Garfield, we tried to point out the vulnerability of the ordinary household to rising energy prices. They do clearly, if crudely, show the impacts that a doubling of energy prices would have on monthly and annual outgoings, and I suggested then that this might well happen by the year 2020, which is why the predicted mortgage costs—in the pre-bust markets—rose for both communities. The price of oil in fact doubled by July 2007. Table 1 shows a sketchy list of what a young professional with a well-paid job might pay per month living a baseline "American Dream" lifestyle in a three-bedroom home in Verrado in April 2007 and what they would have to pay if the price of energy across the board doubled. Altering the cost of the mortgage and doubling the energy and water costs has a profound effect on the income requirements. A required basic salary of $71,488 leaps to $93.756 in what proved to be a devastating difference for many ordinary families in Arizona Valley.

In contrast, the "sustainability" concerns of residents of the Phoenix inner city suburb of Garfield[43] centered around the community-led initiatives to reduce gang crime in their area to which they had considerable success. Garfield inhabitants paid much less for commuting to work and for water, but they had a different additional expense to deal with in 2007: the extra cost of air-conditioning owing to the heat island effect. The climate of Phoenix has been particularly well-studied, and the city has one of the USA's Long-Term Ecological Research (LTER) centers. Thomas Karl[44] demonstrated a clear correlation between growing urban populations and increasing urban temperatures, and there has been extensive subsequent work on the magnitude and workings of the Urban Heat Island (UHI). In 2000, A. J. Brazel et al.[45] published a detailed, comparative study of the UHI in Phoenix and Baltimore and showed that the long-term urban UHI temperatures had increased in both cities in line with Karl's[46] graph, showing a clear correlation between rising UHI temperature and growing populations.

Darren Ruddell et al.[47] built on this work in a study looking at the relationship between health and the UHI. The 1995 heat wave in Chicago had shown that those most likely to suffer and die during such events are the poor, the elderly, and some ethnic minorities.[48] The 2003 heat wave in Europe, during which around fifty-two thousand excess deaths from heat occurred,[49] showed the potential scale of heat related disasters. Ruddell et al. showed that when it gets so hot, the whole city is affected but some areas can provide more cooling protection. Parks, shade, trees, and vegetation play a significant role in reducing temperatures locally. Building

Table 1
Rough Calculation Done in April 2007 of What a Single Young Professional with a Well-Paid Job Might Pay Per Month, and Over a Year, Living in a Three Bedroom Home in Verrado and What They Would Have to Pay if the Price of Energy Doubled Across the Board

Single person	VERRADO 2007		VERRADO 2020	
2020 = x 2 energy prices Without water trucking prices	$ per month	$ per year	$ per month	$ per year
Mortgage	2000	24000	2800	33600
Property Tax	500	6000	500	6000
Water + sewer	80	960	160	1920
Energy	300	3600	600	7200
Car purchase	300	3600	300	3600
Gas / petrol	250	3000	500	6000
House insurance	100	1200	100	1200
Car insurance	100	1200	100	1200
Telephone, internet, mobile	100	1200	100	1200
Cable	50	600	50	600
Health insurance	200	2400	200	2400
Extras	100	1200	100	1200
Groceries	300	3600	300	3600
Credit Cards	200	2400	200	2400
TOTAL	**4,580**	**54,960**	**6,010**	**72,120**
& 30% TAX		16488		21636
Annual salary		**$71,488**		**$93,756**

Source: Sue Roaf.

on such science, ASU academics, including Jay Golden and Harvey Bryan, were able to plot the increase in energy use for air-conditioning, which is due to Phoenix's increasing average summer minimum low night-time temperatures alone. This is the additional energy penalty paid by the ordinary families of Phoenix solely as a result of the city's UHI. Again, the lower-income families who have the cheaper and less efficient cooling systems will pay more in running costs, as they do for driving the cheaper, older, and less-efficient vehicles and living in areas where there are no verdant gardens around their homes.

Table 2 shows that a lower salary will support the minimal "American Dream" standard of life in Garfield but that energy for increased air conditioning use in

Figure 3
Retrospective HVAC Energy Use Analysis (over fifty years) for a 2000 square foot Single Family Home in Phoenix, Arizona. The Increase in Energy Use Shown is Solely Due to Phoenix's Increasing Average Summer Minimum Low Night-Time Temperatures Resulting from the Urban Heat Island

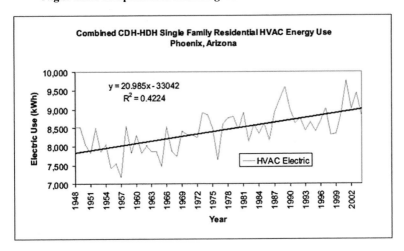

Source: Harvey Bryan and ASU's SMART Program.

Phoenix has the potential over time to drive up the monthly bill in a hot summer. The Arizonan developer has a lot to answer for simply pushing people into larger and larger poorly insulated, often unshaded homes in gardenless lots further from work in the name of profit. Stiglitz[50] highlights the "price of inequality" when he promotes the idea of doing well by others and the "common welfare" as being a prerequisite for improving one's own position. The untrammeled greed of the Arizonan house builders and their friends in power at all levels of Arizonan society actually destroyed their own industry, businesses, and credibility in many cases.

In April 2007, oil prices hovered at a then all time high of around $65 a barrel. At the "Livability" Summit of the Valley Forward organization in Phoenix, I wanted to promote the need to understand oil dependence as a key vulnerability of their development model, and I urged them to consider building smaller homes that run largely on solar energy. I would classify the audience at the event as neutral to hostile. Within a year, Figure 4 shows that the "2020" scenario in Tables 1 and 2 had been reached. By September 2007, oil was at $80, and by July 2008, the price spiked at $147 a barrel. The thirteen-year time frame I had put on the doubling of the price of a barrel of oil occurred in thirteen months. In its wake, the fuel-poor of Arizona fell victim to this energy price shock. The ordinary American dream lifestyle suddenly became a lot more expensive. People had to

Table 2

Rough Calculation Done in April 2007 of What a Single Young Professional with a Well Paid Job Might Pay Per Month, and Over a Year, Living in a Three Bedroom Home in Garfield and What They Would Have to Pay If the Price of Energy Doubled Across the Board

Single person 2020 = x 2 Energy Prices Without heat island rises	GARFIELD 2007		GARFIELD 2020	
	$ per month	$ per year	$ per month	$ per year
Mortgage	1,400	14,400.00	2,000.00	24,000
Property Tax	500	6,000.00	500.00	6,000
Water + sewer	80	960.00	160.00	1,920
Energy	200	2,400.00	400.00	4,800
Car purchase	300	3,600.00	300.00	3,600
Gas / petrol	100	1,200.00	200.00	2,400
House insurance	100	1,200.00	100.00	1,200
Car insurance	100	1,200.00	100.00	1,200
Telephone, internet, mobile	100	1,200.00	100.00	1,200
Cable	50	600.00	50.00	600
Health insurance	200	2,400.00	200.00	2,400
Extras	100	1,200.00	100.00	1,200
Groceries	300	3,600.00	300.00	3,600
Credit Cards	200	2,400.00	200.00	2,400
TOTAL	**3,730**	**42,360.00**	**4,710.00**	**56,520**
& 30% TAX		12,708.00		16,9561
Annual salary		**$55,068.00**		**$73,476**

Source: Sue Roaf.

make choices. They had to keep their children cool in the heat-wave summers of 2007 and 2008; they had to buy gas to get to work, health insurance, food, energy for cooling, credit card bills, and other basic living expenses including water. Further down the line of essential payments were the mortgage repayments. Many residents had leveraged their salaries to be able to buy in aspirational developments, and by the end of 2007, more than six hundred Verrado lots had gone into foreclosure.[51] The 2010 Census in Arizona shows that ninety-one thousand homes in the city of Phoenix were vacant and 261,000 in surrounding Maricopa County. One in seventeen homes in Arizona had been in foreclosure that year. Arizona was the "domino" state in America after which many others

Figure 4
Global Oil Production, Consumption, and Price 2001–2012

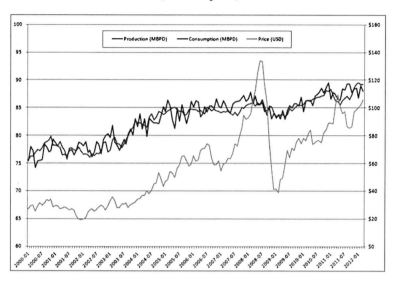

Source: US Energy Information Administration

fell, triggering a US and then a global economic meltdown in 2007/8. Phoenix was one of the worst-hit cities in the global recession with house prices in the city falling 57 percent between 2006 and mid-2011. Only time will tell how much of Verrado eventually does get built, but even now, if one mentions such concerns, there is a tendency among local people to not wanting to be aware, perhaps to protect those people still hanging on in such suburbs or because the reality may shed light on their own housing predicament. Stiglitz may agree here that the sooner the facts are known to the majority, the sooner the adaptive corrections can be made to the system and the safer it will be as a result. The rich can always move state.

Dundee Solar City Case Study

Dundee is a cool, wet city on the east coast of Scotland (Latitude: °N56°45'N), and the fuel poverty that led to the collapse of the Arizonan housing market is prevalent in Dundee too as it is in the rest of the United Kingdom. A UK household is currently said to be fuel-poor if it needs to spend more than 10 percent of its income on fuel to maintain an adequate level of warmth.[52] The adequate standard of warmth for a UK home is usually defined as 21 °C for the main living area, and 18 °C for other occupied rooms. Fuel poverty is therefore based on modelled spending on energy, rather than actual spending. Although the emphasis in the definition is on heating the home, modelled fuel costs in the definition of fuel poverty also include spending on heating water, lights, and appliance usage, and cooking costs.[53]

Table 3 below shows the number of fuel-poor households in England in each year it has been measured. The Department of Energy and Climate Change (DECC)[54,55] gives a range of reasons for the fluctuation in numbers of fuel-poor households year on year in the United Kingdom, but again, the rising cost of domestic energy, oil prices, and the dips in the United Kingdom and global economies, as with Arizona, were seen as key.

Note the clear jump in fuel poverty from 2005 to 2006. Domestic energy prices have been continually rising since then in the United Kingdom, and in 2011, domestic fuel prices rose a further 20 percent pushing more people into fuel poverty. On average, an additional twenty-seven thousand people die in England and Wales a year[56], many of whom will have been suffering from the debilitating effects of fuel poverty. As extreme weather events increase, more deaths are predicted with many being attributable to cardiovascular and respiratory diseases caused by excess cold and fuel poverty.[57]

Significant differences in fuel poverty levels exist across Britain, with one in fourteen households in England, one in nine in Wales, and one in four in Northern Ireland being in fuel poverty by 2008.[58] Recent estimates by the Scottish government put this figure at over a third of Scottish households. In some areas of Scotland, this figure is even higher with 45 percent of rural households deemed

Table 3
Number of Fuel-Poor Households in England

Year	No. of fuel poor households (million)
1996	5.1
1998	3.4
2001	1.7
2002	1.4
2003	1.2
2004	1.2
2005	1.5
2006	2.4
2007	2.8
2008	3.3
2009	4.0
2010	3.5

Source: DECC

to be fuel-poor in 2009 and 20 percent in extreme fuel poverty, while only 30 percent of urban households in Scotland are fuel-poor and 8 percent of those in extreme fuel poverty.[59]

Scotland also has ambitious targets for renewable energy production and for more independence from fossil fuels by 2020, as legislated under the radical 2009 Climate Change Act.[60,61] The Scottish government aims to produce 20 percent of its energy from renewables by 2020 and 80 percent by 2050, but these energy targets are currently not linked to the fuel poverty issue.[62] The key elements in determining whether a household is fuel poor or not are considered to be:

- Income
- Fuel prices
- Fuel consumption (dependent on the lifestyle of the household and the dwelling characteristics)

Households living in older dwellings are also more likely to experience fuel poverty, with around a third of households in Scotland living in dwellings built before 1919 being fuel poor, compared to less than a fifth of households living in dwellings built after 1982.[63] Fuel poverty is a complex issue and government policies to address income poverty, and by extension fuel poverty, have tended to focus on financial subsidies and incentives around children, older people, and the unemployed, whilst deprivation-related policies have focused almost exclusively on regenerating deprived urban areas.

There is a clear need for government policy makers to understand that progress on fuel poverty will be limited unless the government commits to understanding the complex range of issues and drivers in which it operates. One of the initial flawed assumptions is that people live in homes with living rooms at 21 °C. Will Anderson and Vicki White[64] conducted in-depth interviews about real energy habits and found that a wide and diverse range of market behaviors occur as people actively engage in reducing consumption. Nearly half (47 percent) of the low-income households said their homes had been colder and 18 percent said their homes had been much colder than they wanted during the previous winter resulting in various consequences such as controlling systems, damp, or condensation, which would have knock-on effects on health. The low-income households who experienced cold homes reported adverse impacts on their mental health, physical health, and social lives. Nearly half (47 percent) said the cold had made them feel anxious or depressed, 30 percent said an existing health problem had worsened, and 17 percent could not invite friends or family over to the house.

These families were reaching the end of their "adaptive opportunities and capacity" and approaching breaking point. The majority could not, or did not, switch to cheaper suppliers because of ignorance or fear of consequences if it did not turn out well for them. Such families have no tolerance for things to go wrong. If all their actions failed to keep them safe and warm in their own homes, then they have one further conscious action available—to move. Their ecological niche had then collapsed and they could not bounce back in that environment. That is exactly what happened in Arizona. The fracture had occurred.

The Solar Solution for Fuel Poverty in Dundee

Solar Energy is a form of renewable energy whose integration into the fabric of a building, and in turn a city, is the easiest, and it is capable of providing a significant amount of the necessary electricity, heat, and hot water for the comfortable operation of a building over a year. Building-integrated PV and solar thermal (BIPV and BIST) can inform architectural features and aesthetics in forms such as passive solar sun spaces into the design language of the building envelope. They are clean, quiet, robust, energy-neutral to operate and can result in avoided costs of construction if used as rain screen cladding. They may require occasional cleaning over time depending on the climate and air pollution levels where they are located.

People like solar energy in their cities and on their own homes. For many years, we have known this. H. M. Chadwick et al.[65] studied the acceptability of different renewable energy technologies by the local community, as perceived by planners, and found that in comparison to large-scale wind installations and waste incineration, solar energy technologies in the form of passive solar, solar PV, and solar heating are likely to be favorably viewed by local people. The use of solar design in buildings since the earliest times is linked with an embedded cultural appreciation of the power and importance of the sun to human civilizations over millennia.[66] Around 80 percent of people in a recent UK survey said they supported renewable energy for providing electricity, fuel, and heat[67] largely because of the rising price of domestic energy bills.

The Solar City Solution

What happens if we actually try and run whole cities on solar energy? At the turn of the millennium, a new solar movement grew to promote the idea of a city-level integration of solar energy systems. The solar potential of cities lies in not only reducing fossil fuel energy use in, and carbon emissions from, cities but also improving local businesses and the economy, and by doing so while improving the lot of the ordinary citizen. It can be a win-win technology. Roaf and Rajat Gupta[68] demonstrated that with the energy saving measures and low-carbon technologies, BIPV and BIST, around 70 percent of domestic emissions from heating, lighting, and hot water could be eliminated from UK homes at a cost of around £100 billion in 2007. In 2007–2008 alone, the UK government committed some £70 billion to cleaning up and detoxifying existing nuclear power stations in Britain. Since then, the cost of installing a 4 kWp PV system in a high-volume project has fallen by three quarters from around £20,000 a system to £5,000.

That study looked at barriers and influences of solar ordinances, grid configuration issues and markets, planning law, local authority programs, government-level fiscal incentive schemes, and citizen-action programs. The question in Dundee was one of resilience with a focus on questions of the contribution of building integrated solar energy to the resilience of the ordinary citizen in terms of their ability to lead an adequately decent, healthy, and comfortable lifestyle, while living within their means. Could building integrated solar energy systems take citizens out of fuel poverty?

Dundee joined the Solar City movement in 2006 and by 2009 won significant support from the Scottish government for its programs to educate its citizens in energy efficiency and the benefits of using solar systems (www.dundeesuncity. org.uk and www.solarcitiesscotland.org.uk). Scotland patently does not have the best global climate for solar energy, and historically, people have considered the weather as a barrier to solar markets. However, the radiation incidence available is more than adequate and an appropriately sized PV and SWH system on a typical house in Scotland can meet more than 50 percent of the annual needs of a Scottish household for either electricity or hot water. Dundee, like many other Solar Cities around the world, was experimenting with a range of social, fiscal, and commercial incentives to increase the uptake of solar systems in the area. The most effective policy was the UK government's Feed-in Tariff (FIT), a generous rate-based incentive that paid people who installed systems up to 41.5p per kW generated by their systems.

Over 1 GW of PV installations were installed in the United Kingdom as a result of FIT in 2009–2011 alone. The success of the program was responsible for its own demise. Politicians no doubt took stock of the fact that as more people generated their own energy, the need for major new investment in large scale nuclear and a new generation of gas plants diminished. A key objection mooted for its termination was that the poor were subsidizing the middle classes in getting systems fitted, but citizens of the United Kingdom across the board have been heavily subsidizing the clean-up of nuclear plants for generations and that is not considered a reason to stop the program of a new nuclear build in Britain, which will allegedly require £250 billion of support from the UK government over the next forty years. That is enough money to give every home in the United Kingdom £10,000 each for efficiency measures and solar systems, an investment that could preclude the need for new nuclear power stations at all and lay the foundations for a safe renewable energy economy for all coming generations.

With our concern about the increasing numbers of citizens in Dundee in fuel poverty, and the case study of Phoenix Arizona in mind, Andreadis et al.[69] undertook a study for Dundee, a city with very high levels of social deprivation, to explore the potential role for building-integrated solar technologies together with the adoption of some basic energy efficiency measures as a successful and lasting solution for the elimination of fuel poverty in the city.

The Scottish Index of Multiple Deprivation (SIMD) classifies the areas of Scotland according to how deprived the population of areas are. Bailey et al.[70] quote Townsend's definition *"People are relatively deprived if they cannot obtain, at all or sufficiently, the conditions of life—that is, the diets, amenities, standards and services—which allow them to play the roles, participate in the relationships and follow the customary behaviour which is expected of them by virtue of their membership of society. If they lack or are denied resources to obtain access to these conditions of life and so fulfil membership of society, they may be said to be in poverty.*" The term "multiple" has been used to include all the types or domains of deprivation which the SIMD examines which are: Current Income, Housing, Health, Education Skills and Training, Employment, Geographic Access to Services, and Crime.[71,72] The Georgios Andreadis et al.[73] found that the domains that were taken into account are those of income, housing, and employment. They found that 72,329 citizens lived centrally in the city out of a total population in Dundee of 143,090 in 2003. SIMD analysis showed significant concentration of deprivation in the chosen area of 88 data zones in which a third were in the 20 percent most-deprived households in Scotland. It was not possible to identify how many people are under the fuel-poverty threshold; however, they applied some common sense to arrive at his estimate for the fuel-poor of the city. Around 20 percent of the population were income-deprived, 14 percent had no central heating, and 11 percent were out of work.

It was found that there are about 1,300 domestic buildings suitable for solar integration, all of which have tilted roofs with a maximum deviation of ±30° from South. It has also been estimated that there are about three hundred more suitable buildings in the rest of the city. The City Council of Dundee has made a different investigation and concluded that this number is 1,500 for the whole city, and therefore, the two estimations concur. The 1,300 buildings reported here do not correspond to same numbers of households because there were many blocks of flats included, whose roofs seemed suitable for solar integration. The total roof area was found to be 88,313 m^2 which can receive 97,914,848 kWh of the total solar radiation per annum. This gives a 9.6 MW$_p$ total PV capacity and 9,380,242 kWh per year of net electrical energy. The current energy consumption in the city was calculated using the typical annual electricity consumption of Scottish households reported on by Robert Currie et al.[74] being 3,084 kWh for a single person, 4,117 kWh for a working couple and 5,480 kWh for a four-member family (parents at work and children at school). According to another study by DECC,[75] the average annual household electricity consumption in thirty-two zones is 3,601 kWh for the ordinary domestic and 5,697 kWh for the Economy 7 off-peak tariff. If these two averages are combined according to their weights (percentage of the total), the annual average of 4,052 kWh for a dwelling in the city of Dundee is derived. The installed BIPV could thus cover the energy needs of 5,500 four-member households, 12.5 percent of the homes in Central Dundee.

Andreadis et al.[76] also suggested that the unsuitable parts of the roof areas can become usable if solar thermal collectors are installed: another 8 percent of the estimated roof area can be usable for BIST systems bringing in around 2,500,000 kWh of solar hot water and increasing usefully the total solar contribution in Table 5. The total of 11,880,242 kWh of free solar energy potentially installable would provide an additional benefit of 4,443,781 kg of CO_2 emission reduction savings annually from the city.

Energy-efficiency measures were included for five thousand of the worst (most fuel poor) households installing new 24 kW_{th} gas condensing boilers at £725 each; internal insulation of solid walls cost £5,500–£8,500; with the costs of low-energy lighting and retro-fitting selected properties being included, the total average cost per house was around £7,500. The basic cost of this efficiency program would be around £37,500,000. Adding up the total costs for BIPV and BIST solar systems and energy-efficiency measures, the total budget of the suggested solar plan for Dundee would be £67 million. This would then take vulnerable families in Dundee out of fuel poverty, but it would also have a range of additional benefits. Also, £67 million is what will be spent on the new Victoria and Albert Museum waterfront development in Dundee and a quarter of the cost of the new 28-mile Aberdeen Bypass that is being fought by Aberdonians in the courts.

The solar industry in Scotland includes both large scale installers and small local companies who have rapidly grown in numbers in the wake of the FITs bonanza, which ended abruptly in December 2011 when the UK government shut it down with little warning. Energy-efficiency work can again be undertaken by local companies or by larger players in the market. A number of studies have estimated jobs potentially accruing to these "green" industries in the United Kingdom and the United States.

Jonathan Neale[77] argues that the transition to an inclusive green economy will require new metrics that go beyond the prevailing narrow focus on income poverty and gross domestic product (GDP) to a broader way of tracking economic, social, and environmental progress and well-being. Making an inclusive "green economy" work for the poor too is vital for a resilient future, but it needs government commitment. Safeguarding the poor against any adverse impacts during the transition process and functionalizing a system that maximizes the opportunities and benefits for all is key. He sets out the route to building a million new jobs in that new green British economy.

Jacob Kirkegaard et al.[78] in the United States did an extremely useful working paper on issues around "solar employment." They covered in detail the PV industry's recent growth patterns, industry cost structure, trade and investment patterns, government support policies, and employment generation potential. In conclusion, they stressed the need for governments to provide sufficient and predictable long-term support to solar energy deployment. Such long-term frameworks bring investments forward and encourage cost cutting and innovation so that government support can decrease over time. They point out that it is not just the manufacturing jobs in the solar industry that are important but the total number of jobs that could possibly be created including those in research, project development, installation, operations, and maintenance. Their call for a "broader" view of the benefits of a solar economy are welcome in a world where we have seen the prices of energy, food, and water tip systems into freefall. We saw this in the Arab Spring, in the "Occupy" movements around the world and in the violent riots on the streets of Britain in the summer of 2010. We saw it in the Arizonan housing case study and we can clearly calculate the costs, and assess the benefits, of acting locally to build a solar future in a city like Dundee.

Since 2008, the Joseph Rowntree Foundation has published reports on the Minimum Income Standard (MIS) for the United Kingdom, in which they have tracked what members of the public think people need to have a socially acceptable standard of living and how much money different households need to reach this standard. They have achieved this by holding in-depth discussions with people on this subject since 2006 and, although people are aware that times are tougher, people's perception about what this minimum should include has not fundamentally changed. However, there are some instances where people are clearly thinking hard about how they meet their needs in more economical ways like not going out so often or buying cheaper goods and presents.[79] These trends and attitudes are well-captured in the MIS and people can estimate on the MIS website what their own minimum income standards are and test the sensitivity of their MIS to changes in the prices of different good and services. What we can clearly see in the annual MIS reports is the evolution of behaviors and perceptions in a changing economic background. What we don't see is clear evidence of what happens to families when they fall significantly below their own MIS level and what the consequences of that failure might be.

In order to turn the MIS into a standard measure, and indicator, of the resilience of a population, a second measure of how well an individual household meets that standard is needed. The MIS demonstrates the minimum level of income required to pay for a bare minimum standard of living. The second measure required for a typical household figure is how much income is actually coming into households. This represents an individual domestic product (IDP), the total income to one household including multiple wages, benefits, and other incomes. These two measures would provide an alternative finer grained indicator of the relative success or stress of households within different populations of a larger economy providing much more effective policy indicators than the coarse grained instrument of the GDP and average income figures. Stiglitz[80] points out that in the five years after 2007, the top 1 percent seized more than 65 percent of the gain in the US national income, and by 2010, their share was 93 percent. GDP figures and average incomes mask this inequality. The gap between the rich and the poor is increasingly recognized as an indicator of the general success of an economy across multiple factors and populations showing wide gaps fare significantly worse than those in less unequal societies.[81]

The difference between the MIS and the IDP of a population is a proxy for their "adaptive capacity" indicating how much extra strain their own economic system can tolerate. Another modifying factor in a "learning" population would have to be included for the rate of change of the system. Using a MIS/IDP measure, the sensitivities of the system to factors such as the increase in energy or water prices could be tested, and short, medium, and long-term policies developed to avoid system collapse by exploitation of the potentials of adaptive opportunities and maximization of adaptive bandwidths. This need for a revisit on how we measure fuel poverty has recently been addressed in a detailed review of fuel poverty in the United Kingdom by John Hills,[82] who clearly elaborates the need for a more sophisticated approach to the subject than simply defining fuel poverty with the 10 percent of the income figure.

What is clear is that we need a simple calculation for the household income minus household outgoings for individual populations, which gives a clear handle on the exposure and vulnerability of a family, or a population, to the hazard of fuel poverty. The vulnerabilities of a family revolve around variables such as the quality of construction and design of their home, location, climate, household funding model, etc. Secondary vulnerabilities that are in turn influenced by those factors include the price of food, water, and other costs. John Hills[83] recommends that such indicators (he defined one as the LIHC—the Low Income High Costs indicator) should be used as the basis for operational target-setting and to develop policies that a) benefit the individual household not the economy as a whole and b) alert decision makers when populations approach an economic fracture point.

The New American Dream: The Solar Solution for Arizona

Table 4 was drawn up way back in 2007 for a "Valley Forward" audience of business people to show what might have been if, instead of building 4,500 square foot chipboard mansions across the desert, developers had built smaller, more robust, well-insulated, and shaded homes, running on solar energy with a high thermal capacity construction to store heat and cool. Such homes could also have powered solar electric cars, bikes, and scooters that might have taken them to the shaded parking lot of their local branch station of the electric Arizona Rail system. The smaller ecohomes would have cost the same to build with their installed efficiency measures and solar systems as the larger ones, but the running cost is significantly less and these homes have not been susceptible to extreme spiking of energy costs, as once installed, solar energy is not only clean, but also free, and if properly installed, can also provide a further heat insulating barrier for the roof in turn lowering the cooling loads of the building. The impact of a solar lifestyle is the greatest on the expenditure for the home owners of Verrado. The simple point here is that if Arizonans had built such ecohomes[84] with less expensive and carefully zoned air-conditioning systems and PVs, then in the summer of 2007, many of them could have cooled their homes, eaten and paid their mortgages, and remained in their homes, not left on the streets where many of them found themselves, or moved state. Instead, Arizona was a tipping state

Table 4

Table Showing a Rough Estimate of the Impact on Household Bills of an Increased Use of Solar Energy to Run the Home and Transport for Verrado and Garfield with Garfield Residents Walking, Using Bicycles, or the Bus and Verrado Residents Using Electric Vehicles and the Arizona Rail System to Commute. The Figures Include Increased Mortgage and Water Costs and the Original Price Increase with Fossil Fuel Energy is Included in Brackets above the Annual Total

Single Person	GARFIELD 2007		SOLAR GARFIELD 2020		VERRADDO 2007		SOLAR VERRADO 2020	
	per month	per year	per month	per year	per month	per year	per month	per year
Mortgage	1400	14400	2000	24000	2000	24000	2800	33600
Property Tax	500	6000	500	6000	500	6000	500	6000
water + sewer	80	960	160	1920	80	960	160	1920
Energy	200	2400	50	600	300	3600	50	600
Car purchase	300	3600	300	3600	300	3600	300	3600
Gas / petrol	100	1200	0	0	250	3000	100	1200
House insurance	100	1200	100	1200	100	1200	100	1200
Car insurance	100	1200	100	1200	100	1200	100	1200
Telephone	100	1200	100	1200	100	1200	100	1200
Cable	50	600	50	600	50	600	50	600
Health Insurance	200	2400	200	2400	200	2400	200	2400
Extras	100	1200	100	1200	100	1200	100	1200
Groceries	300	3600	300	3600	300	3600	300	3600
Credit Cards	200	2400	200	2400	200	2400	200	2400
	3730	42360	4160	49920	4580	54960	5060	60720
& 30% TAX		12708		14,976		16488		18,216
ANNUAL SALARY		**$55,068**		**($73,476)** **$64,896**		**$71,488**		**($93,756)** **$78,936**

that helped to topple the US and then the global economy system with such dire consequences for us all.

Conclusions

In our search for socioeconomic resilience in a rapidly changing world, both measures and motivations matter enormously. There is a need for a deeper understanding of where populations stand in relation to economic stress, their "adaptive capacity," how much extra economic strain they can tolerate in the real context of their own societies, economies, and environments. A vital factor to determine is at what point their whole economic system fractures, and what factors functionalize the breaking points inherent in those systems.

The MIS/IDP or LIHC indicator approach could be developed to achieve this and to test system sensitivities. When governments need to ratify spending hikes by private and public bodies such as 5–10 percent hikes in transport, water, or energy costs, the figures would be run through the MIS/IDP model to examine the extent to which such price hikes will push a population over the fracture cliff. It is not difficult to see that the Solar City model will fare much better in targeting investment to where it will help reduce economic inequality and take ordinary citizens out of fuel poverty forever. However, we know from Stiglizt[85] and Lazarus[86] that such end points may not be in the interests of the ruling classes, and their lobbyists, in too many countries. But perhaps, this is the exact time when even vested interests must bring about a paradigm shift, developing strategies and measures that enable populations to withstand the growing stresses and strains in our economic systems, in order to ensure the continuation of their own interests.

This exploratory chapter has identified that we need to understand, functionalize, and maximize our adaptive capacity and to identify fracture points in our socioeconomic systems in order to simply survive in an evolving form. It has also highlighted the fact that potentially a solar-powered society is a more resilient society, one with a far larger adaptive capacity to resist the inevitable soaring of fossil fuel prices in the future, as well as its increasingly extreme climates. A solar-powered society can be a safer society, but one in which the "one percent" might not make as much money. It's a choice we all have to make, of who matters to our own best interests in the long run. The real question is who actually makes the decision? *Quis custodiet ipsos custodes*? Who guards the guardians? Perhaps, after all, we all do.

Notes

1. Zolli, A., and A. Healey, *Resilience: Why Things Bounce Back* (London: Headline Publishing Group, 2012).
2. Campbell, Flake C., "Elements of Metallurgy and Engineering Alloys," *ASM International*, 2008, 206, http://en.wikipedia.org/wiki/Resilience (accessed May 1, 2012).
3. Berkes, F., J. Colding, and C. Folke, *Navigating Social-Ecological Systems: Building Resilience for Complexity and Change* (Cambridge, UK: Cambridge University Press, 2008).
4. Meadows, D. H., D. L. Meadows, J. Randers, and W. Behrens, *The Limits to Growth; A Report for the Club of Rome's Project on the Predicament of Mankind* (London: St Martin's Press, 1972)
5. Turner, G., "A Comparison of `The Limits to Growth` with Thirty Years of Reality," *Global Environmental Change* 18, no. 3 (August 2008): 397–411.
6. Meadows, D., *Thinking in Systems* (London: Earthscan, 2009).
7. Cumming, G., G. Barnes, S. Perz, M. Schmink, K. Seiving, J. Southworth, M. Binford, R. Holt, C. Stickler and T. Van Holt, "An Exploratory Framework for the Empirical Measurement of Resilience," *Ecosystems* 8 (2005): 975–87.

8. Crichton, D., "The Risk Triangle," In *Natural Disaster Management*, ed. Ingleton, J., (London: Tudor Rose, 1999).
9. Roaf, S., D. Crichton, and F. Nicol, "Adapting Buildings and Cities for Climate Change," *Architectural Press*, 2nd Edition (In Press, 2009).
10. See Note 3.
11. Wilson, Geoff, A., *Community Resilience and Environmental Transitions* (London: Routledge, 2012).
12. Holling, C. S. ed., *Adaptive Environmental Assessment and Management* (Chichester: John Wiley and Sons Inc., 1978), 105.
13. Stiglitz, J. *The Price of Inequality: The Avoidable Causes and Invisible Costs of Inequality* (London: Allen Lane, 2012).
14. Hardin, G., "The Tragedy of the Unmanaged Commons," *Trends in Ecology & Evolution* 9, no. 5 (1994): 199–224.
15. Hardin, G., "The Tragedy of the Commons," *Science* 162, no. 3859 (1968): 1243–48.
16. See Note 13.
17. Jackson, T., *Prosperity Without Growth: Economics for a Finite Planet* (London: Routledge, 2011).
18. Heinberg, R., *The End of Growth: Adapting to Our New Economic Reality* (Old Saybrook, Connecticut: Tantor Media Inc., 2011).
19. Ayres, R. U., *Turning Point: The End of the Growth Paradigm* (London: Earthscan, 1999). First published 1998.
20. See Note 4.
21. IPCC, *Climate change 2001: Impacts, Adaptation, and Vulnerability: Contribution of Working Group II to the Third Assessment Report of the Intergovernmental Panel on Climate Change* (Cambridge: Cambridge University Press, 2001).
22. Metzger, M. J., M. D. A. Rounsevell, et al., "The Vulnerability of Ecosystem Services to Land Use Change," *Agriculture, Ecosystems & Environment* 114, no. 1 (2006): 69–85.
23. Stadelmann, M., A. Michaelowa, S. Butzengeiger-Geyer and M. Köhler, "Universal Metrics to Compare the Effectiveness of Climate Change Adaptation Projects." Organisation for Economic Co-operation and Development (OECD), 2011, http://www.oecd.org/dataoecd/44/9/48351229.pdf, (accessed August 3, 2012).
24. IPPC., "Definitions of key terms within Climate Change 2007: Working Group II: Impacts, Adaptation and Vulnerability," August 3, 2012. http://www.ipcc.ch/publications_and_data/ar4/wg2/en/frontmattersg.html, 2007.
25. Levina, E., and D. Tirpak, "Adaptation to Climate Change: Key Terms,"*Organization for Economic Co-operation and Development*, (International Energy Agency, 2006)
26. See Note 11.
27. Berrang-Ford, L., J. D. Ford, et al., "Are We Adapting to Climate Change?," *Global Environmental Change* 21, no. 1 (2011): 25–33.
28. See Note 9.
29. Kjell, A., *Peeking at Peak Oil* (Heidleburg: Springer, 2012).
30. Campbell, C. J., *The Coming Oil Crisis* (Brentwood: Multi-Science Publishing and Petroconsultants, 1997).
31. Roaf, S., and R. Gupta, *Optimising the Value of Domestic Solar roofs: Drivers and Barriers in the UK, Sustainable Energy: Opportunities and Limitations: An Introductory Review of the Issues and Choices*, ed. Dave Elliot (Palgrave/McMillan publishers, 2007).
32. Campbell, C. J., *Oil Crisis* (Brentwood: Multi-Science Publishing and Petroconsultants, 2005).
33. See Note 9.

34. Bartsch, U., and Muller, B. *Fossil Fuels in a Changing Climate.* (Oxford University Press. (2000).
35. Busse, K., L. Dirks, K. Kruger, S. Lidberg, T. Miller, D. O'Neill, I. Sakansky, T. Shirmang, B. Stanley, *Verrado's Sustainable Legacy: An Assessment, Global Institute of Sustainability* (Arizona State University, 2007).
36. Roaf, S., A. Horsley, and R. Gupta, *Closing the Loop: Benchmarks for Sustainable Buildings* (London: RIBA Publications, 2004).
37. Glennon, R., *Unquenchable: America's Water Crisis and What To Do About It* (Island Press, 2009).
38. Glennon, R, *Water Follies: Groundwater Pumping and The Fate of America's Fresh Waters* (Washington: Island Press, 2004).
39. Powell, James L., *Dead Pool: Lake Powell, Global Warming and the Future of Water in the West* (University of California Press, 2012)
40. Ross, A., *Bird on Fire: Lessons from the World's Least Sustainable City* (Oxford: Oxford University Press, 2011).
41. See Note 37.
42. Roaf, S., M. Fuentes and S, Thomas, *Ecohouse: A Design Guide*, 4th Edition (London: Earthscan, 2012).
43. Krause, H., C. McGehee, C. Senneville, and S. Swanson, *Measuring Sustainability in the Garfield Community*, Downtown Phoenix, *Global Institute of Sustainability* (Arizona State University, 2007).
44. Karl, T., H. Diaz, and G. Kukla, "Urbanization: Its Detection and Effect in the United States Climate Record," *Journal of Climate* 1, no. 11 (1988): 1099–123.
45. Brazel, A. J., N. Selover, R. Vose, and G. Heisler, "A Tale of Two Climates—Baltimore and Phoenix Urban LTER Sites," *Climate Research* 15 (2000): 123–35.
46. See Note 44.
47. Ruddell, D., S, Harlan, S, Grossman-Clarke, and A. Buyantuyev, "Risk and Exposure to Extreme Heat in Microclimates of Phoenix Arizona," In *Geospatial Techniques in Urban Hazard and Disaster Analysis, Geotechnologies and the Environment 2*, ed. Showalter, P. and Y. Lu, 2010. DOI 10.1007/978-90-481-2238-7_9.
48. Semenza, J., J. McCulloch, W. Flanders, M. McGheehin, J. Lumpkin et al., "Excess Hospital Admissions during the July 1995 Heat Wave in Chicago," *American Journal of Preventive Medicine* 16, no. 4 1999: 260–77.
49. Larson, J., *Setting the Record Straight: More than 52,000 Europeans Died from Heat in Summer 2003* (Earth Policy Institute, Washington, 2006) www.earth-policy.org/updates/2006/Update56.htm, (accessed August 3, 2012).
50. See Note 13.
51. AZCentral, http://www.azcentral.com/business/realestate/articles/2012/02/18/20120218arizona-homes-verrado-development.html#ixzz1wpbjtilt
52. Palmer,G., T. MacInnes, and P. Kenway, *Cold and Poor: An analysis for the Link between Fuel Poverty and Low Income* (London: New Policy Institute, 2008) http://www.poverty.org.uk/reports/fuel%20poverty.pdf (accessed March 18, 2012).
53. DECC, Department for Climate Change Annual Report on Fuel Poverty (National Statistics Publication, 2012b) http://www.decc.gov.uk/assets/decc/11/stats/fuel-poverty/5270-annual-report-fuel-poverty-stats-2012.pdf (accessed August 3, 2012).
54. Ibid.
55. DECC, Department for Energy and Climate Change. Public Attitudes Tracker—Wave 1: Summary of key issues, 2012a, http://www.decc.gov.uk/assets/decc/11/stats/5707-decc-public-att-track-surv-wave1-summary.pdf (accessed August 3, 2012).
56. Hills, J., *Getting the Measure of Fuel Poverty: Final Report of the Fuel Poverty Review*, published by the Centre for Analysis of Social Exclusion (London: London

School of Economics, 2012) www.decc.gov.uk/hillsfuelpovertyreview/ (accessed August 3, 2012).

57. Grasso, M., M. Manera, A, Chiabai, and A. Markandya, "The Health Effects of Climate Change: A Survey of Recent Quantitative Research," *International Journal of Environmental Research and Public Health* 9 (2012): 1523–47, doi:10.3390/ijerph9051523.www.mdpi.com/journal/ijerph.

58. See Note 52.

59. CAB, Citizens Advice Bureau, *The Fuel Poverty Monitor*, 2011. http://www.nea.org.uk/assets/PDF-documents/Monitor-2011-small.pdf (accessed March 18, 2012).

60. Scottish Government, *Carbon Assessment of the 2012–13 Draft Budget*, 2012. Published by the Scottish Government, September 2011, http://www.scotland.gov.uk/Publications/2011/09/21111152/0 (accessed July 7, 2012).

61. Scottish Government, Climate Change (Scotland) Act, 2009a, http://www.legislation.gov.uk/asp/2009/12/contents (accessed August 3, 2012).

62. Scottish Government, Energy Policy Overview, 2008, http://www.scotland.gov.uk/Resource/Doc/237670/0065265.pdf (accessed August 3, 2012).

63. See Note 56.

64. Anderson W., and V. White, *You Just Have to Get by: Coping with Low Incomes and Cold Homes* (Bristol: Centre for Sustainable Energy, 2010). www.cse.org.uk (accessed August 3, 2012).

65. Chadwick, H. M., Batley-White, S. L., and Fleming, P. D., "The UK Planning Process and the Electricity Supply Industry—What Role for Renewables?" In *Proceedings of the Conference on Creating Sustainable Urban Environments: Future Forms for City Living* (Oxford: Christ Church College, 2002).

66. Knowles Ralph, *Ritual House: Drawing on Nature's Rhythms for Architecture and Urban Design* (Washington: Island Press, 2006).

67. See Note 55.

68. See Note 34.

69. Andreadis, G., S. Roaf, and T. Mallick, (Under review), "Tackling Fuel Poverty with Building-Integrated Solar Technologies: The Case of the City of Dundee in Scotland," *Energy and Buildings*, manuscript ENB-D-1200661.

70. Bailey, N, J. Flint, R. Goodlad, M. Shucksmith, S. Fitzpatrick, and G. Pryce, Measuring Deprivation in Scotland: Developing a Long-Term Strategy published by the Scottish Government, 2003, http://www.scotland.gov.uk/Resource/Doc/47176/0025569.pdf (accessed August 3, 2012).

71. See Note 62.

72. Scottish Government, Scottish Index of Multiple Deprivation (SIMD)—Part 4 Employment, 2009b, http://www.scotland.gov.uk/Resource/Doc/933/0096870.xls (accessed August 3, 2012).

73. Ayres, R. U., Turning Point: The End of the Growth Paradigm (London: Earthscan, 1999). First published 1998.

74. Currie, R., B. Elrick, M. Ioannidi, C. Nicolson. Household Electricity Consumption, 2009, http://www.esru.strath.ac.uk/EandE/Web_sites/01-02/RE_info/hec.htm (accessed August 3, 2012).

75. DECC, Department of Energy and Climate Change, Middle Layer Super Output Area Electricity and Gas, 2009, http://www.decc.gov.uk/assets/decc/Statistics/regional/mlsoa_2009/1594-igz-domestic-electricity-scot.xls (accessed August 3, 2012).

76. See Note 69.

77. Neale, J., One Million Climate Jobs, report published by the Campaign against Climate Change (London, 2010). www.climate-change-jobs.org (accessed August 3, 2012).

78. Kirkegaard, J., T. Hanemann, L. Weischer, and M. Miller. "Toward a Sunny Future? Global Integration in the Solar PV Industry," Working Paper 10-6 (Washington: World Resources Institute, 2010).

79. Davis, A., D. Hirsch, N. Smith, J. Beckhelling, and M. Padley, A Minimum Income Standard for the UK in 2012, published by the Joseph Rowntree Foundation, 2012, http://www.jrf.org.uk/publications/MIS-2012 and http://www.minimumincome.org. uk/ (accessed August 3, 2012).

80. See Note 13.

81. Wilkinson, R., and K. Pickett. *The Spirit Level: Why More Equal Societies Almost Always Do Better* (London: Allen Lane, 2009).

82. See Note 56.

83. See Note 56.

84. See Note 42.

85. See Note 13.

86. Lazarus, R. J., *Cornell Law Review* 94 (2009): 1153–234. http://www.lawschool. cornell.edu/research/cornell-law-review/upload/Lazarus.pdf.

8

Energized: The Evolution of the Modern Building

Peter Syrett

Past

The Empire State Building is a self-proclaimed "eighth wonder of the world." The conviction of the building's principal backer, John Jakob Raskob, founder of General Motors, in this statement was so strong that he commissioned stained glass panels for the lobby showing the Empire State Building as a peer to the Seven Wonders of the Ancient World. Raskob had a reason to be boastful, when the Empire State Building was finished in 1931, as it stood 1,453 feet 8 and 9/16th inches tall from the street level to its highest point—over 200 feet higher than its famous neighbor the Chrysler Building.[1]

The Empire State Building held the title of the tallest building in the world for forty-one years, until the twin towers of the World Trade Center took the title in 1972. The Empire State Building's long reign as the tallest building made it an architectural icon. Its profile has been embossed on license plates, it has starred in movies, and it is the most photographed place in the world.[2] The Empire State Building's significance is not limited to its being a popular culture icon, an architectural landmark, or even its stature; it also represents the apex of nineteenth-century building technology.

Pre-war buildings like the Empire State Building, are more similar to late nineteenth-century buildings than those built in the twentieth century because of their low energy consumption. Before the Second World War, buildings relied primarily on "passive" techniques such as high thermal mass to temper interior conditions. Energy was mainly used for heating, and electrical usage was minimal by today's standards. "Active" techniques like mechanical ventilation were

Figure 1
The Empire State Building

a rarity; the opening of windows allowed fresh air to enter and provided for the ventilation of bad air. Daylight was the primary source of illumination, with electrical lighting meeting needs only when it was dark outside. After the war, buildings began to rely on energy-consuming technologies, and now dependence on energy has become so extreme that some buildings are uninhabitable without power. The flooding caused by Hurricane Katrina in 2005 brutally demonstrated this reality. On the day after Hurricane Katrina had hit, doctors, staff, and patients of Charity Hospital in New Orleans began breaking windows that had withstood the hurricane in order to get air.[3] "The hospital was so hot that with no rain or anything, they were better off in the fresh air on the roof," a witness told the Associated Press two days after Katrina's landfall.[4] The patients were evacuated and the hospital was shut down a few days later. Louisiana's governor declared that the hospital would not reopen because flooding had contaminated Charity's air-conditioning and heating systems and that had "affected the core operations of the entire building."[5]

In the first decades of the twentieth century, artificial lighting levels were low. A typical office building in 1913 had illumination levels of 22 to 43 lux owing to the inefficient lighting technologies of that time.[6] (Today, this range of illumination is considered too dark—107 lux is considered a very dark day; lighting levels of over 500 lux are common).[7] Because the quality of artificial light was so poor in earlier office buildings, the rental value increased for spaces with

Figure 2
The Flatiron Building

large windows and high ceilings, which allowed more daylight in. Turn of the century office buildings, like the Flatiron Building in Manhattan, have windows "punched" into 20 to 40 percent of their façades[8] and are narrow and compact so that daylight would reach deep into the interior. The Flatiron Building, typical of that era, used the passive strategy of having deep window niches in conjunction with awnings to let in the winter sun to warm the space, but limited direct gain from the summer sun.

Light and air became a political issue, as cities grew denser. At the end of the nineteenth century, people began protesting the loss of light and air caused by tall buildings. Public health advocates charged that tall buildings made the street dark, dank, and gloomy. In winter, they blocked the sun, and in summer, they acted as "storehouses of heat" that drove up the temperature after sunset "making the once cool and refreshing nights unbearable." Sunlight and wind were as vital to public health as pure water and without them "life would be almost impossible in crowded communities," according to a representative of the Chicago Medical Society in 1891.[9] In response to these protests, the New York State Legislature enacted a series of height restrictions on residential buildings, leading to the Tenement House Act of 1901. Outcry was not limited to residential buildings. When the forty-two-story Equitable Building was completed in 1915, the need for light and air controls for nonresidential buildings became apparent. Rising 538 feet straight up, without any setbacks, the Equitable Building casted a seven-acre shadow over lower Manhattan. This affected the value of those residential structures and resulted in the nation's first comprehensive zoning ordinance in 1916. Cities like Chicago and Detroit passed zoning laws shortly afterward based on the New York model.[10] Inspired by New York's zoning ordinance, in 1922, the US Department of Commerce produced the model zoning framework called the Standard State Zoning Enabling Act (SZEA), which became the basis of thousands of zoning ordinances in smaller communities throughout the country.[11]

The 1916 New York Zoning Resolution assured light and air for everybody by establishing rules that govern the massing (the bulk: width, height, and profile) of buildings. Under the resolution, the stacking of repetitive floors straight up in a compact block was no longer permissible; instead, the stepped "setbacks" were prescribed as seen in the Empire State Building's design with the building growing more slender toward the top. The stepped profile prevented buildings from keeping their neighbors in the dark by pushing the buildings' bulk away from the edge of the property and toward the center. The unintended side effect of the New York City Zoning Resolution increased the energy dependency of buildings because under the new bulk requirements, the lower floors of a tall building were wide and broad like the base tier of a wedding cake—too wide for daylight to fully penetrate and for use of natural ventilation techniques. Setbacks also added area to a building's skin that required more heating energy to be used to compensate for the loss from the additional perimeter. Although earlier twentieth-century buildings were not well-insulated, they benefitted from the use of nineteenth-century building materials. The sparse modern aesthetic "devoid of reminiscent stylistic detail"[12] of the Empire State Building's exterior limestone and metal panels was a marriage of nineteenth century materials and early-twentieth-century aesthetics. These panels, mounted onto a masonry back-up wall with an interior plaster finish, have a high thermal mass that tempers New York City's extreme weather.[13]

The Zoning policy was not the only reason buildings grew dependent on energy in the early decades of the last century. The rapid evolution of technology in the 1920s and 30s meant that interiors could be less reliant on daylight. Artificial lighting levels steadily rose. In 1916, the New York City Department of Health recommended that the lighting levels in offices be between 86 to 97 lux; the levels rose to 108–129 lux in the 1920s and to 269 lux by the 1930s.[14] Other innovations were beginning to change buildings too, and the foremost of these was air conditioning. Before the 1920s, cooling equipment was for industrial applications such as reducing the temperature of engines; then, pioneering engineers began to use mechanical systems as a means to cool people. The J. L. Hudson department store in Detroit, Michigan, was air-conditioned in 1924. In 1925, Carrier Engineering Corporation convinced Paramount Pictures to install air conditioning into their flagship theatre on Times Square in New York City. Both J. L. Hudson and the Times Square theatre quickly became popular attractions, drawing large crowds who wanted to experience the "cool" novelty. The earliest office building to be fully air-conditioned was the Milam Building in San Antonio, Texas, in 1928.[15] The 1932 Philadelphia Savings Fund Society (PSFS) Building by George Howe in association with the Swiss architect William Lescaze was the first architecturally significant building to be air-conditioned. These examples are exceptions; until the depression waned in the late 1930s, the majority of air-conditioning units were simple window mounts for single rooms. As the economy improved, the use of air conditioning began to grow. Manufacturers like General Electric, Carrier, and York all made air conditioners for rooms, which were mostly installed in offices.[16] Fortune Magazine wrote in 1931 that "out of 22,000,000 wired homes in the United States—where real volume sales were expected to blossom—less than 0.25% can yet boast so much as an air-conditioned room."[17] Most buildings before the war still used windows for ventilation and fans when cool air was needed. (The Empire State Building's 2,768,591 square feet[18] relied on 6,514 double hung windows and interior transoms above the doors for ventilation, until tenants began to install air conditioning in their leased spaces.) Architectural historian Henry-Russell Hitchcock observed that it would be another twenty years before other skyscrapers of "the distinguished design of the Philadelphia Savings Fund Society Building were built again in American cities"[19] and it would be equally long before the technological innovation of air conditioning would be installed again in a major American building.

In 1943, Architectural Forum magazine printed a special issue entitled "New Buildings for 194X" on future trends in architecture. The editor Howard Myers invited architects to propose projects for when the war moratorium on civilian construction was lifted. Renowned architects Louis Kahn, William Lescaze, and Mies van der Rohe submitted projects, but it was a lesser-known Pietro Belluschi whose design had the most lasting impact on the energy usage of buildings. Belluschi wrote that his proposed design project was "affected by the peculiar circumstances found in our northwest region—cheap power and a tremendously expanded production of light materials for war use, which will beg for utilization after the emergency."

Mies van der Rohe's 1921 design for the Friedrichstraße in Berlin is widely considered the first vision of the glass skyscraper. His design was far ahead of pre-war technical capabilities and was never built.[20] The war advanced building technology significantly and Pietro Bellushi capitalized on these advancements. His design for an office building liberally used aluminum for cladding, wall-panel frames, air-inlets, louver blinds, and ceiling tiles. He also proposed to maintain internal comfort with unit air-conditioners and radiant heating panels. Bellushi's design was a profound shift away from nineteenth century technology-based buildings like the Empire State Building with high thermal mass exterior walls and naturally ventilated interiors, toward low mass exterior cladding with fully air-conditioned interiors.

After the war, Belluschi realized his Architectural Forum project ideas in the Equitable Savings and Loan Headquarters in Portland, Oregon. The bank's directors shared his vision and were determined that their new building should be one of the most efficient and progressive buildings in the world.[21] When completed in 1948, the Equitable Building was aesthetically and technically different from any other building ever built before.[22] Belluschi went further than Mies and described in detail the systems required to achieve a "crystal and metal tower."[23] Belluschi clad his building in a modular wall system with expansive vision panels made of sealed double glazing with an outer pane of heat-absorbing glass. Working with the pioneering mechanical engineer David Kroeker, he designed the interiors to be heated and cooled by heat pumps that harvested the latent energy of the earth from wells below the building; air-handling units on each floor supplied air from above and returned it low in each room. The modern, sleek, and elegant design of the Equitable Building has been emulated around the world, but according to David Arnold it was:

> "replicated in totally different climates without the availability of renewable energy sources and entirely different geology. The result was the proliferation of clones of this building and many were built without any consideration of the local climate or availability of energy and they could be seen as using profligate amounts of energy and failing to provide comfort for the occupants."[24]

The Equitable Building was the first building to completely break with centuries-old passive technologies and embrace active systems. This changed how buildings were designed, built, and operated.

The bold design of the Lever House, by the architect Gordon Bunshaft of Skidmore, Owings & Merrill (SOM), thrilled the public and critics in 1952. The architecture critic Lewis Mumford quipped, "The public was acting as if the new soap-company headquarters were 'the eighth wonder of the world'."[25] The iconic power of the Lever House came from its light tartan façade of single glazed vision and spandrel panel glass. To achieve this light façade, Bunshaft eliminated operable windows, which would have added more lines to his eloquent grid. Having no operable windows not only helped aesthetically, but also (as it was reasoned at the time) diminished street noise and keep soot out. (Soot was then a major concern in New York because of coal burning combined with the lack of emissions regulations.) From a mechanical engineering perspective, the sealed windows of the Lever House also limited external atmospheric variables and made it easier to control temperature and humidity. The Lever House was the first building that required energy to be habitable; occupants became dependent on energy consuming heating ventilation and air conditioning (HVAC) systems for fresh and tempered air.

The design of the Lever House and its peers such as the United Nations Headquarters (1952) by Wallace K. Harrison quickly became a symbol of modernity. The "international style" buildings (named after a 1932 show curated by Philip

Figure 3
The Lever House

Johnson) also rapidly became building technology models for generations of structures to come. By the mid-1960s, Class A Office Buildings commonly had many technological advances of the early international-style buildings such as integrated HVAC systems, curtain-wall enclosures, and sealed windows. Another significant contribution of the international-style buildings was a shift from the wedding cake massing of their predecessors. Charles Luckman, the president of Lever Brothers in 1952, wanted an iconic building. He gave SOM a program that did not maximize buildable area on the site under the zoning code—it was only half the allowable area;[26] this allowed Bunshaft to design a rectangular building of low density with ample space around it. The design approach, using compact simple massing surrounded by open space, led to a major reworking of New York City's zoning code. The 1961 New York Zoning Law transformed the 1916 "wedding cake" restrictions for a building into floor area ratio (FAR) restrictions for a plot. This change allowed for the development of deep lease spans up the entire volume of the building like the buildings of the pre-zoning resolution era. The 1961 reworking of the bulk requirements in essence codified energy dependence.

As buildings became more and more reliant upon mechanical systems to operate, design approaches became standardized. In 1966, the American Society of Heating, Refrigerating and Air-Conditioning Engineers (ASHRAE), a professional society of mechanical engineers, created a performance criterion called "Thermal Environmental Conditions for Human Occupancy" to replace the outdated 1938 standard. This criterion, cited as the abbreviated name of ASHRAE Standard 55, has been amended numerous times since it was first drafted and is the standard upon which most mechanical designs are based. This standardization of design approach has led to higher energy use. The Center for the Built Environment notes in its 2007 study on operable windows that "comfort standards such as ASHRAE Standard 55 are universally applied across all building types, climates, and populations. This 'one-size-fits-all' approach requires energy-intensive environmental control strategies to deliver stable, consistent temperatures, and can lead to excessive use of air-conditioning."[27]

The energy use of buildings steadily grew in the 1950s and 60s. A study of eighty-six New York City office buildings showed that buildings constructed in the late 1960s have primary energy requirements more than double those constructed in the early 1950s.[28] This increased energy consumption mirrored an overall rise in per capita energy use. In 1949, US energy use per person stood at 215 million BTU and the rate of consumption generally increased until the energy crisis of the mid-1970s. A small dip in usage occurred in the early 1980s in the aftermath of the previous decade's energy crisis, but began to gradually increase again. This overall increase in energy use over five decades is due to greater air conditioning, appliance, and equipment use (plug load). From 1978 to 1993, the number of homes with central air conditioning rose from eighteen million to forty-two million homes; in the South, the growth was even larger.[29]

Another energy factor that has increased energy use is that buildings have grown in size since World War II and accordingly require more energy to operate. The average American new house has more than doubled in size from 983 square feet in 1950 to 2266 square feet in 2000; at the same time, household size has decreased, resulting in a floor area per capita increase of 561 square feet.[30] In the last few decades, commercial floor space also has increased by approximately 260 square-feet per person.[31]

The energy crisis of the 1970s spurred research and building code requirements to improve energy efficiency. The thermal efficiency of building envelopes, light fixtures, and mechanical equipment has improved in the last forty years. In spite of these notable gains in energy efficiency, the total energy use has increased over 30 percent in US residential buildings since 1978, and more than 65 percent in commercial buildings. The amount of energy used for the building sector has increased faster than any other sector of the US economy.[32]

Today

High school physics teacher David Ames wanted his students to create something that would help everybody understand their contribution to global warming; so, he had them build a cube. The cube represents a cubic ton of carbon dioxide (CO_2). Carbon dioxide is a greenhouse gas that is released when fossil fuel (coal, petroleum, and natural gas) is burned. "No-one knows what a ton of CO_2 looks like, and I thought that if people could see it then it might help visualize the issue," Ames told his local newspaper before his students erected their cube.[33] When his students finished their task, the Cohasset High School in Cohasset, Massachusetts, had a 27 feet wide by 27 feet high by 27 feet deep cube on the school's soccer field.[34,35,36] The cube built by Ames' students represents the amount of CO_2 each American emits into the atmosphere every eighteen days, which equals about 20 "cubes" or metric tons of CO_2 a year.[37] If we gather up cubes from everyone in the US for a year, we would have a 27-foot high volume of CO_2, which is the size of the state of California.[38] As of 2008, buildings in the United States are responsible for 8 percent of the world's energy consumption, which is 50 percent higher than in 1980.[39] According to the United Nations' estimates, globally "buildings contribute as much as one-third of total global greenhouse gas emissions, primarily through the use of fossil fuels during their operational phase."[40] The US Energy Information Administration (EIA) found that 38 percent of all CO_2 emissions in the country came from the building sector in 2009. In the EIA study, transportation accounted for another 28 percent and industry for an additional 34 percent of CO_2 emissions.[41] Ed Mazia, the founder of Architecture 2030, a group dedicated to slowing the growth rate of greenhouse gas emissions by having buildings eliminate fossil fuel use by 2030, estimates that buildings account for 48 percent of CO_2 emissions when the embodied energy of the building materials are included in the calculation.[42] Clearly, the size of a building's carbon footprint is huge and is in the range 33–48 percent depending on how and where the calculation is made. In the United States, and based on a more conservative number from the EIA, American buildings release a huge amount of greenhouse gases each year—equivalent to a 27 foot high volume that is little larger in area than the US state of Georgia.

The CO_2 emissions from buildings come from a wide variety of energy sources. In the United States, 76 percent of the energy used by buildings is from fossil fuels, 15 percent comes from nuclear generation, and 8 percent comes from renewable sources like wind. The daily operation of buildings in the United States consumes 74 percent of electricity generation.[43] To put the scale of these numbers in perspective, the US Environmental Protection Agency (EPA) calculates the combined number of commercial buildings and industrial facilities in the United States to be over 5 million, and if the performance of these buildings improved by 10 percent, the result would be $20 billion in savings each year. A 10 percent improvement in greenhouse gas emissions (which does not equate directly to a 10 percent energy savings) from commercial buildings and industrial facilities would be the equivalent of removing about 30 million vehicles from the road for a year, which is the number of registered automobiles in Illinois, New York, Ohio, and Texas combined.[44] These statistics do not cover the energy used to extract, manufacture, and transport the materials that comprise buildings—commonly called the embodied energy. The Department of Energy (DOE) calculates that the embodied energy of building materials is 1.146 MMBTU/SF for new construction and 0.573 MMBTU/SF for renovation.[45] (A BTU—British Thermal Unit—is the approximate amount of energy needed to heat 1 pound of water from 39 to 40 °F).

On an optimistic note, the building sector has the largest potential to appreciably reduce greenhouse gas emissions. The Fourth Assessment Report of the Intergovernmental Panel on Climate Change (IPCC) found that this potential is relatively independent of the cost per ton of CO_2, and there is an equally high probability for greenhouse gas reductions in buildings in both developed and developing countries; this means that with today's technology, the energy use of buildings (both new and existing) can be significantly reduced by an estimated range of 30 to 80 percent. However, if efforts are not made to slow the steady increase in greenhouse gas emissions from buildings, their emissions levels will double by 2030.[46] This will have a profound impact on the planet, and as the IPCC stated:

"Continued greenhouse gas emissions at or above current rates would cause further warming and induce many changes in the global climate system during the 21st century that would very likely be larger than those observed during the 20th century."[47]

In the earliest days of the environmental movement in the 1970s, forward thinkers began to question why our buildings used so much energy, but it was not until the Organization of the Petroleum Exporting Countries (OPEC) embargo oil crisis in 1973 that the general public began to ask the same question. The high cost of gasoline and home heating oil helped US citizens to realize how much the modern world depends on energy to work. As the crisis ebbed, the green building movement began. In 1977, Sir Norman Foster used a grass roof, a naturally light atrium, and mirrored windows in the Willis Faber and Dumas Headquarters project in Ipswich, England, as a means to reduce energy usage. In California, the state commissioned eight energy-sensitive office buildings that had photovoltaic panels and zone climate-control mechanisms. Through the 1980s and 1990s, however, green design and energy-efficient buildings remained on the periphery of the construction industry.[48]

Today, the larger design and construction industry acknowledges that the building industry has an unavoidable responsibility to improve the energy performance of its structures. "Climate change is real" and "architects have solutions," proclaims the American Institute of Architects (AIA) website.[49] The 80,000 members of AIA have endorsed Architecture 2030's challenge of carbon neutrality by 2030 for new and renovated buildings. Numerous other design professional organizations have also adopted the 2030 Challenge, such as ASHRAE, the Royal Architectural Institute of Canada (RAIC), Congress for the New Urbanism, and the American Society of Interior Designers (ASID).[50] However, this endeavor has not been totally successful; only 12.1 percent of the combined design portfolios of the endorsing firms (weighted by gross square footage) are currently meeting the 2010 60 percent energy reduction target of the 2030 Challenge.[51]

An ever-growing number of US states and municipalities are also attempting to address climate change by adopting energy-efficiency codes and green-building initiatives, which target the energy use of buildings. Building energy codes fall primarily under the jurisdiction of state and local authorities, and consequently vary widely across the country. Individual states that allow jurisdictional control to remain at the local level require local codes to exceed a statewide minimum. Some states, like California under its Title 24 Building Standards Code, have statewide mandates that use performance criteria on energy consuming equipment to aggressively reduce energy demand. In California, the per capita building-related energy demand has remained flat since the early 1980s, after the enactment of both a statewide building code and appliance standards.[52] California's latest proposed energy code requires all new residential construction to be zero net energy by 2020 and all new commercial construction to be zero net energy by 2030. Local jurisdictions like New York City have enacted energy policies targeted toward a reduction in energy use. New York estimates that 80 percent of the city's global warming emissions come from the energy used in buildings.[53] New York City's "PlaNYC: A Greener, Greater New York" targets a 30 percent reduction of greenhouse gases by 2030.

Before the OPEC embargo, in the era of "relatively cheap and abundant energy," the Federal government had virtually no energy policy.[54] The embargo changed this scenario that; on October 1, 1977, the Federal government established the Department of Energy to address energy policy and to regulate the safe handling of nuclear material. After an initial flurry of policies targeting energy performance, the DOE has not ventured deeply into policies that require or promote the energy improvement of buildings. Rather, the DOE has focused its attention toward promoting research and technologies that, when applied, would change energy demand. A few notable recent exceptions are the Energy Policy Act of 2005 and the 2009 American Recovery and Reinvestment Act (ARRA), commonly known as the Recovery Act, which have both tax incentives and other initiatives that directly promote the energy improvement of buildings. The Recovery Act invested millions of dollars into the weatherization of existing houses. Meanwhile, another proposed bill called the American Clean Energy and Security Act of 2009 would have advanced a national policy to encourage energy efficiency for all commercial and industrial buildings in the United States by establishing an emissions trading plan similar to the one found in Europe. This bill died in the Senate. The lack of a comprehensive nationwide plan to reduce buildings' energy usage is in sharp contrast to many other western nations. The United Kingdom (UK) has adopted binding emissions reductions of 80 percent below 1990 levels by 2050 with substantial progress by 2030. The UK has created rules to govern the performance of its total building portfolio. In the United States, the Federal government's Executive Order (EO) 13514, under the Federal Leadership in the Environmental, Energy, and Economic Performance program, aims to shift government operations toward greater sustainability. The EO 13514 focuses on high performance and sustainable buildings and includes a requirement that all Federal buildings designed after 2020 achieve zero net energy by 2030.

Over the last decade, beyond legal minimums established by energy codes, by far the biggest influence on the energy performance of buildings in the United States has been Leadership in Energy and Environmental Design (LEED). LEED acts as a third-party, which certifies green buildings and green communities. Developed by the US Green Building Council (USGBC), LEED intends to serve as a universal framework that measures energy savings, water efficiency, indoor air quality, resource stewardship, and site impacts. LEED's simple point system goes from a basic certification to an Olympic medal-like rating of silver, gold, and platinum. LEED's rating system "has given it enormous appeal and made it the most widely accepted program of its kind in the United States."[55] As of September 2011, the USGBC had certified over 22,300 projects (1.538 + billion square feet of commercial spaces).[56] Among the 22,300 certified projects is the Empire State Building, which was awarded a Gold certification under LEED's Existing Buildings rating program on September 13, 2011. The Empire State Building has now reclaimed the tallest building title—it is the tallest building in the United States to receive LEED certification.[57]

In 2007, the USGBC instituted a requirement that all LEED projects achieve an energy reduction of 14 percent over the ASHRAE baseline. This performance requirement followed an eight-point agenda by USGBC leaders to address climate change and buildings. Rick Fedrizzi, USGBC president at the time, said, "Each of the eight specific actions will have an immediate and measurable impact on CO_2 reduction," and "when implemented in concert, they comprise a powerful leadership initiative that sets a high bar for the building industry."[58] The USGBC continues to strive to be a leader in the arena of energy savings in buildings and estimates that if LEED continues to grow at its current rate, it has the potential to save $6 billion in energy expenditures in the next four years.[59]

With this success, nevertheless, the USGBC also has accrued many critics. Henry Gifford, a New York City–based energy consultant, unsuccessfully filed a class action suit against the USGBC for falsely advertising that LEED guarantees energy savings when buildings are awarded LEED certification. Judge Leonard Sand found Gifford's claims to be baseless and "too speculative."[60] In particular, the judge did not concur with Gifford's argument that the USGBC is harming the entire field of green building, and that consumers "will discount all claims of energy saving through design and construction" because of USGBC's false energy performance claims for certified buildings.[61]

A more serious challenge to the USGBC is the assertion that LEED has set the bar too low. Under LEED, a building can receive a platinum certification and still have a high-energy usage per square foot. James Brew, an architect with the Rocky Mountain Institute (RMI), has stated, "LEED Gold is kind of a C right now," and "maybe LEED Platinum is a B or an A-minus."[62]

These critics believe that green buildings, and energy performance in particular, must go well beyond existing standards and that a new approach to building design is required. This has led to a series of new green building initiatives that all try to shift the paradigm toward a more radical change. One of the most far-reaching of these initiatives is the Living Building Challenge (LBC). The Cascadia Green Building Council (based out of Alaska, Oregon, Washington State in the United States, and British Columbia in Canada) created the LBC in November 2006 to augment to LEED, in order to "raise the bar" and to diminish the gap between current limits and ideal solutions."[63],[64] Jason F. McLennan, the CEO of the Cascadia Green Building Council says that LEED

> "nudge(s) us to go a [sic] beyond code, but it's still just being 'less bad.' So what would it take to be 'good,' restorative, truly sustainable? That's what we tried to codify in the Living Building Challenge."[65]

Under the LBC rating system, there are sixteen requirements that are organized into seven general areas. The energy requirement is intended "to signal a new age of design, wherein the built environment relies solely on renewable forms of energy and operates year round in a pollution-free manner."[66] To achieve this goal, the LBC requires projects to obtain 100 percent of the project's energy needs from on-site renewable energy on a net annual basis. Living Building Challenge projects are Net-Zero Energy Buildings (NZEBs).

The standard approach to energy performance is to set goals, typically based on a certain percentage of energy that falls beneath the prevailing code (i.e., California Title 24) or a recognized national standard (i.e., ASHRAE Standard 90.1). The US Green Building Council's LEED rating system bases its energy-efficiency measurement on the ASHRAE Standard or acceptable local codes. This approach has two limitations: the first being that these codes and standards only address a fraction of the energy-using systems in the building and second being that the models are only as good as the assumptions that are made about building operations and management practices.[67] The Living Building System rigorously avoids these problems by requiring twelve months of energy use data after the building is occupied to demonstrate that it is a net zero energy consumer. (Under NREL's NZEB classification system, Living Buildings are NZEB A-Plus, which is the highest level in the 5-tiered system).[68] The Living Building's measurement of actual energy performance is in sharp contrast to LEED's reliance on model energy performance.

As of September 2011, only two projects had achieved full certification under the Living Building rating system: The Omega Center for Sustainable Living in Rhinebeck, New York, and the Tyson Living Learning Center in Eureka, Missouri. There are other NZEB buildings that have not sought to be certified under the Living Building System. As early as the mid-1990s, before LEED existed, some professional practitioners were exploring the idea of minimal or zero energy buildings. From this emerged some pioneering projects, and according to the Center on the Built Environment at the University of California at Berkeley, there are currently forty NZEBs in the United States and another seventeen buildings that are capable of being net zero buildings if photovoltaic panels are added to their roofs.[69] The current portfolio of NZEBs is diverse (nature centers, research buildings, office buildings, homes, and a bank),[70] but all are small scale. There are approximately hundred Living Building projects now in design and construction.[71] The scale and complexity of these projects is growing along with the number of new Living Building projects.

The concept of NZEBs is not limited to the United States. There are numerous projects outside the US that pursue net-zero energy performance such as the Masdar City Headquarters in Abu Dhabi, by Adrian Smith + Gordon Gill Architecture. The Beddington Zero Energy Development (BedZED) is a pioneering net-zero housing development in the Hackbridge District of the Borough of Sutton in London, England, designed by architect Bill Dunster. Completed in 2002, this environmentally friendly housing development has ninety-nine homes and 1,405 square meters of workspace that relies on power generated onsite from solar panels and tree waste. Numerous organizations and international standards promote ZNEBs such as the Swiss MInergie-P system and the Net-Zero Energy Home Coalition in Canada. The program boasting the broadest adoption in Europe is the ultra-low-energy standard of Passivhaus from Germany. Passivhaus—passive house—buildings use design elements with an exceptionally high level of thermal insulation, well-insulated window frames with triple low-e glazing, thermal-bridge-free construction, airtight building envelopes, baseline thermal comfort and ventilation levels, and highly efficient heat recovery systems. More than 22,000 housing units in Europe currently meet the Passivhaus standard. Passive Houses are not zero-energy buildings, but they come very close by using no more than about 1.5 liters of heating oil or 1.5 cubic meters of natural gas (equivalent to 15 kWh) per square meter of living space per year; this equates to energy savings of more than 90 percent over the average existing building. In comparison, a new building code in Germany requires 6 to 10 liters of oil per square meter of living space a year.[72] The Passivhaus homes are now being built in the United States; as of September 2010, thirteen Passive Houses were completed and a dozen more were in the design and construction phase.[73]

The worldwide effort to design and build NZEBs is not limited to the leaders of the design and construction industry. Every two years since 2002, the US Department of Energy holds a competition that challenges collegiate teams to "design, build, and operate solar-powered houses that are cost-effective, energy-efficient, and attractive."[74] The winning team is one that best demonstrates design excellence with optimal energy production and maximum efficiency while being affordable and appealing. In 2011, the Solar Decathlon students did just that. Nineteen different teams from universities in the United States and abroad designed and built homes of 1,000 square feet that use solar energy to generate electricity, hot water, and air with the goal of achieving net-zero energy.

The "grand winner" is the team who designs a house that goes beyond energy balance to be energy positive—generates energy to give back to the grid. The two noteworthy projects at the 2011 Decathlon were the Middlebury College building and the Appalachian State University House. The first is a warm and straightforward modern version of a traditional New England home that utilizes native Vermont materials. The Appalachian design is also based upon local home building traditions with gracious outdoor and interior spaces. NZEBs are achievable and are currently being designed and developed by teachers, artists, and even political scientists, not just professionals traditionally associated with the building sector.[75]

Tomorrow

In 2001, Pamela Mang predicted a revolution in buildings was coming in the form of something called regenerative design. Regenerative design is a shift away from the current sustainable design concept of buildings "doing less harm," and even goes beyond the ecological stasis of net-zero performance toward a new paradigm of buildings that demonstrate a positive ecological footprint. As Mang noted when making her prediction, we "must initiate regenerative processes to replace the degeneration resulting from past practices." This means rethinking how we design, build, and use the built environment.[76]

The physicist Niels Bohr famously quipped, "Prediction is very difficult, especially about the future." Nevertheless, Mang's prediction is becoming a reality. A decade later, there is not a single building that has a total positive ecological footprint, but there are buildings that are beginning to achieve regenerative capacities in some aspect of their performance. The Center for Interactive Research on Sustainability (CIRS) designed by Perkins+Will at the University of British Columbia gives us a glimpse of the regenerative future. Alberto Cayuela, CIRS associate director says, "Given the imperatives of climate change, the new paradigm includes the idea of regeneration in the way we make buildings."[77] Utilizing regenerative design principles, the CIRS project has many innovations, most notably its approach to energy. First, the design team reduced consumption through passive design methods and maximized efficiencies, but then sought to build symbiotic energy relationships with the adjacent buildings. CIRS takes the excess exhaust heat from an adjacent lab building and uses it; then, it exchanges and returns that heat back to the adjacent building to be reused. By working beyond the traditional limits of a project's metes and bounds, this effort lowered the energy usage of the whole campus by 565 MWH per year (172 Tons of CO_2 per year)[78] when the CIRS project went online in September 2011. A building that reduces the carbon footprint for an entire campus is regenerative. The question is: are buildings like the CIRS the next evolutionary step in building design? For this next evolutionary step in building design to become the norm, a broad shift must occur, away from a culture of consumption without consideration, to a culture of thoughtful consumerism.

Notes

1. "The Empire State Building," Glass, Steel, and Stone Website, http://www. glasssteelandstone.com/BuildingDetail/433.php (accessed August 2011).
2. O'Neill, Sean, "10 Most Photographed Places on Earth" *Budget Travel* http://travel. yahoo.com/p-interests-40218938 (accessed August 2011).
3. Mock, Brentin, "Charity Case" *Next American City*, 2008, http://americancity.org/ magazine/article/charity-case/ (accessed August 2011).
4. "Two New Orlean Hospitals Plead for Help: Other Hospitals see gunfire, run out of Medication," *Associated Press*, September 2005, http://www.msnbc.msn.com/ id/9159903/ns/us_news-katrina_the_long_road_back/t/two-new-orleans-hospitals-plead-help/#.Tl-XqXO4Iy4 (accessed August 2011)
5. Burdeau, Cain, "Honere Ex-LA Governor Halted Hospital Reopening," *Associated Press*, July 2009, http://savecharityhospital.com/content/ap-honore-ex-la-governor-halted-hospital-reopening (accessed August 2011).
6. Oldfield, Philip, Trabucco, Dario, Wood, Antony, "Five Energy Generations of Tall Buildings: An Historical Analysis of Energy Consumption in High-Rise Buildings." *The Journal of Architecture* 592 (September 2009).
7. "Illuminance—Recommended Light Levels," The Engineer Toolbox Website, http:// www.engineeringtoolbox.com/light-level-rooms-d_708.html (accessed August 2011).
8. Oldfield, Philip, Trabucco, Dario, Wood, Antony, Ibid.
9. Fogelson, Robert M., *Downtown: Its Rise and Fall, 1880–1950* (New Haven CT: Yale University Press, 2001), 125.
10. New York City Department of City Planning Website, http://www.nyc.gov/html/dcp/ html/zone/zonehis.shtml (Accessed August 2011).
11. Conuel, Thomas, "The Colossus of Zoning: Many of our current land use laws have emerged from a single legal decision that took place in 1926," *Sanctuary: The Journal of the Masschusetts Audubon Society* 11 (Fall/Winter 2011–2012).
12. Hitchcock, Henry—Russell, "Architecture: Nineteenth and Twentieth Centuries," (New York: Penguin Books, 1983), 514.
13. Oldfield, Philip, Trabucco, Dario, Wood, Antony, Ibid 596.
14. Oldfield, Philip, Trabucco, Dario, Wood, Antony, Ibid 595.
15. Pauken, Mike, "Sleeping Soundly on Summer Nights: The First Century of Air Conditioning," *ASHRAE Journal* 40—47 (May 1999).
16. Arnold, David, "Air Conditioning in the Office Building after World War II: The First Century of Air Conditioning," *ASHRAE Journal* 33 (July 1999).
17. "Weathermakers: Carrier Corp," *Fortune*, April 1938, http://features.blogs.fortune. cnn.com/2011/06/05/air-conditioning-history-mass-market-fortune-1938/ (accessed September 2011).
18. http://www.glasssteelandstone.com/BuildingDetail/433.php, Ibid.
19. Hitchcock, Henry—Russell. Ibid, 514.
20. Oldfield, Philip, Trabucco, Dario, Wood, Antony, Ibid, 596.
21. Clausen, Meredith, "Belluschi and the Equitable Building in History," *Journal of the Society of Architectural Historians* L, no. 2 (June 1991): 109–29. http://sah. org/index.php?src=gendocs&ref=JSAH&category=Publications (Accessed August 2011).
22. Arnold, David, "The Equitable Building—The Genesis of Modern Air Conditioned Buildings," 1, www.cibse.org/pdfs/The%20equitable%20building.pdf (Accessed August 2011).
23. Equitable Builds a Leader, Architectural Forum (September 1948), 98–105.
24. Arnold, David. Ibid, 8.

25. Tyenauer, Matt "Forever Modern," *Vanity Fair*, October 2002, http://www.vanityfair. com/culture/features/2002/10/leverhouse200210#gotopage1 (accessed August 2011).
26. Tyenauer, Matt. Ibid.
27. "Operable Windows and Thermal Comfort Project," Center for the Built Environment Website, http://www.cbe.berkeley.edu/research/briefs-opwindows.htm (accessed August 2011).
28. Stein, R. G., "Observations on Energy Use in Buildings." *Journal of Architectural Education* 30, no. 3 (1977): 36–41.
29. Diamond, Richard, "An Overview of the US Building Stock," *Lawrence Berkeley National Laboratory* 8 (January 2001), http://eetd.lbl.gov/ie/pdf/LBNL-43640.pdf (accessed September 2011)
30. Diamond, Rick, Moezzi, Mithra, "Changing Trends: A Brief History of the US Household Consumption of Energy, Water, Food, Beverages and Tobacco," *Lawrence Berkeley National Laboratory*, May 2004, http://eetd.lbl.gov/ie/pdf/LBNL-55011. pdf (Accessed August 2011).
31. Rong, Fang, Clarke, Leon, Smith, Steven, "Climate Change and the Long-Term Evolution of the US Buildings Sector," *Pacific Northwest National Laboratory. US Department of Energy* March 2007, 2.1, http://www.pnl.gov/main/publications/ external/technical_reports/PNNL-16869.pdf (accessed September 2011).
32. Harris, Jeffrey, Diamond, Rick, Iyer, Maithili, and Payne, Christopher, "Don't Supersize Me! Toward a Policy of Consumption-Based Energy Efficiency" *Lawrence Berkeley National Laboratory*, http://eetd.lbl.gov/payne/publications/DontSuper-sizeMe.pdf (accessed August 2011).
33. White, Nancy, "When it's Hip to be Square," *Cohasset Mariner*, April 2007, http:// www.wickedlocal.com/cohasset/local_news/x404593992#axzz1XZWo0Pxb, (accessed September 2011).
34. B. Jay, "Commentary: What does a Ton of CO_2 Look Like," *EnergyRace Website*, March 2008, http://www.energyrace.com/commentary/what_does_a_ton_of_co2_ look_like/ (accessed September 2011).
35. White, Nancy, "It's Good for Go-Blue to Go-Green on Sunday," "When it's Hip to be Square,"*Cohasset Mariner*, April 2007, http://www.wickedlocal.com/cohasset/ local_news/x1201791017#axzz1XZWo0Pxb (accessed September 2011).
36. Please note that Ames' class wasn't the first carbon cube; the first cube appears to have been constructed and displayed in Copenhagen for the 2009 Climate Change Conference. For additional information see http://www.artdaily.com/index.asp?int_ sec=11&int_new=34531&int_modo=2
37. 19.87 metric tons of CO_2 emissions per capita represents the mass of carbon dioxide (CO_2) emitted per person in the US in 2005. Source = World Resource Institute. Accessed September 2011 at http://earthtrends.wri.org/searchable_db/results.php? years=2005-2005&variable_ID=466&theme=3&cID=190&ccID=
38. Cube Calculation:
    1. 27 feet × 27 feet = 729 square feet (area of each cube)
    2. 729 square feet × 19.87 cubes a year = 14,485.23 square feet (annual area of cubes) or 0.000519 square miles
    3. 0.000519 square miles × 312,183,000 Americans = 162,022.977 square miles
39. "Building Energy Data Book," *The United States Department of Energy*, (2010 Edition), Tables 1.1.3 and 1.1.8, http://buildingsdatabook.eren.doe.gov/ChapterIntro1. aspx?1#1 (accessed August 2011).
40. "Buildings and Climate Change: Summary for Decision-Makers," *United Nations Environmental Programme*, 2009, 2, http://www.unep.org/sbci/pdfs/SBCI-BCCSum-mary.pdf (accessed September 2011).

41. "US Energy Information Administration: Emissions of Greenhouse Gases in the United States 2009," *US Energy Information Administration*, March 2011, 5, ftp://ftp.eia.doe.gov/environment/057309.pdf (accessed September 2011).

42. Mazia, Ed., "It's the Architecture Stupid," *Solar Today Magazine*, May/June 2003, 50. http://www.mazria.com/ItsTheArchitectureStupid.pdf, (accessed September 2011).

43. "What LEED Measures," *USGBC Website*. http://www.usgbc.org/DisplayPage.aspx?CMSPageID=1989 (Accessed November 2011).

44. "Fast Facts on Energy Use: Facts about Energy Use in Commercial and Industrial Facilities,"*U.S Department of Energy and the Environmental Protection Agency's Energy Star Program*, http://www.energystar.gov/ia/business/challenge/learn_more/FastFacts.pdf (accessed September 2011).

45. Building Energy Data Book, Ibid.

46. "Buildings and Climate Change: Summary for Decision-Makers," Ibid. 3 and 11.

47. Solomon, S., D. Qin, M. Manning, Z. Chen, M. Marquis, K. B. Averyt, M. Tignor, and H. L. Miller, eds. "IPCC, 2007: Summary for Policymakers," *Climate Change 2007: The Physical Science Basis. Contribution of Working Group I to the Fourth Assessment Report of the Intergovernmental Panel on Climate Change*, Cambridge University Press, Cambridge, United Kingdom and New York, 13. http://www.ipcc.ch/pdf/assessment-report/ar4/wg1/ar4-wg1-spm.pdf (accessed September 2011).

48. "White Paper on Sustainability," *Building Design and Construction*, November 2003, 4, http://www.usgbc.org/Docs/Resources/BDCWhitePaperR2.pdf (accessed September 2011).

49. http://info.aia.org/toolkit2030/ (accessed September 2011).

50. 2030 Challenge Website, http://architecture2030.org/2030_challenge/adopters, (accessed September 2011).

51. Malin, Nadev, "Many AIA Firms Fail To Meet 2030 Commitment," *Green Source Magazine*, June 2011, http://greensource.construction.com/news/2011/06/110601-2030-Challenge.asp (accessed September 2011).

52. Doris, Elizabeth, Cochran, Jaquelin, and Vorum, Martin, "Energy Efficiency Policy in the United States: Overview of Trends at Different Levels of Government," *National Renewable Energy Laboratory*, December 2009, 7, http://www.nrel.gov/docs/fy10osti/46532.pdf (accessed in September 2011).

53. PlaNYC, "A Greener, Greater New York," *The City of New York*, 2007, 101.

54. Fehner, Terrence, Holl, Jack M, "The United States Department of Energy, 1977–1994: A Summary History," *History Division, Executive Secretariat, Human Resources and Administration, Department of Energy*, November 1994, 3, http://energy.gov/sites/prod/files/maprod/documents/Summary_History.pdf (Accessed September 2011).

55. White Paper on Sustainability, *Building Design and Construction*, Ibid, 7.

56. USGBC Website's total project graph (sq.ft. and project number displays). http://www.usgbc.org/ (accessed September 2011).

57. Bloomfield, Craig, *Jones Lang LaSalle Press Release*, September 2011, https://www.usgbc.org/ShowFile.aspx?DocumentID=10266 (accessed September 2011).

58. "USGBC to Boost LEED Energy Performance Standards," *Greenbiz.com*, June 2007. http://www.greenbiz.com/news/2007/06/25/usgbc-boost-leed-energy-performance-standards (accessed 2011).

59. "Government Summit 2011: Fuel for a Clean Economy," *US Green Building Council*, http://www.usgbc.org/DisplayPage.aspx?CMSPageID=1967 (accessed September 2011).

60. Roberts, Tristan, "Gifford Lawsuit against USGBC, LEED Dismissed," *Environmental Building News—Website Posting*, August 2011, http://www.buildinggreen.com/auth/article.cfm/2011/8/17/Gifford-Lawsuit-Against-USGBC-LEED-Dismissed/ (accessed September 2011).

61. Roberts, Tristan, Ibid.

62. Davis, Lisa S. "Structures So Green They Give Back to the Environment." *Plenty Magazine*, August 2008, http://www.mnn.com/money/green-workplace/stories/structures-so-green-they-give-back-to-the-environment (accessed September 2011).

63. Bateson, Shelby, "Move over LEED, is ILBI more green?," *Examiner.com*, November 2005, https://ilbi.org/about/About-Docs/news-documents/pdfs/09-1115%20Examiner%20Move%20Over%20LEED-%20Is%20ILBI%20More%20Green.pdf (accessed October 2011).

64. "Living Building Challenge 2.0—Executive Summary," https://ilbi.org/lbc/Standard-Documents/LBC2-0.pdf (Accessed October 2011).

65. Hitchcock, Darcy, "Beyond LEED: The Living Building Challenge with Jason F. McLennan, Cascadia Green Building Council," *ISSP Insight newsletter*, December 2008, http://www.sustainabilityprofessionals.org/files/Beyond%20LEED.pdf (Accessed October 2011).

66. Ibid, Living Building Challenge 2.0.

67. Brown, Karl, Daly, Allen, Elliot, John, and Higgins, Cathy, "Hitting the Whole Target: Setting and Achieving Goals for Deep Efficiency Buildings," *ACEEE,* August 2010.

68. Pless, Shanti, Torcellini, Paul, "Net-Zero Energy Buildings: A Classification System Based on Renewable Energy Supply Options," *National Renewable Energy Laboratory*, June 2010, http://www.nrel.gov/docs/fy10osti/44586.pdf (accessed October 2011).

69. Author's notes from October 5, 2011 Greenbuild Presentation by HOK Architects, "Achieving Market Rate Zero Emissions Office Building—What Will It Take," Toronto, Canada.

70. "Zen Energy Buildings," *US Department of Energy—Energy Efficiency and Renewable Energy*, http://zeb.buildinggreen.com/ (accessed October 2011).

71. Nelson, Bryn, "The Self-Sufficient Office Building," *New York Times*, October 2011.

72. "Active for More Comfort: The Passive House," *International Passive House Association*, 2010, http://www.passivehouse-international.org/index.php?page_id=70 (accessed October 2011).

73. Zeller, Tom, "Beyond Fossil Fuels: Can We Build in a Brighter Shade of Green?" *New York Times*, September 2010, http://www.nytimes.com/2010/09/26/business/energy-environment/26smart.html?_r=1&ref=earth&pagewanted=all, (Accessed October 2011).

74. US Department of Energy—Solar Decathlon Website. http://www.solardecathlon.gov/about.html, (accessed October 2011).

75. Syrett, Peter, "Witnessing Evolution at the Solar Decathlon." *Solar Decathlon*, September 2011, http://www.metropolismag.com/pov/author/peter (accessed October 2011).

76. Mang, Pamela, "Regenerative Design: Sustainable Design's Coming Revolution," *DesignIntelligence*, July 2001, http://www.di.net/articles/archive/2043/ (accessed September 2011).

77. Cockram, Michael, "The Living Lab: The New Generation of Living Laboratories Fosters Research and Product Development while Providing Educational Tools for Green Building," *Greensource*, September 2011, http://continuingeducation.construction.com/article.php?L=5&C=824 (Accessed October 2011).

78. Energy data given about the CIRS projects is from author's interview with Blair McCarry P.Eng., PE, ASHRAE, LEED AP of Perkins + Will.

9

Sustainable Transport: Managing Auto Dependence through Travel-Time Budgets

Peter Newman and Lee Schipper (post-humous author)

The Issue

Sustainable transport is about reducing the number of cars and trucks and carbon and air quality pollutants in vehicle fuels; this involves addressing a range of measures like carbon taxes and how much we invest in sustainable transport infrastructure. This paper looks at something more fundamental: how cities need to change so that sustainable transport becomes viable and effective.

Urban Form and Transport

Cities are shaped by many historical and geographical features, but at any stage in a city's history, the patterns of land use can be changed by altering its transportation priorities. Italian physicist Cesare Marchetti[1] argued that there is a universal travel-time budget of around one hour on average per person per day. This *Marchetti Constant* has been found to apply in every city, for example, the Global Cities Data Base[2] as well as in data on UK cities for the last 600 years.[3] The biological or psychological basis for the Marchetti Constant seems to be a need for a more reflective or restorative period between home and work, but it cannot go for too long before people become very frustrated owing to the need to be more occupied rather than just "wasting" time between activities.

The Marchetti Constant, therefore, shows how cities are shaped.[4] Cities grow to being "one hour wide" based on the speed with which people can move in them. So far, three city types have emerged:

Walking cities have existed for the past 8,000 years, as no other form of transport was available to enable people to get across their cities other than at a walking speed of around 5–8 km/h. Thus, walking cities were and remain dense, mixed-use areas that are no more than 5–8 kilometers across. These were the major urban form for 8,000 years, but substantial parts of cities like Ho Chi Minh City, Mumbai, and Hong Kong retain the character of a walking city. Kraków is mostly a walking city. In wealthy cities like New York, London, Vancouver, and Sydney, the central areas are predominantly walking cities in character. Many cities worldwide are trying to reclaim the walkability of their city centre, and the cities find that they can't do this unless they have the form of ancient walking cities for their central places.

Transit cities from 1850–1950 were based on trams and trains, which could travel at around 20–30 km/h; this meant they could spread out over 20–30 kilometers, with dense centers and corridors following rail lines and stations. Most European and wealthy Asian cities retain this form, as do the old inner cores in US, Australian, and Canadian cities. Many developing cities in Asia, Africa, and Latin America have the dense corridor form of a transit city, but they do not always have the transit systems to support them, so they become car-saturated. Most of these emerging cities are now building the transit systems that suit their urban form; for example, China is building eighty-two metro rail systems and India is building fourteen. Cities without reasonable densities around train stations are finding that they need to build up the numbers of people and jobs near stations; otherwise, not enough activity is there to support such sustainable transport.

Automobile cities from the 1950s onward could spread over 50–80 kilometers in all directions and at low density because cars could average 50–80 km/h while traffic levels are low. These cities spread out in every direction owing to the flexibility of cars and were given a few buses to support these sprawling suburbs. Canadian, Australian, US, and New Zealand cities that were developed in this way are now reaching the limits of the Marchetti Constant of a half-hour car commute as they sprawl outward. The freeways that service such areas are full at peak times and commuters are unable to keep within a reasonable travel-time budget. This is now a serious political issue as outer suburban residents are demanding fast rail links that can beat the traffic. Also, many people are leaving these areas and moving to better locations where they can live within their travel-time budget. The first evidence is now appearing that these automobile cities are coming back in and reducing their car use.[5]

Cities, like growing megacities or rapidly sprawling ones, are constantly facing the need to adapt their land use or infrastructure to the travel-time budget. They may not realize that is what they are doing, but if the Marchetti Constant is exceeded, then markets and politics invariably ensure that people adapt by moving closer to their work or finding a better transportation option. The search for better options can form the basis of social movements that seek to provide greener transportation.

Land Use for Sustainable Transport

Figure 1 from the Global Cities Database shows the huge range in per capita fuel use that characterizes cities across the world. They all have a combination of these three city types—walking, transit, and automobile cities.

Barcelona uses just 8 GJ per person compared to 103 GJ at Atlanta, yet they have similar per capita levels of wealth. The difference seems to be that Barcelona is substantially a walking city with some elements of a transit city and almost no automobile city, while Atlanta is almost completely an automobile city.

The broader picture is expressed in Figure 2 where travel patterns (as reflected by either per capita car use or passenger transport energy use) are exponentially related to the density of urban activity. Atlanta is six people per hectare and Barcelona is two-hundred per hectare.

The same patterns can be seen across cities where often the centers are walking cities, the middle suburbs are transit cities, and the outer suburbs are automobile cities. This can be seen in Figure 3 where Melbourne and Sydney data are shown covering transport greenhouse gases per person by suburb versus the density of residents and jobs per hectare.

Questions of wealth do not appear to be driving this phenomenon, as there is an inverse relationship between urban intensity and household income in Australian cities—outer suburbs are poorer, yet households in these areas can drive from three to ten times as much as households in the city centre. As the data on

Figure 1
Fuel Use per Person in Cities across the World

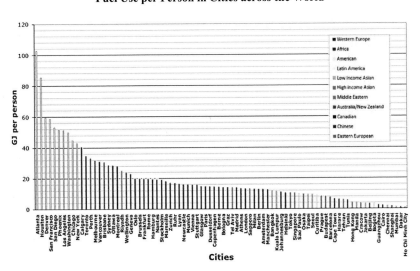

Source: Kenworthy and Laube et al., 1999.

Figure 2
Transport Fuel per Person and Urban Density (people per hectare)

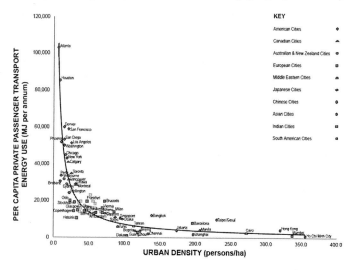

Figure 3
**Transport Fuel (Expressed as Greenhouse Gas per Person) vs. Density
(People and Jobs per hectare) in Suburbs of Melbourne and Sydney**

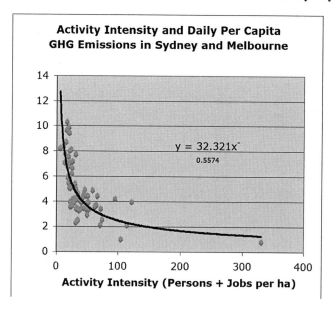

Source: Trubka et al. 2010.

Table 1
Differences in Wealth and Travel Patterns from the Urban Core
to the Fringe in Melbourne

	Core	Inner	Middle	Fringe
Percentage of Households earning >$70,000 pa	12	11	10	6
Car Use (trips/day/cap)	2.12	2.52	2.86	3.92
Public Transport (trips/day/cap)	0.66	0.46	0.29	0.21
Walk/bike (trips/day/cap)	2.62	1.61	1.08	0.81

Source: Kenworthy and Newman, 2001.

Melbourne above (Table 1) indicate, poorer households are driving more, using public transport less, and walking less.

There are obviously factors other than the intensity of activity affecting transport; otherwise, there would be an even stronger relationship within cities between activity intensity and transport patterns; such factors include the network of services provided, income, fuel prices, cultural factors, etc., but all of these can also be linked back to the intensity of activity in various ways. Thus, although many discussions have tried to explain transport in non-land-use terms[6,7], the data suggest that the physical layout of a city does have a fundamental impact on movement patterns. This chapter will now try to take the next step and explain how the relationship between transport and activity intensity works.

What Densities Help Walking and Transit?

From the two density graphs above and from data in areas where viable transit happens and where viable walking happens, there seems to be a density at around thirty-five people and jobs per hectare for transit and around 100 people and jobs per hectare for walking. How can these numbers be understood in terms of guidelines for development to ensure transit and walking are viable options for more people?

A pedestrian catchment area or "Ped Shed," based on a ten-minute walk, creates an area of approximately 220 to 550 hectares for walking speeds of 5 to 8 km/h. Thus, for an area of around 300 hectares developed at 35 people and jobs per hectare, there is a threshold requirement of approximately **10,000 residents and jobs within this ten-minute walking area.** The range would be from 8,000 to 19,000 based on speeds from 5 to 8 km/h. Some centers will have a lot more jobs than others, but the important physical planning guideline is to have a combined minimum activity intensity of residents and jobs necessary for a reasonable local center and a public transport service to support it. Other authors support these kinds of numbers for viable local centers and public transport services[8,9,10,11]. The number of residents or jobs can be increased to the full 10,000, or any combination of these, as residents and jobs are similar in terms of transport demand. Either way, the number suggests a threshold below which transit services become non-competitive without relying primarily on car access to extend the catchment area.

Many new car dependent suburbs have densities more like twelve per hectare and hence have only 1/3 of the population and jobs required for a viable center. When a center is built for such suburbs it tends to just have shops with job densities little higher than the surrounding population densities. Hence, the Ped Shed never reaches the kind of intensity that enables a walkable environment to be created which can ensure viable transit. Many New Urbanist developments are primarily emphasizing changes to improve the legibility and permeability of street networks, with less attention being paid to the density of activity.[12,13] As important as such changes are to the physical layout of streets, it is not surprising when the resulting centers aren't able to attract viable commercial arrangements and have only weak public transport. However, centers can be built in stages with much lower numbers to begin with, provided the goal is to reach a density of at least thirty-five per hectare through enabling infill at higher intensities.

If a walking city center is required then a density of hundred per hectare is needed. This gives an idea of the kind of activity that a Town Center would need: approximately **100,000 residents and jobs within this ten-minute walking area**. The range again is from around 70,000 to 175,000 people and jobs. This number could provide for a viable town center based on standard servicing levels for a range of activities. Fewer numbers than this means services in a town center are nonviable, and it becomes necessary to increase the center's catchment through widespread dependence on driving from much farther afield. This also means that the human design qualities of the center are compromised because of the need for excessive amounts of parking. Of course, many driving trips within a walking Ped Shed still occur. However, if sufficient amenities and services are provided then only short car trips are needed, which is still part of making the center less car dependent. "Footloose jobs," particularly those related to the global knowledge economy, can theoretically go anywhere in a city and can make the difference between a viable center or not. However, there is considerable evidence that such jobs are located in dense centers of activity owing to the need for networking and quick "face to face" meetings between professionals.[14,15] High-amenity, walking-scale environments are better able to attract such jobs because they offer the kind of environmental quality, liveability, and diversity that these professionals are seeking as well as the ease of accessibility for meetings.

Politics of Sustainable Transport

Developing more sustainable transport and land use patterns has been made difficult over recent decades, as car-dependent suburbs have been facilitated by the construction of fast roads. Until recently, it has been hard to do anything for either public transport infrastructure (especially rail) or walking/cycling in centers. Mostly, such changes have been demanded by the public and have been achieved through political intervention.[16]

The reason for this seems to be a professional reticence to push sustainable modes and in particular a deep seated anti-rail sentiment. The idea that "anything a train can do a bus can do better and cheaper" has remained dominant; this has meant that in cities where bus systems are stuck in traffic, the ideology of car dependence has been meekly acceded to. This is seen in organizations such as the International Energy Agency and World Bank. Such analyses suggest car dependence is inevitable and it is no surprise that this has been the major outcome. This is now highly problematic because it is simply not sustainable from either a local or global perspective.

Most citizens who experience car dependence, and have long commutes stuck in traffic, can understand the phenomenon, since they directly feel and bear its economic, social, and environmental consequences. They want to be provided with other options. As cities continue to evolve, the politics of sustainable transport will demand both more livable and less car-dependent options for the future.

The key to this move toward sustainability is a better provision of access to transit that is faster than cars along corridors, and better provision for walking and cycling in local areas, associated with a supportive land use structure of intensive centers with a minimum land use activity intensity of thirty-five people and jobs per hectare. This is due to a fundamental need to ensure that the more sustainable transport modes have a competitive speed advantage for long trips (transit) and for short trips (bike/walk) within centers.

Such change is evolutionary, but it will always require political leadership.

The Demise of the Automobile City?

In 2009, the Brookings Institution was the first to recognize a new phenomenon in the world's developed cities—declines in car use.[17] The data below confirm this trend, especially the global work from Lee Schipper. Peak car use suggests that we are witnessing the end of building cities around cars—at least in the developed world, though it is also possible that they may be peaking as congestion levels are so high that dramatic increases in transit are now happening. These cities also have always been built, in the main, with very high densities and hence have never had the kind of automobile dependence found in New World cities. The peak car use phenomenon suggests that we may now be witnessing the demise of automobile-city development.

The Data on Car Use Trends

Robert Puentes and Adie Tomer[18] first picked up the trend in per capita car use starting 2004 in US cities. They were able to show that this trend was occurring in most US cities, and by 2010, the trend was evident in absolute declines in

Figure 4
Peaking of US Vehicle Miles of Travel (VMT)

U.S. Vehicle Miles Traveled Per Capita, Annualized and Real Gasoline Pump Prices,
January 1991-September 2008

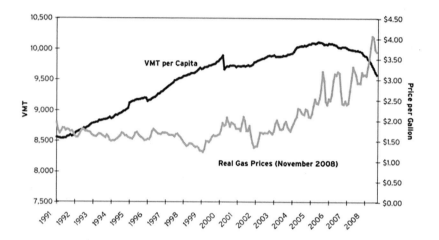

car use. The data are summarized in Figure 4, which also shows the role of fuel prices that reached around $80 a barrel in 2004.

Stanley and Barrett[19] found that a similar trend was obvious in Australian cities. The peak had come at a similar time—2004—and car use per capita at least seems to have been trending down ever since. An updated illustration of Stanley and Barretts' calculation (Figure 5) illustrates this downward trend.

Adam Millard-Ball and Lee Schipper[20] examined the trends in eight industrialized countries that demonstrate what they call "peak travel." They conclude that:

> Despite the substantial cross national differences, one striking commonality emerges: travel activity has reached a plateau in all eight countries in this analysis. The plateau is even more pronounced when considering only private vehicle use, which has declined in recent years in most of the eight countries. . . . Most aggregate energy forecasts and many regional travel demand models are based on the core assumption that travel demand will continue to rise in line with income. As we have shown in the paper, this assumption is one that planners and policy makers should treat with extreme caution.

The Global Cities Database[21,22] has been expanding its global reach since the first data were collected in the 1970s. While the 2005/2010 data are yet to be completed, the first signs of a decline in car use can be gleaned from previous years' data. Cities in the developed world grew in car use per capita in the 1960s by 42 percent, in the 1970s by 26 percent, and the 1980s by 23 percent. The new

Figure 5
Peaking of Car Use in Australian Cities

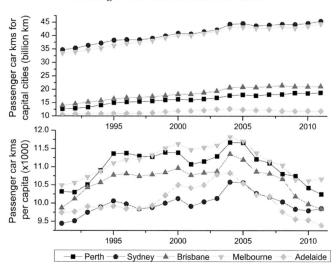

Source: Bureau of Infrastructure, Transport and Regional Economics (2009), *Australian Infrastructure Statistics Yearbook 2013*, Department of Infrastructure, Transport, Regional Development and Local Government, Canberra; Australian Bureau of Statistics (2013), *Regional Population Growth*, Cat. no. 3218.0, ABS, Canberra (Historical Population Estimates by Australian Statistical Geography Standard, 1971 to 2011).

data now show that the period 1995–2005 had a growth in car use per capita of just 5.1 percent, which is consistent with the above data on global peak car use that followed this period.[23]

In an analysis of twenty-six cities that represent the 1995–2005 percentage increase in car vehicle kilometers travelled (VKT) per capita, the data show some cities that have actually declined. Some European cities show this pattern: London has declined 1.2 percent, Stockholm 3.7 percent, Vienna 7.6 percent, Zurich 4.7 percent. In the United States, Atlanta went down 10.1 percent, Houston 15.2 percent (both from extraordinarily high levels of car use in 1995), Los Angeles declined 2.0 percent, and San Francisco 4.8 percent.

Peak car use appears to be happening. It is a major historical discontinuity that was largely unpredicted by most urban professionals and academics. So what is causing this situation?

The Possible Causes of "Peak Car Use"

The following six factors are examined and their overlaps and interdependencies are explored:

Hitting the Marchetti Wall

As outlined above, the travel-time budget matters. Freeways designed to get people quickly around cities have become car parks at peak hours. Travel times have grown to a point where cities based around cars are becoming dysfunctional. As cities are filled with cars, the limit to the spread of the city has become more and more apparent, with the politics of road rage becoming a bigger part of everyday life and many people just choosing to live closer in.

The travel-time budget limit is observable in most Australian and US cities, where the politics of transport has been based on the inability of getting sufficient road capacity to enable the travel-time budget to be maintained at less than one hour. Thus, there has been a shift to providing faster and higher capacity public transport on the basis of the growing demand to go around traffic-filled corridors or to service growing inner area districts. At the same time, the politics of planning in the past decade has turned irrevocably to enabling greater redevelopment and regeneration of suburbs at higher densities closer to where most destinations are located. The automobile city seems to have hit the wall.

The Growth of Public Transport

The extraordinary revival of public transport in Australian and American cities is demonstrated in Figures 6 and 7.

Figure 6
Recent Strong Growth in US Transit Use and Declining Car Use

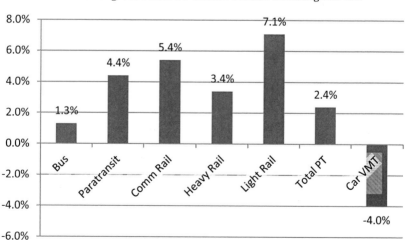

Figure 7
Growth in Transit Use in Australian Cities Since 1999

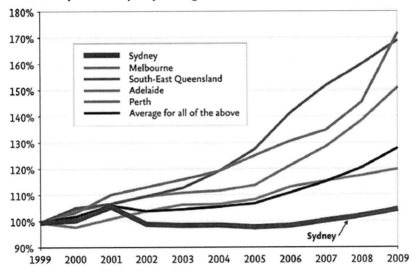

Increase in public transport patronage since 1999

The global cities' data currently being updated show that in ten major US cities from 1995 to 2005, transit boardings grew 12 percent from 60 to 67 per capita, five Canadian cities grew 8 percent from 140 to 151, four Australian capital cities rose 6 percent from 90 to 96 boardings per capita, while four major European cities grew from 380 to 447 boarding's per capita or 18 percent. The growth in transit was always seen by transport planners as a small part of the transport task, and car use growth would continue unabated. However, there is an exponential relationship between car use and public transport use, which indicates how significant the impact of transit can be. By increasing transit per capita, the use of cars per capita is predicted to go down exponentially. This is the so-called "transit leverage" effect.[24,25] Thus, even small increases in transit can begin to put a large dent in car use growth and eventually will cause it to peak and decline.

The Reversal of Urban Sprawl

The turn back of cities leads to increases in urban density rather than the continuing declines that have characterized the growth phase of automobile cities in the past fifty years. The data on density suggest that the peak in decline has occurred and cities are now coming back in faster than they are going out.

Table 2 contains data on a sample of cities in Australia, the United States, Canada, and Europe showing urban densities from 1960 to 2005, which clearly demonstrate this turning point in highly automobile-dependent cities. In the small sample of European cities, densities are still declining owing to *shrinkage* or absolute reductions in population, but the data clearly show that the rate of decline in urban density is slowing down and almost stabilizing as re-urbanization occurs.

The relationship between density and car use is also exponential as shown in Figures 2 and 3 above. If a city begins to slowly increase its density, then the impact on car use can be more extensive than expected. Density is a multiplier on the use of transit and walking/cycling, as well as reducing the length of travel.

Table 2
Trends in Urban Density in Some US, Canadian, Australian, and European cities, 1960–2005

Cities	1960 Urban density persons/ha	1970 Urban density persons/ha	1980 Urban density persons/ha	1990 Urban density persons/ha	1995 Urban density persons/ha	2005 Urban density persons/ha
Brisbane	21.0	11.3	10.2	9.8	9.6	9.7
Melbourne	20.3	18.1	16.4	14.9	13.7	15.6
Perth	15.6	12.2	10.8	10.6	10.9	11.3
Sydney	21.3	19.2	17.6	16.8	18.9	19.5
Chicago	24.0	20.3	17.5	16.6	16.8	16.9
Denver	18.6	13.8	11.9	12.8	15.1	14.7
Houston	10.2	12.0	8.9	9.5	8.8	9.6
Los Angeles	22.3	25.0	24.4	23.9	24.1	27.6
New York	22.5	22.6	19.8	19.2	18.0	19.2
Phoenix	8.6	8.6	8.5	10.5	10.4	10.9
San Diego	11.7	12.1	10.8	13.1	14.5	14.6
San Francisco	16.5	16.9	15.5	16.0	20.5	19.8
Vancouver	24.9	21.6	18.4	20.8	21.6	25.2
Frankfurt	87.2	74.6	54.0	47.6	47.6	45.9
Hamburg	68.3	57.5	41.7	39.8	38.4	38.0
Munich	56.6	68.2	56.9	53.6	55.7	55.0
Zurich	60.0	58.3	53.7	47.1	44.3	43.0

Increase in density can result in greater mixing of land uses to meet peoples' needs nearby. This is seen, for example, in the return of small supermarkets to the central business districts of cities as residential populations increase and demand local shopping opportunities within an easy walk. Overall, this reversal of urban sprawl will undermine the growth in car use.

The Aging of Cities

Cities in the developed world are all aging in the sense that the average age of people living in the cities has been increasing. People who are older tend to drive less. Therefore, cities that are aging are likely to show less car use. This is likely to be a factor, but the fact that all American and Australian cities began declining around 2004 suggests that there were other factors at work than just aging, as not all cities in these places are aging at similar rates. The younger cities of Brisbane and Perth in Australia still peaked in 2004.

The Growth of a Culture of Urbanism

One of the reasons that older aged cities drive less is that older people move back into cities from the suburbs—the so-called "empty nester" syndrome. This was largely not predicted at the height of the automobile city growth phase; nor was it seen that the children growing up in the suburbs would begin flocking back into the cities rather than continuing the life of car dependence.[26] This has now been underway for over a decade and the data presented by the Brookings Institution suggest that it is a major contributor to the peak car use phenomenon.[27] They suggest this is not a fashion but a structural change based on the opportunities that are provided by greater urbanism. The cultural change associated with this urbanism is reflected in the Friends TV series compared to the Father Knows Best suburban TV series of the earlier generation. The shift in attitudes to car dependence is also apparent in Australia.[28]

The Rise in Fuel Prices

The vulnerability of outer suburbs to increasing fuel prices was noted in the first fuel crisis in 1973–1974 and in all subsequent fuel crisis periods when fuel price volatility was clearly reflected in real estate values.[29,30] The return to "normal" after each crisis led many commentators to believe that the link between fuel and urban form may not be as dramatic as first presented by people like us.[31,32] However, the impact of $140 a barrel oil on real estate in the United States dramatically led to the global financial crisis (sub-prime mortgagees were unable to pay their mortgages when fuel prices tripled).

Despite global recession, the twenty-first century has been faced by a consolidation of fuel prices at the upper end of those experienced in the last fifty years of the automobile-city growth. Most oil commentators, including oil companies, now admit to the end of the era of cheap oil, even if not fully accepting the peak oil phenomenon.[33] The elasticities associated with fuel price are obviously going to contribute to reducing car-use growth, though few economists would have suggested that these price increases were enough to cause peak car use, which set in well before the 2008 peak of $140 a barrel.

Implications for Future City Planning

The urban planning profession has been developing alternative plans for automobile cities in the past few decades—often called the polycentric city as in Figure 8—with the rationale of reducing car dependence involving all of the above factors. Few, however, would have thought they would be quite so successful, perhaps because each of the factors had such interactivity and reinforcing effects.

The need for a polycentric city has emerged as the solution to reducing car dependence and creating a new city form that enables people to keep within the Marchetti travel-time budget and create a more sustainable transport system. The polycentric city is demonstrated in Figure 8 showing the need for good public transport across and into the centre of cities as well as the need for density to be increased in a range of centers across the city. These centers would be the key destinations for the upgraded transit system. At these centers, walking and cycling would be the main priority. The economic advantages of the polycentric city have been demonstrated.[34]

Figure 8
The Polycentric City

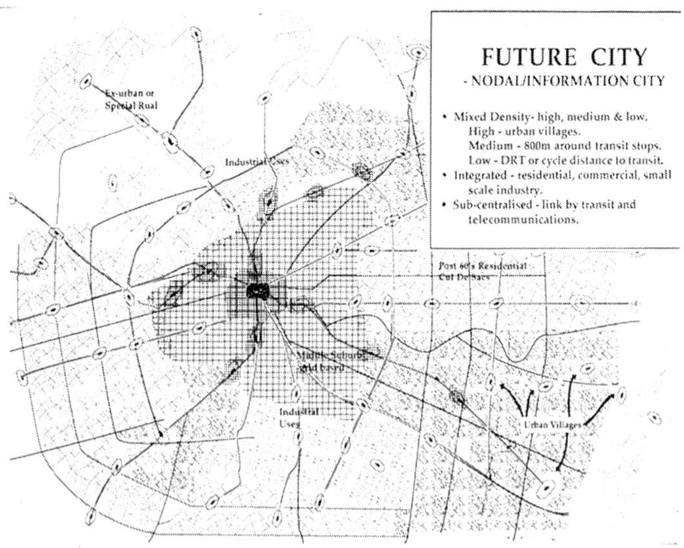

Conclusions

Sustainable transport needs support from land-use planning; it needs more of the transit city and walking city to be made part of the automobile city. This will be the only way to realize the Marchetti travel-time budget in heavily car-dependent areas. It will also significantly decrease car use, and as the phenomenon of peak car use appears to have begun, there are clear signs that sustainable transport is going to grow further in the future. Sustainable transport seems to be in the mainstream.

Notes

1. Marchetti, C, "Anthropological Invariants in Travel Behaviour," *Technical Forecasting and Social Change* 47, no. 1 (1994): 75–78.
2. Kenworthy, J., and Laube, F., *The Millennium Cities Database for Sustainable Transport, Institute For Social Sustainability* (Brussels: Murdoch University, Perth and UITP, 2001).
3. Standing Advisory Committee on Transport, 1994.
4. Newman, P., and Kenworthy, J., *Sustainability and Cities: Overcoming Automobile Dependence* (Washington DC: Island Press, 1999).
5. Newman P, and Kenworthy J., "Peak Car Use: Understanding the Demise of Automobile Dependence," *World Transport Policy and Practice* 17, no. 2 (2011): 32–42.
6. Brindle, R. E., "Lies, Damned Lies and Automobile Dependence—Some Hyperbolic Reflections," *Australian Transport Research Forum* 94 (1994): 117–31.
7. Mindali O, Raveh A, Saloman I., "Urban Density and Energy Consumption: A New Look at Old Statistics," *Transportation Research, Part A* 38 (2004): 143–62.
8. Pushkarev B., and Zupan J., *Public Transportation and Land Use Policy* (Bloomington and London: Indiana Press, 1997).
9. Ewing R., "Transit Oriented Development in the Sun Belt," *Transportation Research Record*, 1552, TRB, (Washington DC: National Research Council, 1996).
10. Frank L., and Pivo G., "Relationships between Land Use and Travel Behaviour in the Puget Sound Region," Washington State DOT, WA-RD 351.1., 1994.
11. Cervero R. et al., "Transit Oriented Development in America: Experiences, Challenges and Prospects," *Transportation Research Board* (Washington DC: National Research Council, 2004).
12. Falconer R, Newman P, and Giles-Corti B. "Is Practice Aligned with the Principles? Implementing New Urbanism in Perth, Western Australia," *Transport Policy* 17, no. 5 (2010): 287–94.
13. Falconer R, and Newman P., *Growing Up: Reforming Land Use and Transport in 'Conventional' Car Dependent Cities* (Germany: VDM Saarbruecken, 2010).
14. Glaezer E., *The Triumph of the City* (London: Macmillan, 2011).
15. Graham, D., *Investigating the Link between Productivity and Agglomeration for UK Industries* (London: Imperial College, 2005).
16. Newman P., "The Transformation of Perth's Rail System," CUSP Discussion Paper (Curtin University, 2011).
17. Puentes, R, and Tomer, A., *The Road Less Travelled: An Analysis of Vehicle Miles Traveled Trends in the US Metropolitan Infrastructure Initiatives Series* (Washington DC: Brookings Institution, 2009).
18. Ibid.
19. Stanley, J, and Barrett, S., "Moving People—Solutions for a Growing Australia," Report for Australasian Railway Association, Bus Industry Confederation and UITP, 2010.

20. Millard-Ball, A, and Schipper, L., "Are We Reaching Peak Travel? Trends in Passenger Transport in Eight Industrialized Countries," *Transport Reviews* (2010): 1–22. First published on 18 November 2010 (iFirst).
21. Kenworthy, J, and Laube, F., *The Millennium Cities Database for Sustainable Transport, Institute For Social Sustainability* (Brussels: Murdoch University, Perth and UITP, 2001).
22. Kenworthy J., Laube F., Newman P., Barter P., Raad T., Poboon C, and Guia B., *An International Sourcebook of Automobile Dependence in Cities, 1960–1990* (Boulder: University Press of Colorado, 1999).
23. These data cover 25 cities in the USA (9), Canada (2), Australia (5) and Western Europe (9) for which per capita car kilometers are consistently available for 1960, 1970, 1980, and 1990 (see Kenworthy and Laube, 1999). The trends in each region and for the average for the whole sample are set out in Table 2.

For the 1995 data in our global cities database the number of cities being monitored and the cities themselves changed, so it is difficult to continue these trends from 1990. However, the update of the data to 2005, which matches with the 1995 data, so far shows that between 1995 to 2005 car vehicle kilometers per capita in US cities rose by only 2.0 percent, in Canadian cities by 2.1 percent, Australian cities by 10.4 percent, and European cities by 5.6 percent, leading to an overall increase across the sample of 5.1 percent (Kenworthy and Laube, 2001; Kenworthy, 2011 unpublished).

End Note Table 1

Car Use Per Capita in Cities in Different Regions from 1960 to 1990 and the Percentage Changes, 60–70, 70–80 and 80–90.

Cities	1960	1970	1980	1990
American	5,489	7,049	8,586	10,710
% change		28.4%	21.8%	24.7%
Canadian	3,482	4,386	6,096	7,913
% change		25.9%	39.0%	21.3%
Australian	2,910	4,466	5,748	6,536
% change		53.5%	28.7%	13.7%
European	1,470	2,755	3,534	4,505
% change		87.5%	28.2%	27.5%
All 25 cities	3,366	4,773	6,000	7,376
% change		41.8%	25.7%	22.9%

Note: The same cities comprise the sample in each year as follows:

US cities: Boston, Chicago, Denver, Houston, Los Angeles, New York, Phoenix, Portland, San Francisco

Canadian cities: Calgary, Winnipeg

Australian cities: Adelaide, Brisbane, Melbourne, Perth, Sydney

European cities: Amsterdam, Brussels, Copenhagen, Frankfurt, Hamburg, London, Munich, Paris, Stockholm.

24. Neff, J. W., "Substitution Rates between Transit and Automobile Travel," Paper presented at the Association of American Geographers' Annual Meeting, Charlotte, North Carolina, April, 1996).

25. Newman, P., Kenworthy J, and Glazebrook, G., "How to Create Exponential Decline in Car Use in Australian Cities," AdaptNet Policy Forum 08-06-E-Ad, 08 July 2008, Also published in *Australian Planner*, 2008).

26. Leinberger, C., *The Option of Urbanism: Investing in a New American Dream* (Washington DC: Island Press, 2007).

27. Puentes, R, and Tomer, A., *The Road Less Travelled: An Analysis of Vehicle Miles Traveled Trends in the US Metropolitan Infrastructure Initiatives Series* (Washington DC: Brookings Institution, 2009).

28. Newman, C. E, and Newman P. W. G., "The Car and Culture." In *Sociology: Place, Time and Division*, ed. Beilhartz, P, and Hogan, T (Oxford: Oxford University Press, 2006).

29. Fels, M. F, and Munson, M. J., "Energy Thrift in Urban Transportation: Options for the Future," *Ford Foundation Energy Policy Project Report*, 1974.

30. Romanos, M. C., "Energy Price Effects on Metropolitan Spatial Structure and Form," *Environment and Planning A* 10, no. 1 (1978): 93–104.

31. Newman, P, and Kenworthy, J., *Cities and Automobile Dependence: An International Sourcebook*, (Aldershot: Gower Publishing, 1989).

32. Newman, P, and Kenworthy, J., *Sustainability and Cities: Overcoming Automobile Dependence* (Washington DC: Island Press, 1999).

33. Newman P., Beatley T, and Boyer H., *Resilient Cities: Responding to Peak Oil and Climate Change* (Washington DC: Island Press, 2009).

34. Trubka R, Newman P, and Bilsborough D., "Costs of Urban Sprawl (1, 2 and 3)," *Environment Design Guide*, 83 (2010): 1–6, 84 (2010): 1–16, 85 (2010): 1–13.

10

High Efficiency Photovoltaics
Lead to Low Energy Cost

Allen Barnett and Xiaoting Wang

Introduction

Photovoltaic solar electric power promises to become one of the major sources of electricity worldwide. Describing a high-value pathway that will achieve this promise is the goal of this work. Since becoming a major electricity source for terrestrial electric energy applications requires achieving a reduced cost of energy, this work studies how the cost of photovoltaic energy is affected by module efficiency.

Selecting a module requires choosing from the many options that are available for the photovoltaic electricity generation device or the solar cell. Choosing between the present and planned technologies for this device is an interesting and important problem, which is difficult to solve, because the solar cell itself appears to be a simple device. Tens of material systems and hundreds of device configurations can, and do generate, electricity from sunlight. Among these, there is not a single choice but a good range of excellent choices. For the energy planner to make these choices requires some type of analytical basis. Developing the understanding that will lead to this analytical basis in order to make near-term decisions will be described in this work.

For decades, it has been assumed that the only requirement for widespread adoption of solar power is a low-cost photovoltaic module. It was further believed that the major goal was achieving grid parity (i.e., equating the cost of photovoltaic solar electric power to the cost of the available grid-generated electricity). However, grid parity has now been achieved in many parts of the world, and it is clear that this relatively new technology not only provides a comparable cost of electricity but also delivers an advantage. Achieving this advantage requires moving in a direction that is driven by clear analysis.

In this work, we will show the value of module efficiency using a rooftop system. This system is an important example of the high-value opportunity provided by generating electricity at the point-of-use. We will identify the components of electricity cost in this system and will then extend the same to outline future opportunities. We will expand this work to other system configurations and new high-value applications. All of this work is designed to give the energy planner an analytical basis for implementing a high-value, long-term, and sustainable photovoltaic energy system.

The Value of Module Efficiency

Overview

The cost of the electricity generated and delivered to the end user is the sum of the cost of the electricity and the cost of the system. The cost of a photovoltaic system includes (1) the hardware costs of the solar generation module (an assembly of solar cells), an inverter to convert the direct current (DC) electricity into alternating current (AC), and the balance of plant (BOP), which is other hardware such as mounting hardware and electrical connections; and (2) the non-hardware soft costs such as engineering design, assembly labor, and permits. These costs are then used to calculate the electricity and efficiency values.[1]

Photovoltaic System Configurations

Photovoltaic (PV) systems can be broadly categorized into three groups on the basis of their application: utility, residential, and commercial. Utility-scale applications are usually ground mounted. In contrast, residential and commercial applications are usually on rooftops. These two fall on the customer (or retail) side of the meter and thus give the electricity a greater value.

Rooftop applications provide two major advantages. First, the electricity can be consumed where it is generated, saving transmission cost and energy loss. Second, no extra land is necessary, thus saving cost and space. The second advantage is especially attractive in countries and areas where the population density is high and open space is not widely available. For example, Japan has 96 percent of its 2011 PV applications installed on rooftops (see Figure 1). Other major PV markets have percentages that are smaller but still significant: Italy 72 percent, Germany 69 percent, United States 57 percent. Even China, whose 2011 PV installations focused more on utility applications (82 percent), has instituted new policy that promotes the development of rooftop PV. According to the Chinese National Climate Change Strategy Research and International Cooperation Center (NCSC), approximately 10 GW of distributive PV is to be commissioned by 2015. On the basis of the recently released twelfth Five-Year

Figure 1
2011 PV Installations by Application in Different Countries (MW)[2]

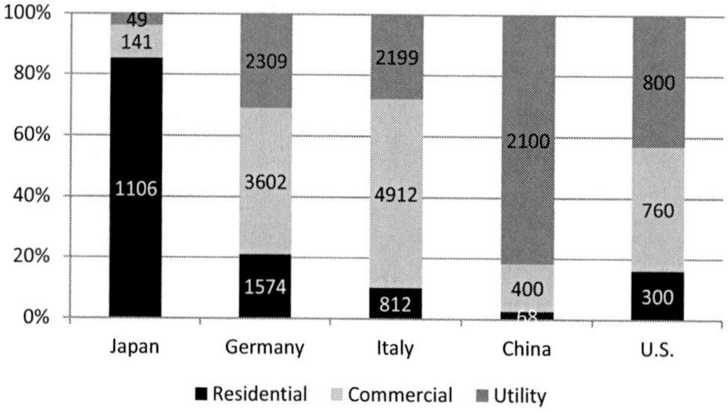

National Development Plan for Strategic Emerging Industries, in which the target of solar electricity projects is set at 21 GW including 1 GW for solar thermal electricity generation, this indicates that residential and commercial applications will constitute 50 percent of the total PV installations by then.

Levelized Cost of Energy (LCOE)

In the past, PV-generated electricity was much more expensive than conventional power. Unlike conventional electricity, which involves major cash input during the entire operating period, PV systems are characterized by high initial capital costs. Thus, to evaluate PV-generated electricity and compare it with other sources, a widely accepted metric is levelized cost of energy (LCOE). Although this metric does not reflect the time value of electricity (e.g., PV electricity fed into the grid has lowered the spot power prices in Germany because its supply peak matches well with the demand peak,[3]) it takes into account the time value of cash and the real generation of energy during the entire operation period. It is defined as "the cost that, if assigned to every unit of energy produced by the system over the analysis period, will equal the total life-cycle cost when discounted back to the base year."[4] Thus,

$$LCOE = \frac{TLCC}{\displaystyle\sum_{n=1}^{N}\frac{Q_n}{(1+d)^n}} = \frac{\displaystyle\sum_{n=0}^{N}\frac{C_n}{(1+d)^n}}{\displaystyle\sum_{n=1}^{N}\frac{Q_n}{(1+d)^n}}$$

where
    LCOE = levelized cost of energy
    TLCC = total life-cycle cost

Q_n = energy output for the nth year
C_n = cash flow for the nth year
d = discount rate
N = analysis period

In this equation, C_n after the installation year is much smaller than the capital cost (C_0). Moreover, a large portion of C_n is the operation and maintenance (O&M) cost and this is proportional to C_0. Accordingly, the numerator can be considered proportional to C_0 (unit of US dollars; \$). In the denominator, Q_n, the annual energy output, can be considered proportional to system capacity (units of Watts). Therefore, LCOE is proportional to the system capital cost (units of \$/W).

With the electricity unit being in kilowatt hours (kWh), this linear relationship between LCOE (¢/kWh) and system capital cost (\$/W) is important in illustrating the value of module efficiency, especially for rooftop applications where the area is restricted. Many system cost items are directly related to the installation area, including structural installation, racking, and site preparation. Thus, for a fixed installation area, the total value of these costs (\$) can be considered a constant. Moreover, a large part of system capital cost comes from business processing, whose total value (\$) for a specific application type and capacity range is also constant. Therefore, with the area-restricted rooftop applications, as module efficiency rises, these area-related and fixed costs can be assigned more capacity so their value (units of \$/W) decreases, producing a lower LCOE.

Grid Parity of Residential PV Applications

To reduce the LCOE and promote the PV industry, many countries have developed supportive policies in the form of incentives that are based on capacity (\$/W) or feed-in tariff (FiT) that are based on energy (¢/kWh). Meanwhile, as the scale of production has expanded, technologies in both new scientific concepts and processing engineering have also experienced significant improvements. Thus, there has been a considerable reduction in the cost of PV system components, such as the module, inverter, and balance of plant (BOP).

Along these lines, an important near-term goal of the developing PV industry is realizing grid parity, that is, having the PV LCOE be equal to or less than the price of purchase from the grid. Even if the system cost (\$/W) is a constant value in all cases, the status of grid parity depends on the location. This is because PV LCOE depends on energy production that in turn is affected by local irradiation, and local electricity prices.

Figure 2 shows the PV grid parity status across the globe in 2011. The graph shows the local irradiation or effective radiation hours and the residential electricity prices for different countries and regions with their potential residential PV markets. It overlays this with a plot of 2011 PV LCOE. Since Italy, Spain, and Hawaii in the United States have local electricity prices higher than the 2011 PV LCOE for their location, they have achieved grid parity. In 2012, the system

Figure 2
Residential PV Grid Parity in 2011.[5] The Circle for Each Country/Region
is Placed According to the Local Irradiation or Effective Radiation Hours (x-axis)
and the Residential Electricity Sales Price (y-axis). The Size of the Circle
Represents its Potential Residential Market (A Reference Circle of 25 GW is
given). The Curve Plots the 2011 PV LCOE as a Function of Local Irradiation.
Above the Curve, Italy, Spain, and Hawaii are the Locations That Have
Achieved Grid Parity.

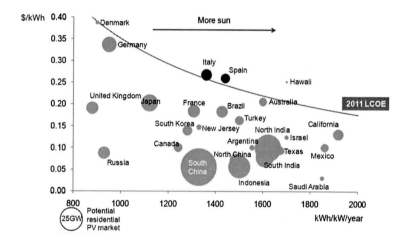

costs have continued to decline, so the LCOE curve has moved downward and
Germany, Australia, and Denmark are entering grid parity.

The Value of Efficiency in Lowering LCOE and Thus Accelerating
Grid Parity

While it is encouraging that some countries/regions have achieved grid parity,
they do share a common feature: a high local electricity price. In other countries,
there is still a remarkable gap between the local electricity price and the PV LCOE.
Among the approaches taken to lower LCOE, an intrinsic and fundamental driver
has been boosting module efficiency. The importance of module efficiency lies
in the fact that the cost of many system components are either proportional to the
installation area or are fixed. In the reference residential PV system illustrated in
Figure 3, these components produce 35.5 percent of the system's total expense.
Details on the cost breakdown are listed below:

- The total system price of $2.21/W is the Q2 2012 average price for
  German systems below 100 kW, as published by German Solar Industry
  Association (BSW).[6]

- The module and inverter prices are average values of the monthly spot price for April, May, and June, as reported by Bloomberg New Energy Finance.[7–9]
- Balance of plant (BOP) includes all hardware involved in PV system and construction except the module and inverter. BOP includes electronic components (such as transformers, cabling, gear switches, and monitoring equipment) and structural components (mainly the aluminum rack). The total BOP price is derived from an estimated distribution of system cost components[10] and the total system price released by BSW. The structural BOP is cited from a research note on the cost breakdown of rooftop systems that was published in January 2012.[11]
- The total price of EPC (engineering, procurement, and construction) comes from the same sources as the total BOP.
- Development costs include applications, licenses, permits, and other business processing costs. The total value comes from the same sources as the total BOP.

The expenses of structural BOP, EPC, and development are either area-related—such as racking, site preparation, and construction—or they are fixed for a certain application (here residential PV) and a certain scale range (here less than 100 kW). Thus, as module efficiency increases, the value of these costs (units of US dollars) remains constant, but the cost of each watt decreases in an inversely proportional manner. Taken together, these three expenses make up 35.5 percent of the total system price of $2.21/W.

Figure 3
Price Breakdown of Reference Residential PV System ($/W)

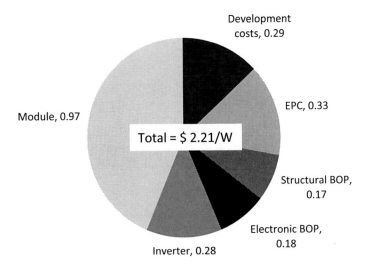

On the basis of these features of the cost items, we analyzed the impact of module efficiency on residential PV LCOE (Figure 4). The five curves correspond to module prices ranging from $0.4/W to $1.2/W. Our derivation is based on the reference system price from Figure 3 and makes the following assumptions: (1) the module price is a parameter; (2) the expenses of the inverter and electronic BOP are proportional to system capacity, so the $/W value is a constant; and (3) the values of the other costs (units of $/W) in Figure 3 are inversely proportional to system capacity. We set the reference system LCOE to 100 percent as the standard (dot in Figure 4). This corresponds to 14.5 percent module efficiency, a typical value for current multicrystalline silicon modules, and to a module price of $0.97/W, the monthly average spot price observed in Q2 2012.[12-14] In this figure, LCOE is considered proportional to the total system cost (units of $/W).

Figure 4 shows two approaches for lowering PV LCOE: reducing module price (units of $/W) as shown by comparing across different curves, and increasing module efficiency, as shown by the drop of each curve. For example, beginning with the reference system, if the module price is decreased to $0.8/W while the module efficiency remains at 14.5 percent, the LCOE decreases by 7.7 percent. Alternatively, beginning at the same point, if the module efficiency is increased to 20 percent while the module price is maintained at $1/W, the system LCOE can be cut by 8.4 percent.

Figure 4
Effect of Module Price ($/W) and Efficiency on LCOE

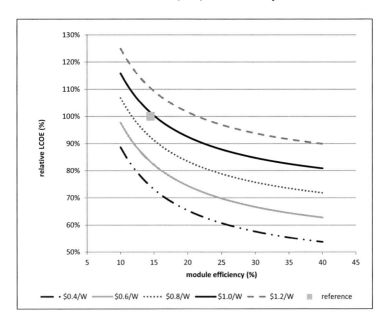

Tracking Systems

So far, we have focused on rooftop systems and have covered their advantages. These systems are mounted in a fixed position that is usually chosen to maximize the total energy collected over a year. They are usually tilted to collect more energy but can also be mounted flat.

One way to collect more energy from a solar cell is to follow the sun. A flat-plate system can collect over 25 percent more energy by following the sun as it goes across the sky. This is called tracking. A tracking technique wherein a system follows the sun from east to west is called 1-Axis tracking. Even more energy can be harvested by the same solar-cell area if tracking follows the sun seasonally (from north to south). This is called 2-Axis tracking.

The options increase further if you consider concentration, the act of magnifying the sun's image onto a solar cell. This process enables the use of a more efficient solar cell. However, this cell can be more expensive. An example of a concentrator is shown in Figure 5. This figure shows a lens that concentrates (magnifies) the sunlight and a solar cell that is placed at the focal point of the lens.

There are several limitations to these concentrator systems. First, since concentrating systems need to be ground-mounted, their use is limited to utility applications. Thus, they are not suitable for rooftops. Second, using a very efficient solar cell requires a concentration of more than 100X or 100 suns. Thus, concentrator systems are usually designed to be 500X to 1000X. Third, 2-Axis

Figure 5
Diagram of the Concentrator Concept Showing the Lens and the Overall Arrangement That Includes the Aperture and the Solar Cell

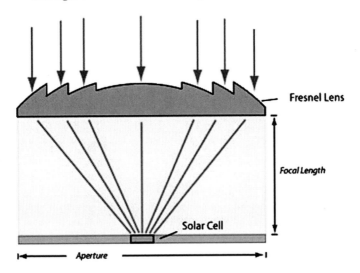

concentrator systems cannot collect energy from scattered sunlight. They can only use direct sunlight, which can be less than 80 percent of the total available energy. However, there are locations where the amount of available direct sunlight is high, so these areas are favorable sites for concentrator systems.

Given all options for a fixed plate, namely 1-axis tracking, 2-axis tracking, and concentrating PV (CPV) systems, the key question is whether the cost of these improvements exceeds the value of the increased energy. Little has been published directly comparing the energy collected by these systems. F. Gomez-Gil et al.[15] published a comprehensive analysis comparing energy collection by a range of systems installed in a sunny region of Spain, Écija, near Seville. Figure 6 illustrates their estimated energy collection by month.

Summing this collection over the course of a year for each configuration gives the total energy values listed in Table 1. This table shows that compared to the fixed flat-plate baseline system we focused on in the previous section, most of the gain in energy collection is achieved by the 1-axis tracking system. The further gain achieved by 2-axis tracking is small. Moreover, the CPV system's gain is less than that of 1-axis tracking, probably because the case that was studied and reported had a reduced amount of direct sunlight compared to total sunlight over the year.

This large gain with 1-axis tracking leads to another opportunity. Once the decision has been made to use 1-axis tracking, a more expensive solar cell can be

Figure 6
Energy Production (EP) Estimates for Fixed, CPV Systems,
1-axis, and 2-axis (bottom to top curves)[16]

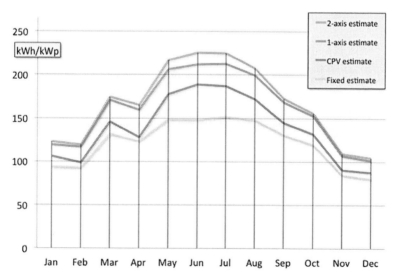

Table 1
**Comparison of Annual EP Estimates for Different PV Congurations
at Écija. The Data Shows the EP Values and Ratios for Different
Congurations; The Fixed Flat-Plate PV is Set as the Reference.[17]**

	Energy production	Ratio (%)
Fixed-flat plate	1452 kWh/kWp	100
1-Axis tracker	1939 kWh/kWp	133.5
2-Axis tracker	2001 kWh/kWp	137.9
CPV	1659 kWh/kWp	114.3

used in a low-concentration configuration. At concentrations of 20X and below, 1-axis tracking can be used to collect all of the available sunlight, not just the direct portion. In fact SunPower does this with their C7 concentrator.[18]

Solar-Cell Operation and Efficiency

Overview

This section briefly describes the solar-cell device and how it works. It then reviews the basis for high-efficiency solar cells and some of the key literature.

The Solar Cell

So far, we have shown that module, or solar cell, efficiency is an important driver of cost and provides an opportunity for achieving increased value. Now, we explain how solar cells achieve their efficiency, and we begin by detailing how solar cells work. A solar cell's greatest strengths are its simplicity and reliability of use. This deceptively simple device is diagrammed in Figure 7, taken from an excellent text on electronics, "PVCDROM," which is available online.[19] The basic steps in its operation are as follows[20]:

- light-generated carriers (electron-hole pairs) are generated in the base
- these are collected to generate a current
- a large voltage is generated across the solar cell at the junction between the emitter and the base (shown by + and -)
- the electrical power is delivered to the load (end-use devices are computers, refrigerators, batteries, etc.)

The sunlight that hits the solar cell is a broad spectrum of energies from ultraviolet and visible through infrared (Figure 8). The percentage of this energy in sunlight that can be converted to electricity is defined as efficiency.

Figure 7
Cross Section of a Solar Cell Showing the Structure and the
Creation of Electrons by the Absorption of Sunlight.[21]

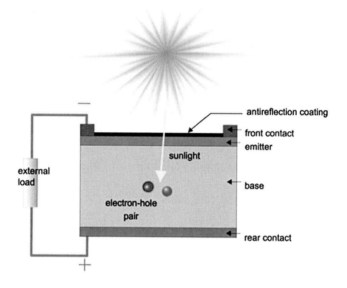

Figure 8
Solar Spectrum Showing Area of Visible Light.[22] **The Bandgap of 1.7 eV**
Falls at the Edge of Visible Light (Arrow). Light at Longer Wavelengths
is in the Near Infrared and Not Visible to the Human Eye.

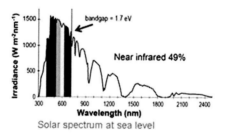

Since a solar cell is a semiconductor device, it has a bandgap. Light (photons) with energy higher than the bandgap is absorbed, and an electron–hole pair is created. Light with energy lower than the bandgap is not converted to electron–hole pairs. Accordingly, energy from these lower-energy photons is not converted to electricity, and for silicon, this reduces the possible energy-conversion efficiency by 19 percent. Meanwhile, photons with higher energy than the bandgap can only convert the energy at the level of the bandgap. Thus, the excess energy is converted to heat, not to electricity. This loss is about 33 percent for silicon. Together, these two effects reduce the possible energy-conversion efficiency for a single solar cell to around 48 percent.

A solar cell can be defined by the following simple equation:

$$I = I_0(e^{qV/kT}-1) - I_L,$$

where I is the current, I_0 is the reverse saturation current (based on the material properties, primarily the bandgap, of the semiconductor junction), q is the charge of an electron, V is the voltage applied across the cell, k is Boltzmann's constant, T is the absolute temperature, and I_L is the light-generated current.

High-efficiency Solar Cells

Since I_0 is determined by the bandgap and is thus the maximum current, the above equation determines the efficiency of the solar cell. I_0 can then be used to determine the voltage. This was first shown by Prince[23] who generated a curve of efficiency versus bandgap. Since all the semiconductor material parameters required to calculate I_0 were not known in 1955, Prince used the observed voltage of silicon solar cells at the time to generate his curve, leading to a calculation of maximum efficiency of around 24 percent. Later Shockley and Queisser[24] used detailed balance to determine the voltage versus the bandgap and they calculated a maximum efficiency of around 30 percent. The plot of efficiency as a function of bandgap for some key solar-cell materials is shown in Figure 9.

Figure 9
Efficiency vs. Bandgap Based on Shockley-Queisser[25] with
High-performance Materials.[26] The Dotted Line Indicates the
Theoretical Maximum Efficiency for Silicon.

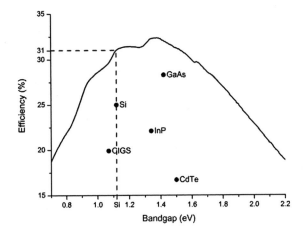

Subsequently Fan[27] and Nell[28]) further refined the parameters that determined voltage and they applied M. B. Prince's equation to get a curve similar to that of Shockley–Queisser. Their major refinement was determining I_0. Interestingly, these investigators determined that the highest diode voltage under sunlight was the bandgap (E_G) minus 0.4 V. Soon after, Tom Tiedje et al.[29] added non-radiative recombination mechanisms to predict the maximum efficiency of silicon solar cells. They recalculated the radiative properties of materials and got a 10 percent increase in the maximum value. This value is used in Figure 9 to plot the maximum efficiency against the bandgap. It is quite similar to what would be achieved using voltage equal band gap minus 0.3 V.

More recently, J. L. Gray et al.[30] calculated solar-cell efficiencies. They implicitly assumed that well-designed solar cells (that primarily used surface passivation and then light trapping) could achieve the highest diode voltage of bandgap minus 0.3 V under sunlight. Indeed, this voltage has recently been achieved for thin, passivated, and light trapped solar cells made from the high-performance semiconductor, gallium arsenide (GaAs).[31] GaAs belongs to a class of very high-performance semiconductors known as III–Vs because the constituent elements occupy groups III and V of the periodic table of elements. Moreover, silicon belongs to Group IV of the periodic table, and a diode voltage of $E_G - 0.37$ has been reported for well-passivated silicon (Si).[32]

The studies we have described so far were focused on establishing the voltage of the best solar cells based on their bandgaps. As knowledge in this field increases and technology improves, this achieved voltage approaches the bandgap. Increasing this voltage offers a leveraging area for solar-cell advancements.

To outline advances in solar-cell technology, we cite several excellent reviews. Some of these are the Solar Cell Efficiency tables by Martin Green et al.,[33] which are updated every six months, an article on solar photovoltaics technology by Larry Kazmerski,[34] an article on the value of high-efficiency photovoltaics by Xiaoting Wang et al.,[35] and an article on the state of the art in silicon solar cells by Martin Green.[36]

Here, we describe these efficient solar-cell materials and designs. The Solar Cell Efficiency Tables have been published in *Progress in Photovoltaics* since January 1993.[37] They report the confirmed efficiencies for a range of solar cells and modules, and are updated every six months. The guidelines for inclusion in these tables are transparent and well-published. The key importance of these listings is that the reported efficiencies are independently confirmed. Within these tables, solar-cell technologies are broadly grouped into crystalline silicon, GaAs (including other III–V compounds), thin films, and novel "emerging" materials and technologies. The improvement in efficiencies over the twenty years of publication demonstrates significant progress in the field. This progress is illustrated by Kazmerski's graphs of record efficiencies that are available online through NREL[38] (example shown in Figure 10).

Figure 10
Progress Over the Last Four Decades of Solar-Cell Efficiencies
for Different PV Technologies[39]

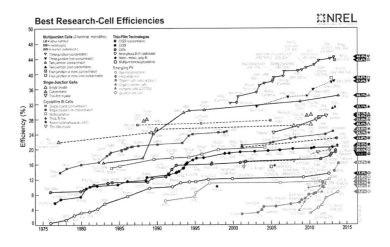

High-efficiency Solar-Cell Technologies

Overview

Solar-cell technologies that show high performance can be divided into several areas: crystalline silicon, GaAs (including other III–V compounds), thin films, and novel "emerging" materials and technologies.

Crystalline Silicon

The dominant commercial solar cell is made of crystalline silicon (c-Si). Laboratory efficiencies for this material have reached 25 percent.[40] The basic design of this 25 percent solar cell is used in a cost-reduced general form by today's mainstream manufacturers and it is shown in Figure 11. The efficiency of this solar cell reaches 80 percent of the theoretical maximum that is shown in Fig 9 (dashed red line). Solar cells are packaged into weather-proof assemblies called modules. These modules have achieved efficiencies of 22.9 percent.[41] This record module efficiency is 92 percent of the record solar-cell efficiency.

Si wafers for PV manufacture are divided into two categories: single-crystal (sc; a single continuous crystal structure, similar to diamonds) and multi-crystalline (mc; multiple crystals each defined by a ground boundary). The grain boundaries in mc-Si cells produce a reduced performance. The current record efficiencies for mc-Si cells and modules are 20.4 percent and 18.2 percent,

Figure 11
Diagram of a 25 percent PERL Solar Cell[42] Illustrating the np Junction,
the Metal Contacts, and a Surface Texture for Improved Light Absorption

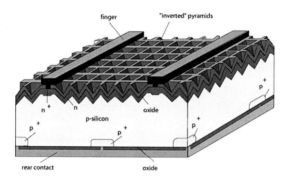

respectively.[43,44] Compared with sc-Si devices, mc-Si devices are 4.6 percent and 4.7 percent (absolute) lower in efficiency at the cell and module levels. At both these levels, the sc and mc devices have the potential to achieve higher efficiencies. However, these record efficiencies are for research solar cells and commercial manufacturers cannot approach these efficiencies.

Since this excellent efficiency requires very high quality material and an extremely sophisticated solar-cell structure, these factors increase cost and can be difficult to achieve in high volume manufacturing. Thus, the present PV market is dominated by Si-based modules that only have efficiencies in the 13–17 percent range. However, even with these constraints, the PV industry has experienced impressive growth in the volume of production, with the market expanding an average of 38 percent yearly over the past two decades.

Although most commercial companies manufacture c-Si modules with efficiencies in the 13–17 percent range, two exceptions, SunPower and Panasonic, have demonstrated commercial-size solar cells with efficiencies of 24.2 percent[45] and 22.9 percent,[46,47] respectively. These high efficiencies are mainly attributed to the long lifetimes of the carriers and the innovative back contact cell designs. Figure 12 shows this design for SunPower's cell. The back contact design avoids the 7 percent shading loss found in standard silicon solar cells with screen-printed contacts. Figure 13 illustrates the Panasonic design. This equally innovative arrangement is based on ultra-thin amorphous silicon layers that provide very high surface passivation. This high surface passivation has led to record high voltages for silicon solar cells of 0.747 V.

GaAs Solar Cells

GaAs has produced single junction solar cells with the highest efficiencies. This is due to its direct bandgap and excellent top surface passivation. However,

Figure 12
SunPower Back Contact Solar Cell

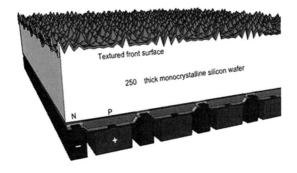

Figure 13
Panasonic HIT Solar Cell

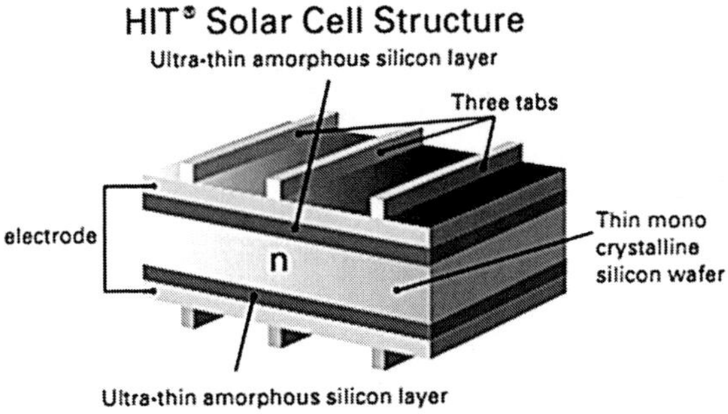

GaAs solar cells have not been considered for wide terrestrial application because the material is very expensive. Recently this problem has been alleviated by the demonstration of "lift-off" technology in GaAs cells.[48] This technology allows the expensive substrate to be re-used after the epitaxial film is removed, thus greatly reducing the material cost. Alta Devices, the commercial company that has deployed R&D on this topic, has not released many fabrication details but the basic concept involves growing a solar-cell structure on an intermediate layer on a GaAs substrate. Similar to the "layer transfer" technique used for thin Si cells, the intermediate layer can then be dissolved leaving a thin wafer of GaAs and a re-usable GaAs substrate. With this technology, Alta Devices has demonstrated a solar-cell efficiency of 28.8 percent.[49] This is the highest efficiency recorded across all types of single-junction solar cells. Furthermore, modules made of thin film GaAs cells by Alta display efficiencies of 23.5 percent, a value that for the first time surpasses the 22.9 percent efficiency of single-junction solar Si modules.[50]

These record efficiencies for cells and modules open another promising avenue for low–cost, high-efficiency PV in terrestrial applications. Future research priorities will include: (1) maintaining the high efficiency of solar cells grown from the re-used substrate; (2) exploring the limit of substrate re-use; (3) raising the growth rate of metal organic chemical vapor deposition (MOCVD) that is currently being used for epitaxial layer growth, or exploring other high-speed growth techniques; (4) producing larger modules with good uniformity; and (5) developing manufacturing equipment and procedures for high volume production.

Thin Film Materials

Another important commercial material for solar cells is cadmium telluride (CdTe). With this material, solar cells can be very thin and still absorb most of the sunlight; on the order of 1 μm (0.001 millimeters) compared with 100 to 200 μm for c-Si. The record efficiencies of CdTe are 16.7 percent and 12.8 percent at the solar cell and module levels, respectively.[51] The dominant CdTe commercial company, First Solar, has produced its FS-390 series with a module efficiency of 12.5 percent, very close to the laboratory record.[52] This company has achieved a total manufacturing capacity of 2.4 GW.[53]

Due to its manufacturing techniques, CdTe has experienced greater market penetration than another type of thin solar cell, copper–indium–gallium–selenide (CIGS). However, CIGS solar cells have been studied since the early 1970s and have properties that lead to higher laboratory efficiencies. Research on these devices has led to improved material deposition methods, better device structure designs, and more cost-effective manufacturing procedures.[54] The current record efficiencies for CIGS are 19.6 percent and 15.7 percent at the solar-cell and module levels, respectively.[55,56]

However, these achievements have been based partially on empirical improvements, and thus the underlying principles are not yet fully understood. For example, it is empirically known that the presence of sodium (Na) in CIGS is beneficial for PV performance.[57] However, the principle underlying Na's effect is not fully understood, although the tentative explanation that Na helps passivation on the grain surface is being investigated.[58–61] Another mystery is the influence of grain size on CIGS cell performance. Unlike c-Si solar cells whose performance directly relates to grain size (due to the correlated recombination), CIGS solar cells display insensitivity to grain size.[62] Although several tentative explanations have been offered, the underlying principles are not yet known.[63–68] Since, as with other solar cells, reducing recombination by improving material quality should be a fundamental R&D direction for CIGS solar cells, modeling of the grain boundary behavior to understand this phenomenon would be greatly beneficial. Moreover, since, unlike c-Si solar cells, multiple types of materials are involved in the fabrication of CIGS, more sophisticated control of material deposition parameters and more diagnostic tools to manage the layer growth process are

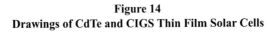

Figure 14
Drawings of CdTe and CIGS Thin Film Solar Cells

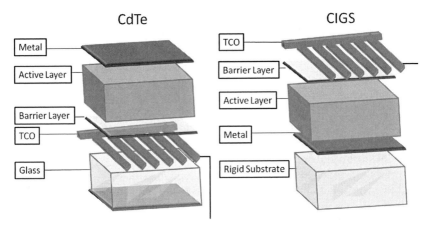

Figure 15
(a) Schematic Drawing of a Multi-junction Three-Solar-Cell Stack.
(b) Overlay of Photon Energy with Absorption in the three Solar Cells[69]

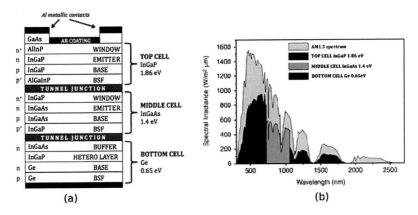

necessary. Thus, developing a comprehensive design of deposition equipment, manufacturing, and diagnostic tools could improve large volume production.

Multi-junction Solar Cells

While single-junction solar cells of various materials are constantly striving for higher efficiencies, multi-junction solar cells have achieved the highest efficiencies. These solar cells, like GaAs cells, are fabricated from the III–V materials.

In addition to the high performance these materials offer in the GaAs single junction solar cell, they provide a wide option of bandgaps and lattice constants for system optimization. However, the high cost of GaAs or germanium (Ge) limits the use of these solar cells to terrestrial applications with concentrating PV (CPV) configurations.

The most commonly adopted III–V multi-junction solar cell has three sub-cells and costs two orders of magnitude more than silicon cells. Thus, for it to compete with silicon cells, it may need to incorporate high-X concentrators. However, since high-X CPV depends on high-accuracy tracking, the cost is significantly greater. To compensate for the high costs of both these solar cells and the trackers, commercial CPV systems on the market are usually around 500X. Triple-junction concentration solar cells at 947X have achieved a record efficiency of 44.0 percent.[70,71]

Triple-junction solar cells adopted in commercial CPV systems typically utilize a monolithic structure with three sub-cells connected in series. Both the design and measurement of these solar cells are based on a reference spectrum, usually ASTM G-173-03 Direct.[72] However, the real spectrum can experience dynamic variation. Thus, although the current match across the three sub-cells, or at least between the top and middle sub-cells, can be achieved for the reference spectrum, it cannot be maintained constantly in real operation. Thus, one focus of CPV research is further splitting the spectrum using four, five, or even six junctions to absorb its broad band.[73] Combining this strategy with efforts to adjust bandgaps to their optimal values could lead to increases in solar-cell efficiency.

However, actual operating conditions suggest that this expansion of spectrum splitting may create more problems than it solves. The value of developing 4-, 5-, or 6-junction solar cells, at least in a monolithic structure for terrestrial application is questionable if there is a high variation in the dynamic spectrum. X. Wang[74] has shown that even for a current matched 3-junction solar cell, the dynamic spectrum variation can show as much as a 10.9 percent loss in the energy generated. Thus, it is reasonable to predict that, as the divisions of the spectrum increase, solar-cell efficiency will be more sensitive to spectrum variation and therefore will experience more energy loss whenever the spectrum deviates from the standard. It remains to be seen if the efficiency gain predicted by increasing the number of sub-cells under the standard spectrum can be realized in field applications where there is dynamic spectrum variation. For example, a switch from three to four sub-cells predicts a theoretical efficiency gain of 8 percent (relative) under the standard spectrum.[75–77] However, in the field, the higher sensitivity to spectrum variation may make the energy production of this four-cell system close to or even lower than that of a three-cell system.

Aside from spectrum variation, another important phenomenon that affects the performance of concentration triple-junction cells is the non-uniformity of illumination. During the design and measurement phases of these solar cells, the illumination is assumed to be uniform across their surface. While this condition is satisfied for flat plate PV, it cannot be guaranteed when a cell is assembled with a concentrator and is installed in the field. High-X CPVs have secondary optical components that are designed to homogenize the rays passing through the primary concentrators, but this design is based on perfect tracking. Thus, if a tracking error occurs during field operations, the uniformity found in ideal tracking is not necessarily maintained.

Lateral Spectrum Splitting

In response to the above problems faced by monolithic solar cells that split the spectrum in a vertical or a series way, another concept has been developed called "lateral spectrum splitting."[78-80] The basic technique entails splitting the spectrum using additional optics before the sunlight arrives at solar cells that are designed for different wavelength ranges. With this lateral spectrum splitting approach, the cells can be fabricated and optimized separately, and so the restrictions of current match or lattice match can be completely or partially lifted. Prototypes of various structures have been demonstrated that have 36.7 percent and 38.5 percent record sub-module efficiencies.[81,82] However, the feasibility of this concept will require more R&D effort on the design of the electronic circuit because this lateral structure creates multiple outputs compared to the monolithic structure's one output. R&D efforts will also be needed to optimize the module structure because the lateral structure has more components than the monolithic one.

New Solar-Cell Opportunities

Overview

This section provides a brief description of some new approaches to further increase silicon solar-cell efficiency. It also introduces the thin GaAs opportunity, briefly touches on new optical concepts, and finally cites a comprehensive overview of research opportunities.

Advanced Silicon Concepts

Since a key driver of solar electric power in the future is expected to be high performance, this section discusses new high-performance opportunities. Since its invention in 1954, crystalline silicon has been the dominant solar cell for terrestrial applications.[83] The existing state-of-the art for this technology has been reviewed by Green.[84] He cites several near-term advances whose origin is the 25 percent efficient PERL solar cell shown in Figure 11. Along with the SunPower and Panasonic solar cells shown in Figures 12 and 13, he also names other active near-term technologies of laser-doped selective emitters that have experienced increases in voltage and increases in current. He cites another new technology that uses the emitter wrap to accomplish the all-back electrical contact approach. This interesting approach eliminates the need for the tabbing step in module assembly, and thus offers the opportunity to further reduce the cost while still retaining high performance.

Tandem Solar Cell on Silicon

Green[85] describes the next level of efficiency improvement in silicon solar cells as building a high bandgap top cell on the silicon that should capture the performance advantages of tandem solar cells while maintaining the cost advantages of silicon (Figure 16). Although silicon is not the ideal choice for a bottom solar cell bandgap, Green shows that this structure has the potential efficiency of 42.5 percent, a value close to the 45 percent maximum that is achievable with ideal bandgaps.

Figure 16
Silicon as the Lowest Cell in a Tandem Stack[86]

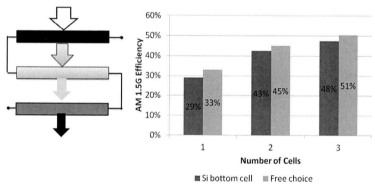

This concept is also being developed by Schmieder et al.[87] They use a SiGe bottom solar cell grown on Si[88] that leads to lattice matching for the two solar cells. This matching promises to lead to high performance in the top solar cell. Schmieder graphically shows that this lattice-matched combination has the potential to achieve efficiencies over 40 percent. The structure and analysis are shown in Figure 17.

Ultra-thin Silicon

Another interesting approach to improving module efficiency is the thin silicon wafer grown by epitaxy on porous silicon that in turn is formed on a re-usable silicon wafer. One advantage of thin-film cells is the cost saved on the material. Another important advantage is that thin solar cells offer the potential of producing higher voltages that will lead to greater solar-cell efficiencies. This

Figure 17
a) Diagram of GaAsP on SiGe on Si[89] (b) S-T Model of Tandem Cell with Lattice-match Points for GaInP (Top Left) and GaAsP (Middle Left).[90]

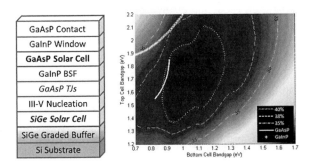

higher efficiency results from the thinner bulk region creating less bulk recombination. However, this concept requires excellent surface passivation and light trapping. For light trapping, silicon oxide and silicon nitride have been used.[91] Other innovations that have led to improvements can be found in the literature.[92]

The idea of this thin wafer is based on the concept of "layer transfer."[93,94] This approach allows the thin solar cells to be lifted off the substrate wafer that can then be used multiple times. As shown in Figure 18, the innovative component that allows the layer transfer is the mesoporous double layer between the substrate wafer and the thin solar cell. Currently, this technology has fabricated a 43-micron-thick wafer on thin Si solar cells that has a record efficiency of 19.1 percent.[95,96]

Figure 18
Schematic of Thin-film Si Growth Structure Using "Layer Transfer" Technology. The Breaking Layer Has a Thickness of 350 nm and a Porosity of 55 percent; The Seed Layer Has a Thickness of 1200 nm and a Porosity of 20 percent.[97] Diagram is not to scale.

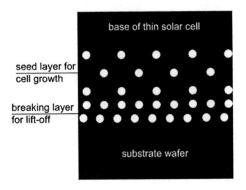

A technology used to improve the performance of an ultra-thin, silicon solar cell is the high-performance back junction and all-back contact that has been demonstrated by Solexel[98] in a wafer that is 35-microns thick. In a commercial size wafer, the reported efficiency is 20.6 percent. However, these epitaxially grown wafers are very fragile. Thus, one approach to overcoming this fragility is bonding them to steel. This steel substrate approach is being pursued by AmberWave.[99]

Thin GaAs and Multi-junctions

The thin GaAs solar cell by Alta[100] is an excellent example of the efficiency gains of having thin bulk layers (it is 3 microns), excellent surface passivation, and light trapping. These multi-junction solar cells that have been described previously meet the performance goals, but need a low-cost implementation.

New Optical Concepts

Recently, Polman and Atwater[101] have proposed new approaches to "light management that systematically minimize thermodynamic losses" that can "enable ultrahigh efficiencies previously considered impossible."

Research Overview

The future opportunities in the solar technology field are best summarized in a recent overview by Kazmerski[102] (figure is reproduced in Figure 19). These opportunities are divided into three types: accelerated evolutionary (3–5 years), disruptive (5–15 years), and revolutionary (15 years and beyond).

Figure 19
Description of the Technology Pathways That Could Lead to the Next Generation of High-Performance Solar Cells[103]

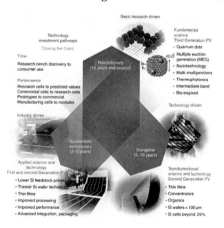

Summary and Conclusion

Solar-cell efficiency can be an important driver in the achievement of low-cost solar electricity. This potential can be demonstrated through analytical methods. These analytical approaches will lead to the most promising pathways to new generations of high-performance solar cells.

Notes

1. Wang X, Kurdgelashvili L, Byrne J., and Barnett A., "The Value of Module Efficiency in Lowering the Levelized Cost of Energy of Photovoltaic Systems," *Renewable and Sustainable Energy Reviews,* 15 (2011): 4248–54.
2. Woodward, T., "Japan's 50 cent Solar FiT: Straight to the Bank," Research Note of Bloomberg New Energy Finance, June 6, 2012.
3. Czajkowska, A., "Aces and Faults in First Year of German 'Turnaround'," Research Note of Bloomberg New Energy Finance, July 9, 2012.
4. Short, W., Packey, D. J., and Holt, T., "A Manual for the Economic Evaluation of Energy Efficiency and Renewable Energy Technologies," NREL/TP-462-5173, 1995.
5. Chase, J., Key Issues in the Top 10 PV Markets of 2012 and 2013, 17 May 2012. Presentation slides of Bloomberg New Energy Finance.
6. http://www.solarwirtschaft.de/en/start/english-news.html
7. Simonek, M., "April 2012 solar spot price index," Analyst Reaction of Bloomberg New Energy Finance, April 20, 2012.
8. Kim, A., "May 2012 Solar Spot Price Index," Analyst Reaction of Bloomberg New Energy Finance, May 16, 2012.
9. Simonek, M., "June 2012 Solar Spot Price Index," Analyst Reaction of Bloomberg New Energy Finance, June 22, 2012.
10. Chase, J., Simonek, M., Kim, A., Radoia, P., Wang, M., DeSilva, R., Bhushan Agrawal, B., Goldie-Scot, L., Woodward, T., and Bhavnagri, K., PV Market Outlook, Q3 2012, Market Outlook of Bloomberg New Energy Finance, August 7, 2012.
11. Radoia, P., "Are small PV Costs Falling Off the Roof?" Research Note of Bloomberg New Energy Finance, January 5, 2012.
12. See Note 7.
13. See Note 8.
14. See Note 9.
15. Francisco Javier Gómez-Gil, Xiaoting Wang, and Allen Barnett, "Energy Production of Photovoltaic Systems: Fixed, Tracking, and Concentrating," *Renewable and Sustainable Energy Reviews* 16 (2012): 306–13.
16. Ibid.
17. Ibid.
18. SunPower C7 www.sunpowercorp.com.
19. C. Honsberg and S. Bowden, PVCDROM http://pveducation.org/pvcdrom.
20. Ibid.
21. Ibid.
22. www.Photonics.com.
23. Prince MB, "," *J. Appl. Phys.* 26 (1955): 534.
24. W. Shockley, and H. J. Queisser, "Detailed Balance Limit of p-n Junction Solar Cells," *J. Appl. Phys.*, 32 (1961): 510.
25. Ibid.
26. K. J. Schmieder, A. Gerger, M. Diaz, Z. Pulwin, M. Curtin, C. Ebert, A. Lochtefeld, R. Opila, and A. Barnett, "Progression of Tandem III-V/SiGe Solar Cell on Si Sub-

strate," *Proc. 27th European Photovoltaic Solar Energy Conf.*, (Frankfurt, September 2012).

27. J. C. C. Fan, B. Y. Tsaur, and B. J. Palm, "Optimal Design of High Efficiency Tandem cells," In Conf. Rec. 16th IEEE Photovoltaic Specialists Conf, 1982, 692–701.

28. Nell M. E, Barnett A. M, "The Spectral p-n Junction Model for Tandem Solar-Cell Design," *IEEE Transactions on Electron Devices* ED-34, no. 2, (February 1987).

29. Tiedje T., Yablonovitch E., Cody G. D., and Brooks B. G., "Limiting Efficiency of Silicon Solar Cells." *IEEE Transactions on Electron Devices* 31 no. 5 (1984): 711–16.

30. J. L. Gray, A. W. Haas, J. R. Wilcox, and R. J. Schwartz, "Efficiency of Multijunction Photovoltaic Systems," *Proc. 33rd IEEE Photovoltaic Specialists Conf.*, (San Diego, May 2008).

31. http://www. techmedia.com/articles/read/stealthy-alta-devices-next-gen-pv-challenging-the-status-quo/

32. Kinoshita T., Fujishima D., Yano A., Ogane A., Tohoda S., Matsuyama K., Nakamura Y., Tokuoka N., Kanno H., Sakata H., Taguchi M., and Maruyama E., "The Approaches for High Efficiency HIT Solar Cell with Very Thin (<100 mm) Silicon Wafer Over 23%." *26th EUPVSC Proceedings*, 2011, 871–74.

33. Green, M. A., Emery K., Hishikawa Y., Warta W., and Dunlop, E. D., "Solar Cell Efficiency Tables (Version 41)," *Progress in Photovoltaics: Research and Applications* 21, no. 1–11 (2013): 10.

34. Kazmerski, L.L., "1.03 - Solar Photovoltaics Technology: No Longer an Outlier." In *Comprehensive Renewable Energy*, edited by Ali Sayigh (Oxford: Elsevier, 2012), 13-30, ISBN 9780080878737, http://dx.doi.org/10.1016/B978-0-08-087872-0.00101-3.

35. Wang X., Byrne J., Kurdgelashvili L., and Barnett A., "High Efficiency Photovoltaics: On the Way to Becoming a Major Electricity Source," *WIREs Energy Environ* 1 (2012): 132–51. doi: 10.1002/wene.44.

36. Green, M. A., "Radiative Efficiency of State-of-the-Art Photovoltaic Cells," *Prog Photovolt: Res.Appl.* 20 (2012): 472–76

37. Green, M. A, Emery K., Hishikawa Y., Warta W., "Solar Cell Efficiency Tables (version 33)." *Progress in Photovoltaics: Research and Applications* 17 (2009): 85–94.

38. http://www.nrel.gov/ncpv/images/efficiency_chart.jpg

39. Ibid.

40. Green, M. A., "The Path to 25% Silicon Solar Cell Efficiency: History of Silicon Cell Evolution," *Prog. Photovolt: Res. Appl.* 17 (2009): 183–89.

41. See Note 33.

42. See Note 40.

43. See Note 35.

44. See Note 38.

45. Cousins, P. J., Smith, D. D., Luan, H. C., Manning, J, Dennis, T. D., Waldhauer, A., Wilson, K. E., Harley, G., and Mulligan, G. P., "Gen III: Improved Performance at Lower Cost," 35th IEEE PVSC (Honolulu, HI, June 2010).

46. See Note 32.

47. H. Sakata, T. Nakai, T. Baba, M. Taguchi, S. Tsuge, K. Uchihashi, and S. Kiyama, 28th IEEE PVSC, 2000, 7–12.

48. Kayes, B. M., Nie, H., Twist, R., Spruytte, S. G., Reinhardt, F., Kizilyalli, I. C., and Higashi, G. S., "27.6% Conversion Efficiency, a New Record for Single-Junction Solar Cells under 1 Sun Illumination," Proceedings of 37th IEEE PVSC, 2011.

49. See Note 33.

50. See Note 33.

51. See Note 33.

52. http://www.firstsolar.com/~/media/WWW/Files/Downloads/PDF/Datasheet_s3_ NA.ashx (accessed March 24 2012).

53. http://en.wikipedia.org/wiki/First_Solar (accessed March 24, 2012).

54. Shafarman, W., Siebentritt, S., and Stolt, L., *Chapter 13 of Handbook of Photovoltaic Science and Engineering, Second Edition* (Chichester, West Sussex: Wiley), 2011.

55. See Note 33.

56. Repins, I., Contreras, M. A., Egaas, B., DeHart, C., Scharf, J., Perkins, C. L., To, B., and Noufi, R., "19.9% Efficient ZnO/CdS/CuInGaSe2 Solar Cell with 81.2% Fill Factor." *Progress in Photovoltaics: Research and Applications* 16 (2008): 235–39.

57. Hedstrom, J., Ohlsen, H., Bodegard, M., Kylner, A., Stolt, L., Hariskos, D., Ruckh, M., and Schock, H. W., "ZnO/CdS/Cu(In,Ga)Se$_2$ Thin Film Solar Cells with Improved Performance," Proceedings of the 23rd IEEE PVSC, 1993, 364–71.

58. Kronik, L., Cahen, D., and Schock, H., "Effects of Sodium on Polycrystalline Cu (In,Ga)Se$_2$ and Its Solar Cell Performance." *Advanced Materials* 10: 1998: 31–36.

59. Niles, D., Al-Jassim, M., and Ramanathan, K., "Direct Observation of Na and O Impurities at Grain Surfaces of CuInSe2 Thin Films," *Journal of Vacuum Science & Technology A* 17 (1999): 291–96.

60. Boyd, D., and Thompson, D., *Kirk-Othmer Encyclopaedia of Chemical Technology, 3rd Edition.* (John Wiley & Sons, Inc., 1980), 807–80.

61. Kessler, F, Herrmann, D, and Powalla M., "Approaches to Flexible CIGS Thin-film Solar Cells," *Thin Solid Films* 480/481, 2005: 491–98.

62. See Note 54.

63. Hetzer, M. J., Strzhemechny, Y. M., Gao, M., Contreras, M. A., Zunger, A., and Brillson, L. J., "Direct Observation of Copper Depletion and Potential Changes at Copper Indium Gallium Diselenide Grain Boundaries," *Applied Physics Letters* 86 (2005): 162105.

64. Lei, C., Li, C. M., Rockett, A., and Robertson, I. M., "Grain Boundary Compositions in Cu(InGa)Se$_2$," *Journal of Applied Physics* 101 (2007): 024909.

65. Yan, Y., Noufi, R., and Al-Jassim, M., "Grain-Boundary Physics in Polycrystalline CuInSe$_2$ Revisited: Experiment and Theory," *Physical Review Letters* 96 (2006): 205501.

66. Seto, J., "The Electrical Properties of Polycrystalline Silicon Films," *Journal of Applied Physics* 46 1975: 5247.

67. Siebentritt, S., and Schuler, S., "Defects and Transport in the Wide Gap Chalcopyrite CuGaSe$_2$," *Journal of Physics and Chemistry of Solids* 64 (2003): 1621–26.

68. Castaldini, A., Cavallini, A., and Fraboni, B., "Deep Energy Levels in CdTe and CdZnTe," *Journal of Applied Physics* 83 (1998): 2121–26.

69. Fraunhofer Institiute for Solar Snergy, www.ise.fraunhofer.de/en.

70. See Note 33.

71. Solar Junction, http://www.sj-solar.com/.

72. http://rredc.nrel.gov/solar/spectra/am1.5/ASTMG173/ASTMG173.html (accessed March 24 2012).

73. Dimroth, F., Baur, C., Bett, A.W., Meusel, M., and Strobl, G., "3–6 Junction Photo-voltaic Cells for Space and Terrestrial Concentrator Applications," Proceedings of the 31st IEEE PVSC, 2005, 525–29.

74. Wang, X., and Barnett, A., "The Effect of Spectrum Variation on the Energy Production of Triple-junction Solar Cells," *IEEE Journal of Photovoltaics* 2, no. 4 (2012): 417–23.

75. Ibid.

76. Torrey, E., Ruden, P., and Cohen, P., "Performance of a Split-spectrum Photovoltaic Device Operating under Time-varying Spectral Conditions," *Journal of Applied Physics* 109 (2011): 074909.

77. Torrey, E., Krohn, J., Ruden, P., and Cohen, P., "Efficiency of a Lateral Engineered Architecture for Photovoltaics," Proceedings of the 35th IEEE PVSC, 2010, 002978–002983.
78. Vincenzi, D., Busato, A., Stefancich, M., and Martinelli, G., "Concentrating PV System Based on Spectral Separation of Solar Radiation," *Physica Status Solidi (A)* 206 (2009): 375–78.
79. Stollwerck, G., and Sites, J., "Analysis of CdTe Back Contact Barriers," Proceedings of the 13th EU PVSEC, 1995, 2020–22.
80. Allen Barnett, Douglas Kirkpatrick, Christiana Honsberg, Duncan Moore, Mark Wanlass, Keith Emery, Richard Schwartz, Dave Carlson, Stuart Bowden, Dan Aiken, Allen Gray, Sarah Kurtz, Larry Kazmerski, Myles Steiner, Jeffery Gray, Tom Davenport, Roger Buelow, Laszlow Takacs, Narkis Shatz, John Bortz, Omkar Jani, Keith Goossen, Fouad Kiamilev, Alan Doolittle, Ian Ferguson, Blair Unger, Greg Schmidt, Eric Christensen, and David Salzman, "Very High Efficiency Solar Cell Modules," *Progress in Photovoltaics*, 17, no. 1 (January 2009): 75–83.
81. Wang, X., Waite, N., Murcia, P., Emery, K., Steiner, M., Kiamilev, F., Goossen, K., Honsberg, C., and Barnett, A., "Lateral Spectrum Splitting Concentrator Photovoltaics: Direct Measurement of Component and Submodule Efficiency," *Progress in Photovoltaics: Research and Applications* 20 (2012): 149–65.
82. McCambridge, J., Steiner, M., Unger, B., Emery, K., Christensen, E., Wanlass, M., Gray, A., Takacs, L., Buelow, R., McCollum, T., Ashmead, J., Schmidt, G., Haas, A., Wilcox, J., Meter, J., Gray, J., Moore, D., Barnett, A., and Schwartz, R., "Compact Spectrum Splitting Photovoltaic Module with High Efficiency," *Progress in Photovoltaics: Research and Applications* 19 (2011): 352–60.
83. Chapin, D. M., Fuller, C. S., and Pearson, G. L., "A New Siliocn p-n Junction Photocell for Converting Solar Radiation into Electrical Power," *J. Appl. Physics* 25 (1954): 676–77.
84. Green, Silicon solar cells: State-of-the-art, to be published.
85. Ibid.
86. Ibid.
87. See Note 26.
88. Yi Wang, Andrew Gerger, Anthony Lochtefeld, Lu Wang, Chris Kerestes, Robert Opila, and Allen Barnett, "Design, Fabrication and Analysis of Germanium:Silicon Solar Cell in a Multi-Junction Concentrator System," *Solar Energy Materials & Solar Cells* 108 (2013): 146–55.
89. See Note 26.
90. See Note 26.
91. Tobias, I., Canizo, C. D., and Alonso, J., *Chapter 7 of Handbook of Photovoltaic Science and Engineering, Second Edition* (Chichester, West Sussex: Wiley, 2011).
92. Aberle, A. G., "Surface Passivation of Crystalline Silicon Solar Cells: a Review," *Progress in Photovoltaics: Research and Applications* 8 (2000): 473–87.
93. Brendel, R., "A Novel Process for Ultrathin Monocrystalline Silicon Solar Cells on Glass." Proceedings of 14th EU PVSEC, 1997, 1354–57.
94. Petermann, J. H., Zielke, D., Schmidt, J., Haase, F., Rojas, E. G., and Brendel, R., "19%-Efficient and 43 Um-Thick Crystalline Si Solar Cell From Layer Transfer Using Porous Silicon," *Progress in Photovoltaics: Research and Applications* 20 (2012): 1–5.
95. See Note 33.
96. See Note 94.
97. See Note 59.

98. Moslehi, M. M., Kapur, P., Kramer, J., Rana, V., Seutter, S., Deshpande, A., Stalcup, T., Kommera, S., Ashjaee, J., Calcaterra, A., Grupp, D., Dutton, D., and Brown, R., "Worldrecord, 20.6% Efficiency 156mm_156mm Full-Square Solar Cells Using Low-Cost Kerfless Ultrathin Epitaxial Silicon and Porous Silicon Lift-Off Technology For Industry Leading High-Performance Smart Pv Modules," PV Asia Pacific Conference (APVIA/PVAP), October 24, 2012.
99. AmberWave http://www1.eere.energy.gov/solar/pdfs/incubator_7_awardees_2012.pdf.
100. See Note 31.
101. Polman, A., and Atwater, H. A., "Photonic Design Principles for Ultrahigh-Efficiency Photovoltaics," *Nature Materials* 11 (2012): 174–77.
102. See Note 34.
103. See Note 34.

11

The Need for a Storage Revolution for a Green Energy Economy

Bryan Yonemoto, Gregory Hutchings, and Feng Jiao

The future use of green electrical energy greatly depends on the capability to store that electricity in a cheap and efficient way, because the generation of electricity from renewable and sustainable sources (e.g., solar panels, biomass, wind, and wave energy) is intermittent.[1] The electricity must be stored during excess production and delivered when it is demanded. Battery technology has been used to store electrical energy chemically for more than 160 years. Nowadays, batteries are being developed to power a wide range of applications, such as mobile phones, laptops, electric vehicles, power tools, and stationary power storage. Although some recent improvements have made rechargeable batteries ubiquitous in today's portable electronic devices, the basic design of batteries has not changed substantially since their invention.

Transformational changes in battery technologies are critically needed to extend the operation time of powerful mobile electronics, to enable the effective use of alternative energy sources such as solar and wind, to allow the expansion of hybrid electric vehicles (HEVs) to plug-in HEVs (PHEVs) and all-electric vehicles (EVs), and to assist in utility load-leveling. For these applications, batteries must store more energy per unit volume and weight, and they must be capable of being cycled for thousands of times. Depending on the applications, the required characteristics and costs of energy storage devices could be very different; therefore, it is logical that future energy storage technologies will be highly application oriented. This chapter explores such trends, with the discussion divided into three major sections based on the scale of energy storage. Also explored are current energy storage technologies that are already in the market, as well as those now being developed in research laboratories. The goal is to assess in detail their current technical capability and future development for a wide range of applications, from electric vehicles to wholesale energy services.

Small-Scale Energy Storage

Small, portable electronic devices have undergone significant technological improvements since the 1990s. In many ways, this technological revolution towards smaller, sleeker rechargeable electronics has been dependent upon the introduction of lithium ion (Li-ion) batteries in 1992 by Sony. Here, a comparison of Li-ion batteries to other established battery technologies is presented to help explain why Li-ion batteries have come to dominate the portable electronics market. Also explored is Li-ion use in electric or hybrid electric vehicles, followed by a discussion about proton exchange membrane fuel cells, which is an alternative energy storage technology for electric vehicles.

Consumer Electronics

Batteries can be broken into two general types: primary (single use) and secondary (rechargeable) batteries. In general, primary batteries cost more per kilowatt-hour (kWh) than their rechargeable counterparts. For rechargeable batteries, the four most common commercialized battery chemistries are lead-acid, nickel metal hydride (NiMH), nickel/cadmium (NiCd), and Li-ion. As the batteries discharge, releasing the stored energy, electrons flow from the anode to the cathode. Important characteristics for the four battery chemistries are presented in Table 1.[2,3,4]

Table 1
Comparison of Different Battery Chemistries Available for Use in Small, Portable Energy Storage Devices such as Mobile Phones or Laptops. Values Presented in the Table are Estimates, and are Provided to Compare the Different Systems.

	Lead-acid	NiCd	NiMH	Li-ion
Cell Voltage (nominal)	2V	1.25V	1.25V	3.6V
Weight Energy Density	20–40 Wh/kg	30–50 Wh/kg	50–110 Wh/kg	80–165 Wh/kg
Volume Energy Density	45–95 Wh/L	40–110 Wh/L	200–280 Wh/L	135–400 Wh/L
Specific Power	1–500 W/kg	10–950 W/kg	25–1,150 W/kg	5–9,000 W/kg
Cost per kWh ($US)	$8.50	$11.00	$18.50	$24.00

V = Volt, Wh = Watt-hour, kg = kilogram; L = liter

As seen in Table 1, Li-ion batteries are more expensive per kWh of energy storage than the other three battery chemistries. Manufacturers install Li-ion, despite the higher costs, because the energy by weight and energy by volume that can be stored using Li-ion batteries is significantly higher. High energy densities are very attractive for portable electronics because longer operation times are desired without substantial increases in battery size.

To understand why Li-ion batteries offer higher energy density over other commercial battery options, a basic knowledge about the cells' construction is needed. The schematic diagram for a typical Li-ion cell is presented in Figure 1. Li-ion cells differ significantly from the other battery chemistries because an organic electrolyte, instead of a water based electrolyte, is placed inside the battery. This organic electrolyte allows the nominal cell voltage to go as high as 4.5 V in Li-ion cells. Beyond 4.5 V, most organic electrolytes will begin to decompose, which is unsafe and ruins the performance. For most aqueous based electrolytes, voltages above 1.25 V will cause decomposition of the electrolyte because electrolysis of the water begins to occur. The equations to calculate power and energy density are as follows:

$$Power = Voltage \times Current$$

$$Energy\,Density = \frac{Power \times Discharge\,Time}{Weight\,or\,Volume}$$

Figure 1
Schematic of Li-ion cell. During Discharge, Electrons and Li-ions Flow from the Anode to Cathode. During Recharge, the Electron and Li-ion Motion Reverses. The Separator Stops the Anode and Cathode from Touching, Which Would Short the Cell.

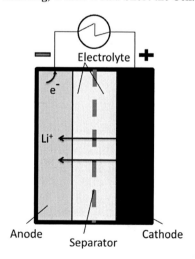

Both terms are calculated by multiplying the voltage of the cell, so the Li-ion voltage of 3.6 V versus the 1.25 V of aqueous electrolyte batteries results in significant energy storage improvements.

The downside to an organic electrolyte is the additional safety concerns that it creates for Li-ion batteries. When a Li-ion battery overheats, either from an internal reaction/short or external heat source, the flammable organic electrolyte becomes very dangerous. In the most basic scenario, a short circuit will cause a spark that ignites the electrolyte. In other cases the heat will build, causing the electrolyte to vaporize from the liquid phase to the gas phase. During the vaporization the pressure builds, until eventually the battery containment explodes from the high pressure. To help prevent these problems, many small electronics such as mobile phones use Li-ion polymer batteries. Li-ion polymer batteries are unique because the organic electrolyte is replaced with a solid polymer, resulting in safer, slightly cheaper batteries. While some safety concerns continue to exist, the Li-ion polymer batteries can usually take up less space because less safety protection is needed in the battery. Unfortunately, the solid polymer electrolyte can only operate with very low currents, so higher power electronics (i.e., laptops) usually need a traditional Li-ion battery installed. Due to the safety advantages, research is currently underway to improve the polymer electrolytes so they perform better in high power conditions, such as electric vehicles.

Electric Vehicles

As gasoline prices rose in the early 2000s, carmakers and consumers began to seriously look for new technologies and fuels that could supply the energy currently provided by gasoline. Technical and cost challenges for fuel cells and biofuels has put most of the commercial focus on battery powered hybrid electric vehicles (HEVs), plug-in hybrid electric vehicles (PHEVs), and fully electric vehicles (EVs). HEVs, PHEVs, and EVs provide improvements in greenhouse gas emissions and energy efficiency when compared to a conventional internal combustion engine (ICE) automobile.[5,6,7] Most electric vehicle reports focus on light duty automobiles, so care should be taken if extrapolating to larger vehicles such as trucks.

Hybrid electric vehicles offer improved fuel economy and lower emissions per car because the rate of gasoline consumption is reduced by about 30 percent.[8] Typical HEVs have the lower cost NiMH battery installed over Li-ion batteries because the ICE is always running; meaning gasoline provides the stored energy needed to drive the car. The NiMH battery is installed in HEVs to improve energy management by capturing excess energy and storing it for later use. A significant concern for HEVs is the time it takes to break even financially—in terms of the economic cost of purchasing and operating the vehicle—versus a conventional car because there is usually a higher initial price tag from the battery installation. A study by researchers at the U.S. National Renewable Energy Laboratory (NREL) in 2006 suggested the payback time would be around ten years near term, and 4 years long term if gas prices were $3/gallon and no additional cost breaks were provided.[9] The calculated ten year breakeven time seems to be a conservative value moving forward, because lower costs from mass production and gas prices higher than $3/gallon will decrease the breakeven time significantly.

Plug-in hybrid electric vehicles are a technological stepping stone between gasoline powered automobiles and fully electric vehicles. These cars operate fully electric until the energy stored within the battery is significantly depleted, and then a backup ICE activates to supply energy until a user can recharge the battery by plugging into the power grid. When the ICE is online, these cars operate very similarly to a HEV. Normally an all-electric range of 40–50 miles is targeted, because it is estimated that 70 percent of drivers travel less than 50 miles a day.[10] To store the battery energy, most PHEVs will use Li-ion batteries because of the high weight and volume energy density described previously. While running only on electric power, no greenhouse gas emissions are produced by PHEVs. This does not make the car zero emission, however, because the pollution has just been shifted from the tailpipe to a power plant. The overall economics for PHEVs and their environmental and energy impact can be very difficult to determine, because power costs and power plant energy sources vary from region to region. The 2006 NREL article suggested PHEVs would never reach breakeven costs in the near term, and would require six to eight years in the long term compared to conventional vehicles.[11] These near term estimates are probably too high because no tax incentives were built into the model estimates; gas prices will likely be higher than $3/gallon, and the battery cost is decreasing.

Battery electric vehicles run entirely on electricity stored from the onboard Li-ion battery. For a fully electric car, volume (and weight) energy density from the battery needs to be as large as possible because once the car runs out of battery power, there is no backup fuel source to give drivers additional miles until they can plug in the vehicle for a recharge. Current recharge times for EVs and PHEVs are on the order of hours, so commuters would have to be certain they do not need to drive more distance than the battery can allow. These cars will be marketed as "zero emission" since the individual car will not emit greenhouse gases, but just like PHEV the emissions are coming from power plants instead of tailpipes. This pollution shift by PHEVs and EVs from tailpipe to power plant may be advantageous, because regulation and carbon capture will be easier to implement. The economics for these cars rely almost entirely upon the initial purchase price of the vehicle since electric power is considerably cheaper than gasoline. The battery is the most significant cost for electric vehicles, so the U.S. Advanced Battery Consortium has set a battery price target of $200–$300/kWh for EVs (and PHEVs) to become commercially viable, and a minimum goal for long term commercialization at less than $150/kWh.[12,13] Currently, the battery costs are estimated between $500–$1,700/kWh, but estimates suggest $300/kWh or lower at full production volume is possible.[14,15] These Li-ion battery costs and price goals extend to PHEV commercialization as well.

For PHEVs and EVs to have long term success in the market place, battery costs must decrease closer to the cost targets, or the price of gasoline will need to significantly increase. It is believed that increasing production quantities of Li-ion batteries for vehicles can substantially reduce the cost of cells through an "economy of scale" effect.[16] The idea here is that improved production efficiency and bulk material ordering will lower costs. This will certainly help lower the cost, but it seems unlikely that the economy of scale effect can provide all the cost reduction required to meet the U.S. Advanced Battery Consortium cost goals. More than 75–80 percent of a cobalt based Li-ion battery cost comes from the purchase of raw materials to make the battery, with the cathode accounting for 30–70 percent of the raw material cost.[17]

In Figure 2, the relative costs for a battery pack and the raw materials are presented.[18,19] The cell material cost has historically been calculated using cobalt as the active cathode metal, so switching to other transition row metals such as nickel, manganese, iron, or vanadium could substantially reduce the calculated costs. Cobalt is the traditional cathode metal oxide, and offers very high energy densities. Li-ion polymer batteries have traditionally used manganese as a cathode, but the energy density is slightly smaller. To reduce the cost, manganese, nickel-manganese, and nickel-manganese-cobalt oxide cathodes are being researched as possible alternatives. Another alternative cathode to cobalt is an iron phosphate material being commercialized by A123 Systems, Inc. The iron phosphate cathode is safer, and has significantly decreased costs compared to

Figure 2
a) Proportion of Cost in Manufacturing Vehicle Li-ion Battery Packs.
b) Breakdown of Raw Materials Costs. c) Relative Cost for
Common Cathode Metals Cobalt (Co), Nickel (Ni), and Manganese (Mn)

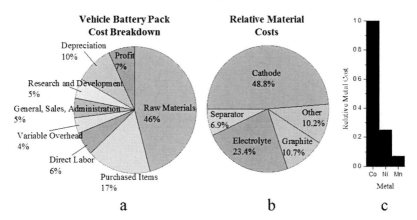

cobalt. One drawback is the iron phosphate cathode has lower energy density compared to some metal oxide cathodes. While it is unclear what cathode will eventually reign supreme in the electric vehicle market, using a cobalt oxide cathode is unsustainable and will ensure that the battery price targets are never achieved. The raw material cost to make a cobalt cathode can be more than $200/kWh and historically has cost as much as $700/kWh.[20] PHEV, EV, and to some extent HEV battery cost is the major technical hurdle that must be overcome for electric vehicles to replace the conventional automobile.

PEM Fuel Cells

The polymer electrolyte membrane fuel cell (PEMFC), instead of the Li-ion battery, was originally considered the technology that would lead to electric vehicles. Figure 3 shows the schematic for a PEMFC. Technically, the important metric for fuel cells is power density, not energy density, because the energy storage is provided by a fuel that feeds into the reactor, just like gasoline feeds into an ICE. For comparison purposes to batteries, however, energy density will be explored, with PEMFC providing at least an order of magnitude larger energy storage than Li-ion batteries. Unfortunately, some technical hurdles relating to the fuel cell catalyst and fuel source hindered its deployment to automobiles. Research is still underway to improve the system, and if the cost and technical challenges can be solved, fuel cells could probably replace Li-ion batteries as the energy system used to drive electric cars.

Figure 3
Fuel Flow to the Anode Catalyst That Splits Hydrogen or Methanol into Protons and Electrons. The Protons Travel through the Polymer Electrolyte Membrane to the Cathode. At the cathode, a catalyst reacts oxygen from the air with protons to make water.

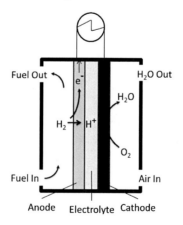

The catalyst for PEMFC is platinum because the noble metal is needed to efficiently separate the hydrogen into protons and electrons. Wild price fluctuations can occur in the platinum market ($2,280–$782/ounce in 2008), and it is estimated that 20–25 percent of the world platinum supply would be needed if 50 percent of all cars in 2050 were powered by PEMFC.[21,22] Most cost studies use approximately $1,000 as the platinum price, so higher prices can significantly change the economic feasibility. Besides the cost, platinum becomes inactive if carbon monoxide is present, so unless the fuel source is pure hydrogen additional noble metals must be installed, increasing the cost, to treat the carbon monoxide. For the catalyst, most research focuses on finding new, cheaper catalysts that could be used instead of platinum.

Hydrogen gas or methanol is the fuel source most likely to be used in PEMFC. The main technical problem for hydrogen gas is that it is a flammable gas. Large heavy containment cylinders must be installed into the car to hold the pressurized hydrogen gas. The extra weight and gas phase fuel has a detrimental effect upon energy density, making it difficult to drive a similar distance to a conventional automobile. The hydrogen tank also raises safety concerns, so researchers began to look for solid materials that could bind hydrogen to improve the safety and energy density. So far, these materials are not practical, making a hydrogen powered PEMFC car a risky alternative.

The other fuel option, methanol, has a real possibility to go mainstream. The original thought was to take methanol and treat it in a chemical reformer to convert the methanol to CO_2 and H_2, which would then flow into the fuel cell. The reformer adds additional cost, lowers the fuel efficiency, and will produce small amounts of unreacted carbon monoxide that must be dealt with by another rare metal catalyst before it reaches the platinum. The new idea is to make direct methanol fuel cells. In this system, methanol flows directly into the fuel cell and reacts on the platinum catalyst. To protect the platinum, other heavy metal catalysts must be present, but it eliminates the energy consuming reformer. Since methanol is a liquid at room temperature, it could easily be introduced into the current gasoline infrastructure. A major holdup is protecting the platinum catalyst from carbon monoxide and methanol crossing through the polymer electrolyte membrane to the cathode.

Just like PHEVs, a PEMFC vehicle's effectiveness at reducing energy use is dependent upon the energy source. The well-to-wheel energy efficiency for a PEMFC vehicle is higher than a conventional vehicle, and if the fuels are generated by renewables, a substantial decrease compared to PHEVs is possible.[23] This is because the fuels, hydrogen and methanol, will probably need to be made through solar water electrolysis and biofuels, respectively. While the possibility of low emissions using PEMFC is very exciting, safer hydrogen storage materials and biofuels are not fully developed technologies, so deployment could need multiple technology breakthroughs before mass commercialization of PEMFC vehicles is worthwhile. This lag, as the technologies for PEMFC develop, gives time for other alternatives, like next generation batteries, to become commercially possible.

For Li-ion batteries to meet the $/kWh price goals, technological improvements to raise the energy density are required. From the energy density equation, the voltage, the discharge time at a given current (referred to as capacity), or weight/volume can be changed. Weight/volume can be optimized for a given system, but it will be difficult to get the materials much lighter moving forward. In current batteries, the working capacity is usually lower than the theoretical capacity because the materials become structurally unstable, particularly with cathode oxides. To stabilize the material, researchers are making composites of active material and highly stable, inactive material. Another strategy for capacity is raising the theoretical value through new chemistry mechanisms or by using multiple metal oxidation states, meaning more Li can be stored at little to no increase in weight or volume. The final option is to try and manipulate the cell voltage to higher values. Any technological improvements to the energy density must be accompanied by stable operation in the electrolyte and be capable of thousands of charge/discharge cycles. For some perspective in this regard, a 400 Watt-hour (Wh)/kg battery was recently announced by Envia Systems, which is more than double the common energy density for Li-ion batteries as reported in Table 1. Envia also asserts that the battery costs can be $125/kWh, though this estimate is yet to be proven commercially.[24]

Li-ion battery EVs will have a difficult time being adopted in the U.S. because the battery packs may not store enough energy to alleviate range anxiety for most consumers, where drivers are worried about being stranded. This sentiment could change if charge times were reduced to below ten minutes, or if a battery leasing program similar to the one being developed in Norway is economically feasible for implementation in the U.S.[25] In Norway, consumers "rent" the battery. Accordingly, if the battery depletes while the vehicle is being operated, drivers may simply head to a service station and exchange the battery for one that is fully charged.

Realistically, due to range anxiety among consumers, PHEVs or HEVs are more likely to see widespread adoption. Since PHEVs and HEVs have both an ICE and advanced battery onboard, the additional installation causes a price mark up of approximately $2,000–$6,000 (for HEVs) or $4,000–$18,000 (for PHEVs) over a conventional automobile.[26] From an energy consumption standpoint, the well-to-wheel energy usage is very similar for HEVs and PHEVs; both provide some 30 percent less total energy consumption compared to a conventional vehicle, though gasoline reduction could be 30 percent greater with PHEVs.[27,28]

The really interesting comparison between HEVs and PHEVs entails estimates for greenhouse gas emissions. A 2010 U.S. Argonne National Lab report shows that, over the entire U.S., only a small reduction in emissions occurs by using PHEV technology over HEV technology. However, the power plant fuel source significantly affects the carbon footprint, so locations that generate electricity from cleaner fuels or technologies will see significant emission reductions by using PHEVs over HEVs.[29] For example, compared to HEVs, PHEVs in California have lower emissions, but PHEVs in Illinois have higher emissions.[30] In many ways, EV and PHEV deployment depends upon the local power grid and how much investment is expected to upgrade power generation from conventional to renewable sources. If the regional power grid is very "dirty" and little renewable energy investment is expected, HEVs probably provide more environmental gain than PHEVs and EVs.

Medium-Scale Energy Storage

For long distance and heavy duty electric vehicles, current state-of-the-art Li-Ion technologies become both prohibitively large and expensive. The primary limit for Li-ion is the amount of lithium that may be inserted into the crystalline cathode material, which for transition metal-based electrodes, is estimated to cap at approximately 500 Wh/kg, or around 1,000 Wh/L. New chemistries are required to achieve the breakthrough in specific energy required to power vehicles over longer ranges and provide effective, medium-scale grid energy storage.

Two candidates for the next generation of lithium-based batteries are lithium-air (technically lithium-oxygen, or Li-O$_2$) and lithium-sulfur (Li-S). Both technologies rely on direct reactions with either oxygen or sulfur, increasing the amount of lithium that can react per formula unit. In the case of lithium-air, the oxygen required for reaction can be obtained from the atmosphere rather than contained within the battery pack, further increasing the specific energy of the charged battery. More importantly, the electrodes may be engineered to support a large number of charge/discharge cycles, allowing for long-term installations.

Both lithium-air and Li-S technologies are in the research and development stage, with lithium sulfur forecasted for commercialization in five to ten years, and lithium-air forecasted for ten to twenty years. The materials of construction and final design are still under development, and so direct comparison of specific energies to those of commercialized Li-ion batteries is difficult. For Li-S, energy storage is calculated in a similar manner to that of Li-ion cathodes (by discharged mass). The theoretical specific energy is 2,567 Wh/kg, with practical estimates ranging from 350 Wh/kg in current prototypes to 600 Wh/kg in the near future.[31]

The specific energies of lithium-air systems are much more complex to determine, as the theoretical values range from 3,460–5,220 Wh/kg, depending on the choice of battery chemistry (aqueous vs. non-aqueous, catalytic cathode material properties, etc.). Practical estimates are more difficult to determine, but range from 500–2,200 Wh/kg based on electrode materials alone.[32] For a non-aqueous lithium-air battery producing Li$_2$O$_2$ as a discharge product, it is estimated that the complete system could achieve a specific energy of 900 Wh/kg. Assuming that one driven mile in a car takes approximately 300 Wh of energy with a 200 kg onboard battery, this corresponds to a range of 600 miles, which would be more than sufficient for the majority of personal transportation. In contrast, an "economy" class EV Li-ion can currently supply a range of about 100 miles.

In terms of cost, both battery technologies could boast considerably lower prices than current Li-ion technologies. In the case of lithium-air, the main cathode material is carbon, for holding oxygen obtained from the atmosphere and storing discharge reaction products, along with a catalyst for rechargeability. Ideally, the catalytic material would be an inexpensive transition-metal oxide and present in very small amounts compared to the carbon, which would keep costs low. For Li-S, there is no need for a catalyst, but there would be an additional cost of sulfur, which is abundant and inexpensive. Common to both technologies are lithium anodes, which may need to be protected for safety reasons, as well as electrolytes and battery pack construction materials. All of these components could be optimized for cost and efficiency.

Active Research

The Li-S system consists of a lithium anode, an organic electrolyte for Li^+ conduction, and solid sulfur loaded into a carbon support. The net reaction is $2Li^+ + S + 2e^- \leftrightarrow Li_2S$, with a voltage of 2.2 V vs. Li metal, though the reaction proceeds by slowly breaking apart an S_8 species from Li_2S_8 to the final, maximum Li^+ storage (Li_2S). Both S and the discharge product Li_2S are insoluble, and S is insulating. A diagram of this cell is shown in Figure 4(a).

Recently, advances in nanostructured and mesoporous carbon materials have allowed for new approaches to controlling the Li-S reactions. However, the most difficult research challenge that remains is related to compounds present in the discharged state. These compounds, specifically lithium polysulfides, are the desired reaction products in the battery. However, they dissolve easily and migrate away from the controlled cathode area, preventing the cells from sustaining a large number of cycles. Current research is focused on controlling this mechanism, as it is the single greatest source of degradation.

Figure 4
Li-S and Lithium-Air (Li-O2) Schematics

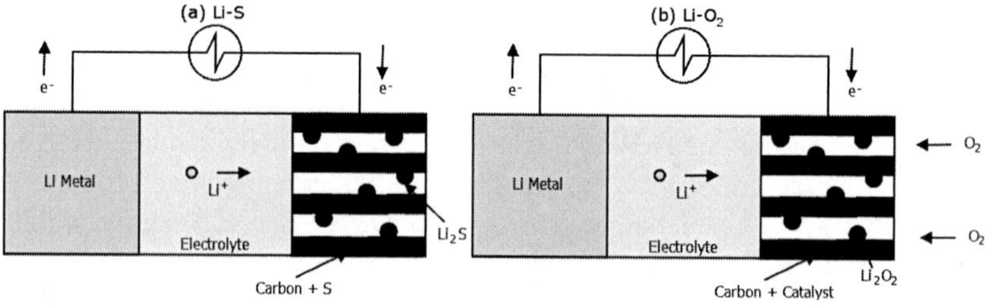

(a) Li-S   (b) Li-O$_2$

Lithium-air research, particularly in terms of rechargeable lithium-air cells, is at an even earlier stage of development. The earliest research was focused on the primary (non-rechargeable) system, and most advances in electrolytes, membranes, and carbon supports were optimized towards this direction.[33] In 2006, the secondary (rechargeable) system was discovered in the presence of a catalyst, and since then the focus has been on this system.[34]

A schematic of the Li-O_2 system is shown in Figure 4(b). The cell is similar to the Li-S system, and consists of a lithium anode, an aqueous or organic electrolyte to conduct lithium ions, and an oxygen cathode. The oxygen cathode consists of a porous, electroconductive carbon support, some form of catalyst for cell recharge, and, in the realized system, oxygen from the atmosphere that would not need to be carried with the battery system before discharge. For the non-aqueous system, the desired net reaction is either $2Li^++O_2+2e^-\leftrightarrow Li_2O_2$, which gives a theoretical operating voltage of 3.0 V vs. Li metal, or $4Li^++O_2+4e^-\leftrightarrow Li_2O$, which has a voltage of 2.9 V vs. Li metal, but, would also be more difficult to reverse. For the aqueous system, the reaction is $2Li^++\frac{1}{2}O_2+H_2O+2e^-\leftrightarrow 2LiOH$, which could operate at 3.5 V vs. Li metal. In both the non-aqueous and aqueous systems, the products have limited solubility in the solvents, which significantly complicates the cathode design.

Catalysts are required to achieve long-term cycleability, and many options have been explored. The primary focus has been on inexpensive, first-row transition-metal oxides, which can be optimized for either the discharge or charge reaction. Some work has been done on precious metal catalysts with promising results, but the cost of using such materials in mass-produced systems would be far too high. Ideally, the catalyst materials could be used in very small quantities and engineered alongside the carbon material, allowing for complete control of the reaction system.

Additionally, side reactions with many commonly-used cell components hinder optimization of the cathode materials. As of now, many issues related to electrolyte selection and improved materials of construction remain unresolved. Research is also delving into lithium-protection coatings and oxygen-permeable membranes. These would allow the cells to be operated in the ambient atmosphere, which is ultimately desired in commercial applications.

Commercialization Outlook

As considerable research challenges remain to be overcome, there are—as of the spring of 2012—no complete prototypes of lithium-air cells, let alone mass-produced battery packs. Commercial activity is mostly limited to research and development of new materials, which is complementary to fundamental studies conducted in the academic setting. It is likely that a few companies producing first-generation lithium-air cells at the device scale will arise in the coming years, once some of the more pressing challenges for lithium-air batteries are resolved.

While also still in the development phase, workable prototypes for Li-S cells have been showcased in recent years. Of note is Sion Power, which developed a weight-optimized system with an approximate specific energy of 350 Wh/kg. These cells were showcased in the QinetiQ Zephyr unmanned aircraft, which was able to set a world record with a 336-hour flight duration in 2010. Even these cells suffered from the degradation mentioned earlier, and until it can be solved, market availability will likely be a few years away. However, Li-S is much closer to production than lithium-air, and it is hoped that the former will serve to power a whole new generation of portable devices.

Large-Scale Energy Storage

Energy storage at a large scale (above 100 kW) has been considered as a valid approach to improve grid reliability and integrate with renewable energy sources.[35,36] Currently, three major energy storage technologies, that is, mechanical, chemical, and electrochemical, are available for large-scale applications. Mechanical energy storage includes compressed air energy storage (CAES), flywheel energy storage, hydroelectric energy storage and gravitational potential energy storage, while the chemical energy storage technologies are based on energy storage in chemical bonds, such as hydrogen gas, biofuels, hydrocarbons, and liquid fuels. In terms of installed energy generation capacity, pumped hydro has a capacity of 127,000 MW (approximately 99 percent of the total installed capacity), far more than the total energy storage capacity of all other available storage technologies. For the remaining 1 percent, 440 MW is from CAES, 316 MW from sodium/sulfur batteries, 35 MW from traditional lead-acid batteries, 27 MW from nickel/cadmium batteries, 25 MW from flywheels, 20 MW from Li-ion batteries, and less than 3 MW from redox flow batteries (see Figure 5).[37]

Figure 5
Installed Energy Storage Capacity (MW) for Various Technologies

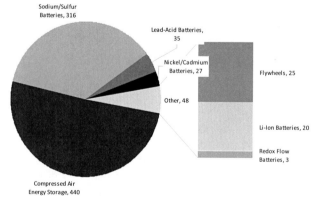

Although mechanical and chemical energy storage technologies currently have large installed capacities for grid-scale energy storage, electrochemical energy storage has attracted increased attention in the past few years because it shows great potential in smart grid applications, such as peak load shifting and frequency regulation. A smart grid will improve energy efficiency, and could help integrate renewable energy generation technologies like solar, wave, and wind into the power grid. Therefore, this section will present current state-of-the-art electrochemical energy storage technologies for large-scale applications, identify the major challenges, and discuss the future perspectives.

Different from energy storage at small and medium scales, energy storage at a scale above 100 kW entails its own distinctive set of criteria. Before discussing detailed technologies, however, it seems prudent to clarify the applications that will be discussed in this section. They include retail applications, transmission and distribution system applications, and generation and system level applications (see Figure 6). These applications require energy storage at a scale from 100 kW to 100 MW. More specifically, a backup power supply for home customers usually requires a capacity of 100 kW to meet the critical requirements of reliability.

Electrochemical energy storage can usually be divided into three types: batteries, flow batteries, and fuel cells.[38] Among them, fuel cells require external fuels to be fed into the cells and they are not rechargeable. Fuel cells are commonly utilized for back-up power applications, but the technology has limited application for smart grids that require frequent recharging. In contrast, both battery and flow battery technologies are very attractive because of their capability to store energy when excess electricity is generated and to release the energy as needed. The key characteristics of a wide range of technologies that may be employed for large-scale energy storage applications are presented in Table 2.[39]

Figure 6
Applications for Energy Storage at Various Scales/Sizes

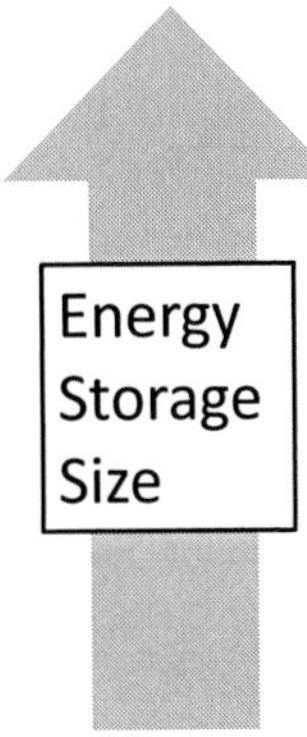

		Application
Energy Storage Size	Generation & system-level applications	Wholesale energy services
		Renewable integration
	Transmission and distribution applications	Stationary storage for peak load shift
		Stationary storage for frequency regulation
		Distributed energy storage
	Retail applications	Back-up power supply for industrial customers
		Back-up power supply for home customers

Table 2
Key Characteristics of Various Technologies for Large Scale Energy Storage

System	Cost ($/kWh)	Energy Efficiency (%)	Power (MW)	Capacity (MWh)	Maturity
Li-ion batteries	1,000–5,000	>90	1–100	0.25–25	Demo
Na/S batteries	450–550	75	50	300	Commercial
Fe/Cr redox flow batteries	300–400	75	1–50	4–250	R&D
V redox flow batteries	620–830	65–75	1–50	4–250	Demo
Pb/acid batteries	650–3,500	75–90	1–100	0.25–400	Demo
CAES (above ground)	390–430	N/A	50	250	Demo

Li-ion batteries, as one of the major battery chemistries, are being considered for small-scale distributed energy storage.[40,41] Lithium iron phosphate and titanate batteries (2 MW, ~250 kWh) have demonstrated the capability of frequency regulation, which is necessary for a smart grid. Commercial systems based on Li-ion batteries are already available. A123 Systems, using a nanophosphate Li-ion battery, currently provides several energy storage solutions for smart grid applications. The key parameters for the individual module commercialized by A123 Systems are 480V 3-phase AC voltage, DC voltage of 960V, 2 MW of power, 500 kWh of energy, 90 percent efficiency, operating temperature between –30 to +60°C, and recharge time of 15 minutes. The potential benefits of installing a Li-ion storage system such as the one manufactured by A123 include increases in power quality, higher reliability of the electric grid, better integration of renewable sources, and improvements to power plant efficiency. However, in order to succeed at the wholesale scale the cost of Li-ion batteries must be further reduced by at least a factor of 3, and a much higher energy density is required.

The high large scale cost for well-developed small scale battery chemistries such as Li-ion or lead-acid have generated strong interest in sodium/sulfur batteries and redox flow batteries, because the potential generation capacity and cost are very attractive for large scale applications.

Redox Flow Battery

A typical redox flow cell consists of two parallel electrodes, which are separated by an ion exchange membrane (see Figure 7). The separated electrolyte

Figure 7
A Schematic Diagram of a Redox Flow Battery

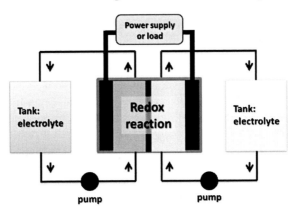

solutions contain different soluble redox couples, which are circulated through an independent electrolytic circuit. The energy in redox flow batteries is stored in the electrolyte, which is charged or discharged by either applying or withdrawing current, respectively. The redox flow battery energy density is mainly determined by the concentration of redox couples in the electrolytes.

The first generation of redox flow batteries was developed by NASA in the 1970s to store electrical energy generated from photovoltaic panels.[42] The battery is based on the Fe^{2+}/Fe^{3+} and Cr^{2+}/Cr^{3+} redox couples. Such redox couples provide a one-electron process, resulting in simple charge transfer and low overpotential on the surface of carbon electrodes. However, the Cr^{2+}/Cr^{3+} experiences slow kinetics and the open circuit voltage of this battery is low (~1.0 V). Since the first generation Fe/Cr redox couple, flow batteries based on other redox couples have been developed. The most successful one is based on two vanadium redox couples (i.e., V^{2+}/V^{3+} and V^{4+}/V^{5+}).[43] Advantages of the vanadium redox flow battery include fast kinetics compared to the Cr^{2+}/Cr^{3+} redox couple, less damage when the electrolytes are mixed together accidentally, and use of a single element with four oxidation states.

As mentioned previously, in order to increase the value of renewable energy generation, an energy storage system must be utilized because renewable sources such as wind, wave, and solar systems produce power intermittently. The redox flow battery is an excellent candidate for utility scale energy storage because the amount of battery capacity is flexible, and determined primarily by the electrolyte tank size. The ability to store large amounts of energy means the redox flow battery can deliver stored energy during peak demand and use off-peak power for charging. This stabilizes operations and provides the flexibility to store renewable energy even when generated at off-peak times.

Compared with many other battery technologies, the current generation redox flow battery has already demonstrated many advantages such as a high power rating, long energy storage time, and excellent response time.[44] It is possible to deliver full power in a few seconds. Such characteristics make this technology highly competitive in the distributed energy storage market because the fast response times and high power rating can help improve grid regulation and stability.

Flow batteries still need some technological improvements before these systems can see widespread installation for smart grid applications. Scaled-up, effective control systems, optimized flow geometries, better material resistance to degradation, and a larger temperature window are all under investigation to make flow batteries an attractive commercial energy storage system.[45] For the vanadium redox battery, the most expensive component is the ionic exchange membrane used to separate the two electrolyte solutions, so significant research efforts are underway to reduce the cost and, ideally, improve the membrane technical specifications.[46] If some of these technical challenges are overcome, the redox flow battery is well poised for installation into the current power grid because its large energy storage and fast response times make it ideal for intermittent renewable energies and smart grid energy management.

Sodium/Sulfur Batteries

Sodium/sulfur batteries are considered as another alternative for grid scale energy storage.[47,48] This type of battery operates between 300–350°C using sodium as the anode material and sulfur as the cathode material to provide a 2 V battery cell (see Figure 8). Such an operation temperature is needed to achieve high ionic conductivity within the solid sodium β-alumina ($NaAl_{11}O_{17}$) electrolyte, which approaches the ionic conductivity of an aqueous electrolyte (e.g., sulfuric acid) at these temperatures. Also, sulfur is in its liquid phase at this temperature, which helps minimize the interfacial contact issues.

During discharge, sodium is oxidized at the anode to form sodium cations that diffuse through the solid sodium β-alumina ($NaAl_{11}O_{17}$) electrolyte until reaching the cathode. At the electrolyte/cathode interface, sodium ions react with sulfur and initially form Na_2S_5. At elevated temperature, the discharge product Na_2S_5 and sulfur form a two-phase liquid. Further sulfur reduction leads to the formation of sodium polysulfides, such as Na_2S_4 and Na_2S_3. At 100 percent discharge, the final product composition is Na_2S_3 (see Figure 8). During charging, the sodium polysulfides decompose to reform a mixture of Na_2S_5 and sulfur. The half-cell and full-cell reactions are described as follows:

$$\text{Cathode: } xS + 2Na^+ + 2e^- \leftrightarrow Na_2S_x \tag{1}$$
$$\text{Anode: } 2Na \leftrightarrow 2Na^+ + 2e^- \tag{2}$$
$$\text{Overall: } 2Na + xS \leftrightarrow Na_2S_x \tag{3}$$

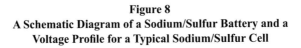

Figure 8
A Schematic Diagram of a Sodium/Sulfur Battery and a
Voltage Profile for a Typical Sodium/Sulfur Cell

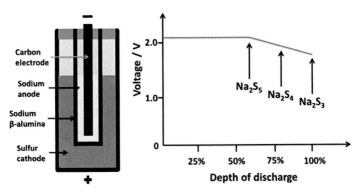

In 2002, the sodium/sulfur battery was commercialized in Japan. This technology subsequently has been employed in utility-scale frequency regulation and peak load shift. The cells are assembled into series-parallel configurations as modules, which are further connected in series–parallel to form a battery pack. This design minimizes the impact of failure of an individual cell.

After substantial development, the current generation of sodium/sulfur batteries have high cycle efficiency and good cycle life, which are very attractive for grid-scale applications. Still, technological improvements are being investigated to lower the material cost, enhance safety, and improve performance.[49] Developing methods to reduce sulfur formation at the cathode/electrolyte interface will reduce unwanted internal resistances and improve rechargeability. In addition, significant research efforts are focused on reducing the battery operating temperature, which could improve material durability and reduce cost by eliminating the installation of thermal components. Reducing the operating temperature can only be considered, however, when a solid electrolyte capable of fast, low temperature sodium ion transport is discovered.

All large scale electrochemical energy storage systems, including the already commercialized sodium/sulfur battery, need lower installation and operating costs before significant market penetration can be achived. Research to accomplish this feat is underway. These systems, by their underlying electrochemistry mechanisms, are well suited for implementation in smart grids and to store excess power generated via intermittent renewable energy sources.

Future Prospects

There is no doubt that electrochemical energy storage is playing, and will continue to play, an important role in sustainable energy development. As markets experience rapid growth for mobile electronics and electric vehicles, the Li-ion battery is one of the most important high-value technologies and will likely continue to dominate the mobile energy storage market. With significant investment in research and development, the Li-ion battery is expected to have improved lifetime, enhanced energy density, better reliability, lower cost, and increased safety. These features will assure its success in future applications, particularly plug-in hybrid electric and fully electric vehicles, the biggest market for mobile energy storage in the next ten to twenty years.

For grid scale energy storage, the Li-ion battery, redox flow battery, and sodium/sulfur battery may have new opportunities in the near future because the reliability and quality of electricity is becoming more and more important. Also, in order to fully integrate renewable energy sources into the existing electric grid, electrochemical energy storage is the critical technology to mitigate intermittency. Novel technologies, such as the rechargeable lithium-air battery and Li-S battery, are in the early stage of development and it is still too early to estimate their potential impact on the energy storage market. Yet their impacts over time merit close observation, as technological advances supportive of alternative energy futures will impact the sustainability of our society in significant ways, and shape the direction of critical industries and infrastructure in ways that influence us all.

Notes

1. Dunn, B., Kamath, H., and Tarascon, J. M., "Electrical Energy Storage for the Grid: A Battery of Choices." *Science* 334 (2011): 928–35, doi:10.1126/science.1212741.
2. "Battery University—Cost of Power from Batteries," *Cadex Electronics Inc.*, February 2012, http://batteryuniversity.com/lern/article/cost_of_power.
3. Amirault, J. et al., "The Electric Vehicle Battery Landscape: Opportunities and Challenges," (# 2009.9.v.1.1, Center for Entrepreneurship and Technology, University of California Berkeley, December 2009).
4. Technology Roadmap: Electric and plug-in hybrid electric vehicles, *International Energy Agency* (2011).
5. Peterson, S. B., Whitacre, J. F., and Apt, J., "Net Air Emissions from Electric Vehicles: The Effect of Carbon Price and Charging Strategies," *Environmental Science and Technology* 45 (2011): 1792–97.
6. Elgowainy, A. et al., "Well-to-Wheels Analysis of Energy Use and Greenhouse Gas Emissions of Plug-In Hybrid Electric Vehicles," (ANL/ESD/10-1, Argonne National Laboratory, US Department of Energy, Argonne, IL, June 2010).
7. Eberhard, M., and Tarpenning, M., "The 21st Century Electric Car." *Tesla Motors Inc.* (2006).
8. Markel, T., and Simpson, A., "Cost-Benefit Analysis of Plug-In Hybrid Electric Vehicle Technology (NREL/JA-540-40969)," *World Electric Vehicle Journal* 1 (2006).
9. Ibid.
10. Gonder, J., Markel, T., Simpson, A., and Thornton, M., "Using GPS Travel Data to Assess the REal-World Driving Energy Use of Plug-In Hybrid Electric Vehicles (PHEVs)," (NREL/CP-540-40858, National Renewable Energy Laboratory, US Department of Energy, Transportation Research Board 86th Annual Meeting, Washington D.C., January 2007).

11. See Note 8.

12. Anderson, D. L., and Patino-Echeverri, D., *An Evaluation of Current and Future Costs for Lithium-Ion Batteries for use in Electrified Vehicle Powertrains*, Masters Degree thesis, Duke University, 2009.

13. USABC Goals for Advanced Batteries for EVs, *US Advanced Battery Consortium*, February 2012, http://batt.lbl.gov/files/Goals_for_Advanced_EV_Batteries-21.pdf.

14. Cheah, L., and Heywood, J., "The Cost of Vehicle Electrification: A Literature Review," (MIT Energy Initiative Symposium, Massachusetts Institute of Technology, Cambridge, MA, April 2010).

15. Miller, T. J., *Electrical Energy Storage for Vehicles: Targets and Metrics* (US DOE Advanced Reseach Projects Agency-Energy, 2009).

16. Gaines, L., and Cuenca, R., "Costs of Lithium-Ion Batteries for Vehicles" (ANL/ESD-42, Argonne National Laboratory, US Department of Energy, Argonne, IL, May 2000).

17. Ibid.

18. Ibid.

19. Nelson, P. A., Santini, D. J., and Barnes, J. *International Battery, Hybrid and Fuel Cell Electric Vehicle Symposium* (Stavanger, Norway, 2009).

20. See Note 16.

21. Anderson, A. F., and Carlson, E. J. "Platinum Availability and Economics for PEMFC Commercialization," DE-FC04-01AL67601, Tiax LLC Report to US Department of Energy, December 2003, http://www1.eere.energy.gov/hydrogenandfuelcells/pdfs/tiax_platinum.pdf.

22. James, B. D., and Kalinoski, J. A., "Mass Production Cost Estimation for Direct H2 PEM Fuel Cell Systems for Automotive Applications: 2008 Update," (v.30.2021.052209, Directed Technologies Report to US Department of Energy, March 2009).

23. An, F., and Santini, D., "Assessing Tank-to-Wheel Efficiencies of Advanced Technology Vehicles," (2003-01-0412, Argonne National Laboratory, US Department of Energy, Argonne, IL, 2003).

24. "Envia Systems-Innovation," Envia Systems, Newark, CA, http://enviasystems.com/innovation/ (Accessed March 2012).

25. "The electric car—a green transport revolution in the making?," *European Environment Agency*, February 2012, http://www.eea.europa.eu/articles/the-electric-car-2014-a-green-transport-revolution-in-the-making.

26. See Note 8.

27. See Note 6.

28. See Note 8.

29. See Note 6.

30. eGRID2010 Version 1.1 Year 2007 Summary Tables, *US Environmental Protection Agency*, May 2011, http://www.epa.gov/cleanenergy/documents/egridzips/eGRID2010V1_1_year07_SummaryTables.pdf (Accessed March 2012)

31. Hamlen, P., and Atwater, T. B., In *Handbook of Batteries*, David Linden, and Thomas Reddy (McGraw-Hill, 2001).

32. Christensen, J. et al., "A Critical Review of Li/Air Batteries," *Journal of the Electrochemical Society* 159 (2012): R1–R30, doi:10.1149/2.086202jes.

33. Abraham, K. M., and Jiang, Z., "A polymer electrolyte-based rechargeable lithium/oxygen battery." *Journal of the Electrochemical Society* 143 (1996): 1–5, doi:10.1149/1.1836378.

34. Ogasawara, T., Debart, A., Holzapfel, M., Novak, P., and Bruce, P. G. "Rechargeable Li2O2 electrode for lithium batteries." *Journal of the American Chemical Society* 128 (2006): 1390–93, doi:10.1021/ja056811q.

35. See Note 1.
36. Yang, Z. G. et al. "Electrochemical Energy Storage for Green Grid." *Chemical Reviews* 111 (2011): 3577–613, doi:10.1021/cr100290v.
37. "Electrical energy storage technology options" (Report 1020676, Electric Power Research Institute, Palo Alto, CA, December 2010).
38. Winter, M., and Brodd, R. J., "What are batteries, fuel cells, and supercapacitors?," *Chemical Reviews* 104 (2004): 4245–69, doi:10.1021/cr020730k.
39. See Note 36.
40. Arico, A. S., Bruce, P., Scrosati, B., Tarascon, J. M., and Van Schalkwijk, W. "Nanostructured materials for advanced energy conversion and storage devices," *Nat. Mater.* 4 (2005): 366–77, doi:10.1038/nmat1368.
41. Armand, M., and Tarascon, J. M., "Building Better Batteries." *Nature* 451 (2008): 652–57, doi:10.1038/451652a.
42. Bartolozzi, M. "Development of Redox Flow Batteries—A Historical Bibliography," *J. Power Sources* 27 (1989): 219–34, doi:10.1016/0378-7753(89)80037-0.
43. Skyllaskazacos, M., Rychcik, M., Robins, R. G., Fane, A. G., and Green, M. A., "New All-Vanadium Redox Flow Cell." *J. Electrochem. Soc.* 133 (1986): 1057–58, doi:10.1149/1.2108706.
44. Skyllas-Kazacos, M., Chakrabarti, M. H., Hajimolana, S. A., Mjalli, F. S., and Saleem, M., "Progress in Flow Battery Research and Development." *J. Electrochem. Soc.* 158 (2011): R55–R79, doi:10.1149/1.3599565.
45. Ibid.
46. Ibid.
47. Dustmann, C. H., "Advances in ZEBRA Batteries," *J. Power Sources* 127 (2004): 85–92, doi:10.1016/j.jpowsour.2003.09.039.
48. Lu, X. C., Xia, G. G., Lemmon, J. P., and Yang, Z. G., "Advanced Materials for Sodium-Beta Alumina Batteries: Status, Challenges and Perspectives," *Journal of Power Sources* 195 (2010): 2431–42, doi:10.1016/j.jpowsour.2009.11.120.
49. Ibid.

12

Hydrogen Fuel Cells: Current Status and Potential for Future Deployment

Ajay K. Prasad

Introduction

Concerns about dwindling fossil fuel reserves and climate change caused by rising greenhouse gas emissions have generated tremendous public interest in renewable energy for buildings, alternative fuels for automobiles, energy conservation, and environmental protection. At the same time, fuel cells have become well known in the popular media as clean and efficient energy-conversion devices that are powered by hydrogen. Hence, it is not surprising that the past decade has also witnessed a tremendous increase of interest in the science and technology of fuel cells. One of the pivotal announcements was made by President George W. Bush in his 2003 State of the Union address when he launched the Hydrogen Fuel Initiative to propose funding to develop clean, hydrogen-powered automobiles. While research in fuel cells was already beginning to intensify in the last decade of the twentieth century, this initiative helped to further accelerate research and development in academic, government, and industrial laboratories. Significant advances have subsequently resulted in fuel cell performance, durability, and cost-effectiveness. Several successful products have been introduced that have enjoyed commercial success.

Fuel cell research and development continues to be vigorous today as researchers and manufacturers strive to overcome significant challenges in order to meet the 2015 target metrics specified by DOE for the various types of fuel cells. This report will highlight the current status of the fuel cell industry in three major market segments–transportation, portable power, and stationary power–and their prospects for future growth.

A Brief History

A major milestone in the demonstration of the fuel cell as a practical device was achieved in 1958 by Francis Bacon who fed hydrogen and oxygen to nickel electrodes on either side of a potassium hydroxide electrolyte.[1] The success of Bacon's design attracted the attention of Pratt and Whitney who found a customer in the US space program. In addition to extremely high reliability, reducing weight and volume is also of great importance in on-board power systems in spacecraft; it is therefore impressive that fuel cells found rapid acceptance within the space program and became integral components in the Gemini and Apollo missions, and eventually the Shuttle Orbiter.

In the 1960s the Apollo missions employed an alkaline fuel cell similar to Bacon's original design weighing 250 lbs that produced 1.5 kW; the device was fed with hydrogen and oxygen and the crew drank the product water.[2] Two decades later, the Shuttle Orbiter also employed an alkaline fuel cell of almost identical weight, but with an increased power output of 12 kW.[3] Today, automotive fuel cell systems of similar weight produce 100 kW. These numbers demonstrate the one-hundred-fold improvement in power density that has resulted from substantial investments in fuel cell R&D.

Looking at the remarkable advances in fuel cell technology from the 1960s till today, it is tempting to ask if there is an equivalent of Moore's Law that predicts a ten-fold increase in fuel cell power density (kW/kg) every twenty years! Unfortunately, it is the opinion of experts that while future improvements are inevitable, we are unlikely to witness such dramatic leaps in performance and weight reduction. Instead, the breakthroughs that will make a real impact in the future will be through the discovery of new component materials and manufacturing methods that will greatly reduce fuel cell cost while improving reliability and durability.

Fuel Cell Basics and Applications

In its simplest implementation, a fuel cell is an energy conversion device that operates on hydrogen gas as its fuel (Figure 1). For terrestrial applications, the oxygen that is required to complete the fuel cell reaction is drawn directly from the air. Unlike an internal combustion engine (ICE), the fuel is not burned in the presence of oxygen to create heat; the fuel cell does not contain pistons and cylinders, in fact, there are no moving parts within the fuel cell "stack". Instead, the cell contains an electrolyte and electrodes, and the hydrogen and oxygen combine electrochemically such that the chemical energy contained within the fuel is converted directly to electrical energy. Fuel cells are not heat engines and therefore they are not limited by the Carnot thermodynamics, which place an upper bound on the efficiency of any heat engine. As a result, fuel cells can be two to three times more efficient than ICEs. The lack of moving parts implies

Figure 1
A Schematic View of the Fuel Cell

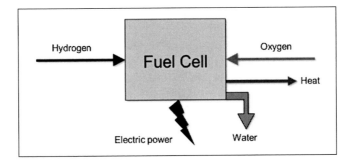

Figure 2
Cross-Sectional View of a Polymer Electrolyte Membrane Fuel Cell

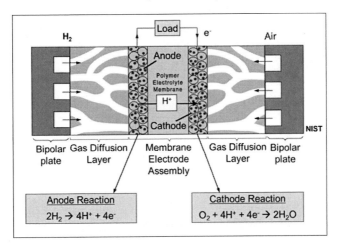

that fuel cells can run more quietly as well. Finally, as shown in Figure 1, the only product of the reaction is water making it a zero emission device.

In order to illustrate the operating principle of the fuel cell, we have chosen the polymer electrolyte membrane (PEM) fuel cell whose cross-section is shown in Figure 2. The cell is shown to consist of seven layers. The central layer is the polymer electrolyte membrane. This membrane is very thin (25–50 microns) and is coated on either side with electrode layers. The electrodes are about twenty microns thick and are black in appearance due to the presence of carbon black powder. Importantly, the electrodes contain platinum nanoparticles, which catalyze the anode and cathode reactions. The membrane electrode assembly is sandwiched between two gas diffusion layers, which are porous carbon papers with a thickness of about three hundred microns. Finally, bipolar end plates with 1 mm channels cut in them distribute gases to the electrodes and serve as the terminals, while also helping to compress and seal the cell.

Within the fuel cell, the hydrogen supplied to the anode diffuses through the anode gas diffusion layer and is transported to the anode electrode. The hydrogen dissociates to form protons and electrons (see Figure 2). The PEM has a special property that it conducts protons while blocking electrons. Hence, the electrons are forced to travel through the external circuit and drive the load. The electrons then return to the cathode, and combine with the protons arriving through the PEM and the oxygen permeating through the cathode gas diffusion layer to complete the reaction and form water.

The ideal voltage produced by the hydrogen–oxygen reaction at room temperature is 1.23 V. In practice, several loss mechanisms contribute to curtail this voltage to a much lower value. Raising the operating temperature greatly reduces losses and improves performance. However, even at a typical operating temperature of 60 °C, a PEM cell will only produce about 0.7 V. Generally this voltage is much too small to be useful, so typically several cells are placed in series to boost the voltage to useful values. When arranged in series, the cells resemble a stack, hence the name. Stacks can feature a hundred or more cells in series.

In addition to the stack, a number of other components are required to make up the fuel cell's balance-of-plant. These include mechanical devices such as an air compressor, hydrogen recirculation pump, humidifiers, heat exchangers, a coolant pump and radiator, as well as electrical devices such as a boost converter, inverters, and other power conditioning equipment.

It is important to remember that because a fuel cell produces electric power, it can be easily integrated into a vehicle chassis that employs an electric drivetrain. Typically, automotive stacks are paired with batteries or ultracapacitors to provide hybrid operation. Similarly, small fuel cells can be easily mated with portable electronic devices to either completely replace Li-ion batteries, or work in conjunction with them in a hybrid configuration. Finally, fuel cells for stationary power can directly supply electric power to their clients and also feed power to the grid.

Efficiency of Fuel Cells

The efficiency of an energy conversion device is defined as the useful work obtained from the device, divided by the calorific value of the fuel that is put into the device. In the case of a fuel cell, this could be expressed as the electrical work produced by the fuel cell stack, divided by the higher heating value (HHV) of the hydrogen fuel consumed by it over a given period of time. Defined this way, the ideal thermodynamic efficiency of a hydrogen fuel cell operating at room temperature is 83 percent. Often, efficiency is defined on the basis of the fuel's lower heating value (LHV), giving an ideal efficiency of 98 percent. Moreover, the fuel cell stack efficiency is often quoted in terms of the gross power produced by the stack, whereas it is more useful to express it in terms of the net power (gross power minus the parasitic power consumed by the fuel cell's balance-of-plant). Due to the multitude of available efficiency definitions, one must exercise caution when making comparisons between different energy conversion devices.

Just as an internal combustion engine can never approach the ideal Carnot efficiency, the ideal thermodynamic fuel cell efficiencies mentioned above can never be realized in practice due to a variety of loss mechanisms. One of the principal fuel cell losses arises from activation losses incurred at the anode and/or cathode due to sluggish electrode kinetics. Activation loss results in a steep drop in the cell's voltage from the ideal value even at very modest current draws. The use of an effective catalyst is vitally important in reducing these losses. Operating at higher temperatures also reduces activation losses. The second major loss, termed the ohmic loss, occurs due to the fuel cell's internal resistance that causes the voltage to drop linearly (albeit less steeply) with the current drawn from the cell. Ohmic losses can be reduced by improving the electronic and ionic conductivities of the fuel cell materials, and reducing the path length traversed by the electrons and ions within the cell. Ohmic losses are more apparent over the mid-range values of the cell's current. Finally, at high current draws, the fuel cell suffers from mass transport losses to due fuel starvation. Mass transport losses result in a rapid drop in cell voltage at high currents and limit the maximum current that can be drawn from the cell. This loss mechanism can be mitigated by improving the gas and liquid water transport within the cell by better internal design and material choices.

The consequence of these losses is that practical fuel cell efficiencies are much lower than the ideal thermodynamic values. In the case of PEM fuel cells, practical stack efficiencies (based on the HHV of hydrogen) are about 50 percent. Direct methanol fuel cells suffer from higher activation and ohmic losses than PEM fuel cells and can therefore produce practical efficiencies of around 30 percent. Solid oxide fuel cells exhibit low activation losses due to very high operating temperatures resulting in efficiencies of around 60 percent. In contrast, a typical (non-hybrid) gasoline engine operates at an efficiency of only about 20 percent.

Current Status of the Fuel Cell Industry

Over the past decade, fuel cell research and development in universities, industry and government laboratories has grown substantially. As a result, the number of fuel cell patents granted worldwide has increased from 403 in 2000 to 1,801 in 2010.[4] In 2010, Japan was the leader in the number of fuel cell patents granted (617) followed closely by the United States (598), with Germany (187) and Korea (177) in the next tier.[5] The top ten companies receiving fuel cell patents in 2010 were Samsung (140), Honda (135), General Motors (130), Toyota

(90), Panasonic (58), Nissan (46), Hitachi (26), Delphi Tech. (24), Toshiba (21), and Canon (20).

The automotive industry has invested heavily in PEM fuel cell technology, and therefore it is not surprising that automotive original equipment manufacturers (OEMs) figure prominently in the list of fuel cell patent recipients in 2010 with Honda in the lead, followed by General Motors, Toyota, and Nissan. However, Toyota had the highest number of fuel cell patent *applications* in 2010 with 381, followed by Honda (103), General Motors (103), Daimler AG (75), and Nissan (52).[6] Apart from automotive OEMs, the names of companies involved in consumer electronics (Samsung, Panasonic, Toshiba, Canon) figure prominently in the list of fuel cell grantees in 2010.

The 2010 Fuel Cell Technologies Market Report[7] produced by the Breakthrough Technologies Institute for the USDOE gives a comprehensive assessment of the fuel cell industry. Some of the key findings of this report are summarized below.[8]

- A study by McKinsey & Co. concluded that fuel cell vehicles are "the best low-carbon substitute" in medium- and large-car segments, which account for 50 percent of cars and seventy 5 percent of CO_2 emissions. In addition, it stated that fuel cell vehicles would be cost-competitive by 2020 or 2025.
- Fuel cell units shipped from North America quadrupled between 2008 and 2010.
- Japan unveiled a plan to sell two million fuel cell vehicles by 2025, and install 1,000 hydrogen fueling stations to support them.
- Ballard Power Systems surpassed one million membrane electrode assemblies produced. AC Transit's fuel cell bus fleet passed a durability milestone with one bus logging 7,000 hours of operation on a single fuel cell stack.

The report also noted that commercial deployments of fuel cells continued to grow, particularly in the forklift, primary power, back-up power, and auxiliary power sectors. In fact, it is interesting to note that several of these products are becoming cost-competitive with conventional technologies.[9] In terms of 2010 gross revenues, FuelCell Energy ($70 million) and Ballard Power Systems ($65 million) accounted for the bulk of the revenues from North American companies. Cumulative global investment in fuel cells and hydrogen totaled $630 million between 2008 and 2010, with US companies accounting for about half of the global total. The total number of worldwide shipments of fuel cell systems doubled from about 7,500 in 2008 to just under 15,000 in 2010. Shipment by MWs also increased significantly from 52 MW in 2008 to 87 MW in 2010. In 2008, transportation, portable and stationary shipments accounted for 62, 12, and 26 percent of the global total, respectively. In 2010, the corresponding fractions were 34 percent, twenty one percent, and 45 percent, respectively. Shipments of stationary systems overtook transportation systems due to increases in residential and back-up power (primarily due to the Japanese residential fuel cell program). The transportation sector owed its growth primarily to fuel cell powered forklifts, while the growth in portable applications came from external battery chargers, military demand, and remote monitoring.[10]

The 2010 Fuel Cell Technologies Market Report[11] concludes that despite the rapid growth in market penetration and success of fuel cells in recent years, a number of challenges remain. High cost continues to be an issue for commercialization, and improvements in performance and durability are necessary for fuel cells to compete with incumbent technologies. The cost of hydrogen production and storage must also be reduced, especially hydrogen production methods with a low carbon footprint. Safety regulations and product standards need to be updated and made uniform across jurisdictions. The report also mentions the need for public outreach and education to promote awareness and acceptance of fuel cells and hydrogen among the general population. Government support is seen as critical in all of these endeavors. Despite these challenges, on the basis of the current progress on key indicators, the outlook for the fuel cell industry is seen to be very positive.

Polymer Electrolyte Membrane (PEM) Fuel Cells

Operating Principles, Materials and Architecture

The basic seven-layer architecture of a PEM fuel cell consisting of one membrane, two electrodes, two gas diffusion layers, and two bipolar plates has already been illustrated in Figure 2, and the associated discussion there introduced the PEM fuel cell's component materials and fuel cell reactions. At the heart of the PEM fuel cell is a proton-conducting polymer film, typically a perfluorosulfonic acid (PFSA). DuPont's Nafion® is an example of this type of membrane. As stated earlier the PFSA membrane conducts protons while blocking electrons. The ionic resistance of the electrolyte membrane is a major contributor to the cell's internal resistance and voltage loss, therefore, it is very important to minimize its thickness. However, reducing the thickness below twenty-five microns can compromise the membrane's mechanical integrity and lead to the formation of pinholes and microtears, which are generally fatal to fuel cell operation.[12,13] In addition, the polymer must be well hydrated in order to conduct protons, hence in most applications, the anode and cathode gas feeds are humidified prior to their inlet to the cell. Since water is a byproduct of the fuel cell reaction, water is usually captured from the cathode exhaust and recycled to the inlet gases. However, the presence of excess water within the cell (which can occur at high current draw) can cause the opposite problem of flooding, which can saturate the porous layers with liquid water and block the path of the reactant gases. Hence, the water content within the cell must be maintained within a narrow range in order to promote membrane hydration while also avoiding flooding; this issue is termed "water management."[14] In cold climates, liquid water within the cell can freeze when the fuel cell is turned off. Freezing can lead to a host of problems such as delamination of the electrodes from the membrane, which compromises the lifetime of the cell. Hence, manufacturers must employ special features to prevent water accumulation and degradation in cold climates; this issue is termed "freeze tolerance".[15]

The PFSA membrane is currently a significant cost-driver of PEM fuel cells. Furthermore, it can only operate at temperatures below 80 °C in order to maintain mechanical integrity, while also ensuring good hydration and preventing dry-out. This is a serious drawback because higher operating temperatures would greatly speed up reaction kinetics and reduce losses. Hence, vigorous research is ongoing to identify and test novel membranes that can operate at higher temperatures and low relative humidities in order to enhance performance and reduce the overhead associated with water management.[16,17]

The anode and cathode layers employ Platinum (Pt) catalyst supported on carbon black. While Pt is a precious metal and greatly adds to the cost of the fuel cell, there is no viable alternative that can catalyze the fuel cell reaction as effectively. One of the major advances in PEM fuel cells over the past decade is the reduction in Pt loading. Today, it is possible to obtain a performance of around 700 mW/cm^2 with a catalyst loading as low as 0.2 mg/cm^2. However, the search for cheaper catalysts continues,[18,19] as also methods to further reduce Pt loading.[20,21] Platinum also suffers from CO poisoning, therefore the hydrogen that is supplied to PEM fuel cells must be of sufficiently high purity. Hydrogen is typically produced by steam reforming natural gas. The reformate stream contains substantial amounts of CO, which must be removed by the shift reaction to make the hydrogen suitable for PEM fuel cells.

Another contributor to PEM fuel cell cost is the bipolar plate. Bipolar plates must be light, thin, rigid, impermeable to gas, and excellent conductors of heat and electricity. The fuel cell environment is fairly corrosive, so the bipolar plate material must exhibit a high degree of corrosion resistance. Typical materials include graphite, stamped metal, and electrically conducting plastics.[22,23]

The cumulative cost of the component materials such as the membrane, the catalyst, and the bipolar plate poses a major challenge for fuel cell commercialization today. A second major challenge arises from limited fuel cell durability. Although the fuel cell stack does not experience catastrophic failures, its performance does decline gradually over time due to degradation mechanisms such as Pt dissolution and carbon corrosion within the electrodes, loss of the membrane's mechanical integrity due to stresses generated by repeated temperature and humidity cycling, development of interfacial defects leading to delamination between the membrane and the electrode, etc. Vigorous research is ongoing to understand these degradation mechanisms and to propose solutions with improved materials, manufacturing methods, and operating procedures.

Apart from the stack itself, the fuel cell system requires a number of additional mechanical components to supply the gases at the appropriate temperature and humidity and recycle water such as an air compressor, hydrogen recirculation pump, humidifiers, and heat exchangers. A stack cooling system consisting of a coolant pump and radiator is required to maintain the stack temperature. Electrical components include power conditioning equipment, sensors, and control electronics. All of these together are termed the balance-of-plant. Each component of the balance-of-plant must be selected for high reliability and adds to the overall cost.

Hydrogen storage is also a major challenge.[24,25] Hydrogen is the lightest of gases, so its storage density is low even when it is pressurized to 700 bar (10,000 psi) or stored as a cryogenic liquid. A passenger car would require 5 kg of hydrogen for adequate range. At a typical gravimetric storage efficiency of 3 percent, the weight of the corresponding hydrogen tank would be over 160 kg. The volumetric efficiency of hydrogen storage is also poor, implying that current hydrogen storage tanks are rather bulky. Furthermore, such high-pressure tanks, while extremely safe, are quite expensive.

Finally, despite rapid gains in the past decade, PEM fuel cell manufacturing is still in its infancy. In contrast, a century's-worth of evolution of the internal combustion engine has led to enormous sophistication of its manufacturing industry and large production volumes of a highly optimized product. Hence, it is not surprising that fuel cell costs have remained high despite the spurt of interest in recent years. As the technology matures and production volume increases, it is expected that costs will come down.

Current Market and Future Prospects

The strong commitment to PEM fuel cell research and development by the automotive OEMs, as evidenced by the number of patents granted to them in 2010, underscores their desire to bring fuel cell vehicles to commercialization by 2015. Audi, Daimler AG, General Motors, Honda, Hyundai, Kia, Nissan, Suzuki, Toyota, and VW have all produced prototypes.[26] Companies such as GM, Daimler, Honda, and Toyota have already leased fleets of about a hundred cars to customers in chosen locations.

The US Department of Energy hosts the Hydrogen and Fuel Cells Program and Vehicle Technologies Program Annual Merit Review (AMR) every year in May in Washington DC. All organizations that receive DOE funding (universities, national laboratories and businesses) are required to report on their progress during the previous year. From the 2011 AMR Meeting, DOE compiled data collected from 180 fuel cell vehicles that had logged over 500,000 trips for a cumulative travel distance of 3.6 million miles. Data compiled from twenty-five hydrogen refueling stations included about 33,000 fillings with over 152,000 kg of hydrogen produced or dispensed.[27] Based on these data, DOE published its targets for fuel cell efficiency, durability, fuel economy, vehicle range, refueling rate, and fuel cell cost, along with the current progress toward meeting these targets.[28] These metrics are summarized in Table 1. Significant results include reductions in fuel cell costs by 30 percent since 2008 and more than 80 percent since 2002. These cost reductions are attributed to R&D efforts that reduced Platinum content, increased power density, and simplified the balance-of-plant.[29] However, continued R&D investments are required to drop the stack cost to the 2015 target of $30/kW.

Table 1

DOE Targets for Fuel Cell Vehicles and Current Progress toward These Targets Compiled from Operating Data from 180 Vehicles and Twenty-Five Hydrogen Refueling Stations[30]

Metric	Goal	Current progress
Fuel cell efficiency	60% for commercialization	59%
Fuel cell durability	5,000 operating hours for commercialization	Demonstrated more than 2,500-hour (75,000 miles) durability of fuel cell systems in vehicles operating under real-world conditions, with less than 10% degradation.
Fuel cell power density	650 W/L by 2017 850 W/L by 2020	400 W/L
Fuel cell specific power	650 W/kg by 2017 850 W/kg by 2020	400 W/kg
Fuel economy	Undefined	"Window-sticker" fuel economy range of 43–58 miles/kg hydrogen
Vehicle range	300 miles for commercialization	Driving range of more than 250 miles between refueling with one vehicle achieving up to 430 miles on a single fill.
Refueling rate	Similar to conventional fueling (about 1 kg/minute)	Median fueling time is 0.77 kg/minute, which includes data from 70 MPa fueling for the first time. Refueling rates are close to being acceptable.
Fuel cell cost	$30/kW by 2015	$49/kW in 2011 (more than 80% reduction since 2002), based on projections to high-volume manufacturing. Cost reductions are due to the development of durable membrane electrode assemblies with low Platinum content. DOE anticipates further reductions in cost from improvements in manufacturing methods.

(Continued)

Table 1 (*Continued*)

Metric	Goal	Current progress
Cost of hydrogen delivery	$1/gge	Reduced from $5/gge in 2003 to a projected $3/gge, assuming high-volume production of 500 units at 1,500 kg of hydrogen per day.
Hydrogen cost at the station	$3/gge for delivered hydrogen	Hydrogen from natural gas is $7.70–$10.30/kg. Hydrogen from electrolysis is $10.00–$12.90/kg. DOE is revising the target cost from $3.00 to $6.00 in 2020 primarily due to the higher anticipated cost of gasoline in 2020.

Table 1 reveals that certain vehicle metrics like efficiency, hydrogen refueling rate, and driving range have reached acceptable levels. Fuel cell durability has shown good improvement as well. However, fuel cell cost continues to remain a challenge due to high material and manufacturing costs. The cost of delivered hydrogen is also higher than the target, which in fact, was revised from $3/gge to $6/gge. The main reason for this upward revision was that the original target was based on a gasoline cost of $1.30/gallon, whereas the cost of gasoline in 2020 was projected to be $4.50.

Fuel cell buses are an excellent application of PEM fuel cells, especially for urban transit. The low average vehicle speed in urban transit buses implies that the average power expended is quite low. As a result, the fuel cell power plant can be greatly downsized with significant cost savings. All fuel cell vehicles are electric hybrids, meaning that the fuel cell is usually paired with a battery bank to complete the electric drivetrain. In passenger car applications, the hybrid platform is designed such that the fuel cell is expected to be "load-following" and provide the bulk of the propulsive power, with the battery pack providing a boost in high-demand situations. Manufacturers have converged on a typical gross stack power of about 100 kW for cars. On the other hand, for urban transit buses, the battery bank can be made much larger with the fuel cell serving as a range extender. All of the power demand is then supplied by the battery and the role of the (smaller) fuel cell is to replenish the battery's state-of-charge. For example, the University of Delaware's first fuel cell bus (22 ft long, 22 passengers) employs just a single 20 kW Ballard Mark9 SSL stack.[31] This type of hybrid platform is called a battery-dominant series hybrid. The other major advantage with fuel cell buses is that the maintenance and refueling operations can be centralized, which reduces the infrastructural costs. Moreover, fuel cell buses are an excellent tool for public outreach and education. Bus routes chosen for high visibility and maximum ridership can effectively convey the message that fuel cells represent a safe, reliable and clean alternative for transportation.

The number of fuel cell buses currently in operation in the United States is about twenty. In contrast, Europe has seen major fuel cell bus demonstration programs such as the current CHIC (Clean Hydrogen In European Cities) project, which builds upon previous programs such as CUTE (Clean Urban Transport for Europe) and HyFleet:CUTE. The 2010 Winter Olympics in Whistler, Canada, witnessed the operation of twenty fuel cell buses. The International Fuel Cell Bus Collaborative maintains a website that lists all the current fuel cell bus demonstrations worldwide.[32] Although the current market for fuel cell buses is rather small, fuel cell buses are perfectly suited for low-emission urban transit needs. Greater market penetration is expected as costs reduce and durability improves.

Direct Methanol Fuel Cells

Direct methanol fuel cells (DMFCs) are well suited for low-power applications where a high premium is placed on long, uninterrupted periods of operation such as portable power for laptops and other electronic devices. The advantage of the DMFC is that it employs a liquid fuel, methanol, which has very high energy density and is extremely easy to handle in contrast to compressed or liquefied hydrogen. Consequently, DMFCs can operate for long durations on methanol stored in simple plastic tanks. Similar to PEM fuel cells, DMFCs operate at fairly low temperatures (around 50 °C) and hence their start-up and shutdown procedure is quick and easy. However, DMFCs suffer from low power density owing to higher activation losses at their electrodes compared to PEM fuel cells. Despite this drawback, DMFCs can compete well with Li-ion batteries in terms of energy density (Wh/liter) for many practical applications.

Operating Principles, Materials and Architecture

The architecture of a DMFC is very similar to the PEMFC. As shown in Figure 3, it also features a central proton-conducting membrane (typically a PFSA) coated with catalyst layers, sandwiched between gas diffusion layers and bipolar plates. The major difference is that the fuel supplied to the anode is methanol. Hence the reaction on the anode is different, and requires a mix of two precious metal catalysts, platinum and ruthenium. The oxygen reduction reaction on the cathode is identical to the PEM fuel cell, and hence, platinum alone suffices there. Due to poor kinetics on both electrodes, a far higher catalyst loading is required in DMFCs. Furthermore, in order to prevent methanol crossover through the membrane, the membrane thickness has to be increased (to about 200 microns), which in turn increases ohmic losses. All of these factors lower the DMFC's performance to a typical value of only 100 mW/cm^2. As a result, DMFCs are not suitable for high power density applications like automobiles. Instead, they

Figure 3
Architecture and Operating Principle of a Direct Methanol Fuel Cell

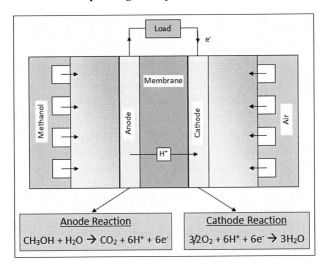

are well-suited for portable electronics applications that require low power. In addition, since the methanol fuel has a high energy density, it is possible to run the DMFC for long durations on a single "charge" of methanol. When the methanol is consumed, it is very straightforward to replenish the liquid fuel. Since the primary role of the membrane is to conduct protons while blocking electrons, the same polymer membrane as in PEM fuel cells is employed, albeit of much greater thickness.

As shown in Figure 3, the anode reaction breaks methanol down to produce protons and electrons with the simultaneous release of carbon dioxide gas, which must be vented out. As in the case of PEM fuel cells, the protons travel across the membrane while the electrons travel through the external circuit and perform work. The electrons then return to the cathode and combine with the protons and oxygen to form water. The methanol must be greatly diluted before supplying it to the anode, hence it is necessary to capture the product water at the cathode and resupply it to the anode stream.

The balance-of-plant in a DMFC would include a methanol pump to drive the fuel through the anode, and a methanol sensor to ensure that the methanol is being fed at the right concentration. In small systems, air is drawn into the cathode in a passive manner; larger systems may need active components to drive the air. Similarly, small systems can be cooled by natural convection, whereas larger systems may need a cooling fan. Hence small systems are simple, light, and offer weight and volume advantages when compared with lithium-ion batteries especially when the required duration of operation is long.

Current Market and Future Prospects

Starting around 2005, companies such as Casio, IBM, Samsung, Sony, Toshiba, etc., announced prototype DMFC solutions for portable electronics applications. The launch of the Toshiba Dynario in 2009 remains the most recent announcement; since then, portable fuel cell announcements have been scarce.[33] DMFCs release considerable heat and water vapor during operation, and it appears that companies have found it difficult to overcome miniaturization and system integration challenges. Integration problems can be avoided by developing external fuel cell chargers, which have also been introduced for the lucrative portable electronics market. Horizon Fuel Cell Technologies' MiniPak is a pocket-sized DMFC charger that can deliver 2 W continuously using a standard USB port, and uses refillable fuel cartridges able to store up to 12 Wh of net energy.[34] While DMFCs still show great promise for the huge market represented by laptops, tablet computers, and cell phones, this market has not realized the commercial success that was initially predicted, and some companies appear to have ceased development activities in DMFCs. In contrast, in December 2011, Apple Computers announced patent applications for fuel cells that envision portable electronics operation lasting "weeks without refueling".[35]

On the other hand, sales of DMFC auxiliary power units have reached tens of thousands since 2007. The industry leader in direct methanol fuel cell APUs (up to 100 W) is SFC Energy with cumulative sales of almost 17,500 units between 2007 and 2010.[36] The bulk of these sales have been in the recreational market. SFC Energy also produces a range of DMFC products for the military. Soldiers have to carry large quantities of batteries to power portable computers, communication devices, laser rangefinders, night vision and navigational equipment, and any device that can provide portable power with a reduced weight and volume would be in high demand. Another opportunity for DMFC APUs comes from

Table 2
Technical Targets for Portable Power Fuel Cell Systems (10–50 W)[37]

Characteristic	Units	2011 status	2013 targets	2015 targets
Specific power	W/kg	15	30	45
Power density	W/L	20	35	55
Specific energy	Wh/kg	150	430	650
Energy density	Wh/L	200	500	800
Cost	$/W	15	10	7
Durability	hours	1,500	3,000	5,000
Mean time between failures	hours	500	1,500	5,000

the trucking industry, which must look for a low-noise and low-emission alternative to the idling of heavy engines while parked overnight. DOE has published targets for portable fuel cells for three ranges: <2 W, 10–50 W, and 100–250 W. Of these, the target for the mid-size fuel cells (10–50 W) are presented below (Table 2). It should be noted that these targets do not specify the type of fuel cell or fuel employed.

Solid Oxide Fuel Cells

Solid oxide fuel cells (SOFCs) operate at very high temperatures (around 800–1,000 °C). This represents a major departure from PEMFC and DMFC discussed earlier. As a result, the architecture of the SOFC, component materials, operating principles, and applications are also substantially different. Several benefits are gained by operating at such high temperatures: the anode and cathode reaction kinetics are extremely facile, so much so that high performance is achieved without the need for precious metal catalysts. The fuel cell shows high tolerance to impurities such as CO, which occur in the reformate stream. In fact, CO actually constitutes a fuel for the SOFC producing a voltage very similar to the hydrogen reaction. It is also possible to exploit the high temperature exhaust from the SOFC to drive a gas turbine or a steam turbine (commonly called a "bottoming cycle") and obtain even higher combined efficiencies. Most importantly, it is possible to feed natural gas directly to the SOFC. Due to the presence of water and high temperatures, natural gas is steam-reformed within the SOFC anode to H_2 and CO, which are then combined electrochemically with O_2 to produce H_2O and CO_2. The conversion of natural gas to H_2 and CO within the SOFC is called direct internal reforming. The major benefit of this feature is that a natural gas distribution network already exists all over the United States. Hence, the bottleneck imposed by the lack of a hydrogen infrastructure is completely circumvented in this case.

The high operating temperatures imply that SOFC systems require an elaborate start-up and shut-down procedure, which generally precludes their use in automobiles and portable electronics applications. Instead, they are typically designed for stationary power applications and can provide continuous, reliable power for buildings. Second, the high temperatures also pose materials challenges for SOFC. The electrolyte, anode and cathode are all fabricated from ceramic materials. Maintaining effective sealing is also difficult due to the high temperatures.

Operating Principles, Materials and Architecture

Figure 4 shows the construction of the SOFC and the relevant reactions. Here, the depicted anode feed gas is H_2, so the anode and cathode reactions pertain to the conversion of H_2 to H_2O. The electrolyte is a solid ceramic material composed of yttrium-stabilized zirconia (zirconia doped with a small amount of yttrium),

Figure 4
Architecture and Operating Principle of a Solid Oxide Fuel Cell

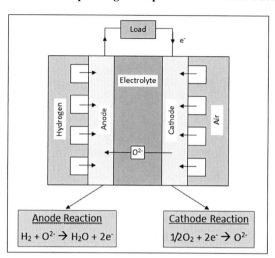

which renders it an oxide-ion conductor at high temperatures. The anode is a porous nickel-zirconia cermet, and the cathode is a porous p-type semiconductor composed of strontium-doped lanthanum manganite.

As shown in Figure 4, at the cathode, oxygen accepts electrons arriving from the external circuit to form oxide ions. The oxide ions dissolve in the YSZ electrolyte and diffuse through it to arrive at the anode, where they combine with H_2 to form H_2O with the simultaneous release of electrons. Due to the high temperatures, the produced water is exclusively in the gas phase, which simplifies its interaction with the solid porous matrix. While Figure 4 depicts the hydrogen reaction, it is also possible to use CO as a fuel at the anode. The corresponding anode reaction is $CO + O^{2-} \rightarrow CO_2 + 2e^-$, while the cathode reaction is unchanged. As stated earlier, natural gas (CH_4) can be fed directly to the SOFC anode. In the presence of water, CH_4 under goes a reformation reaction given by: $CH_4 + H_2O \rightarrow CO + 3H_2$. Both CO and H_2 serve as fuel inputs to the SOFC and their respective electrochemical reactions produce nearly identical voltages.

A number of SOFC designs exist. The planar design resembles the construction depicted in Figure 4 with plate-like anode, electrolyte, and cathode layers manufactured by plasma spraying or electrochemical vapor deposition. Depending on the manufacturing method, the thickness of the various layers can range from tens of microns to about 1 mm. Again, it is critical to reduce the thickness of the layers to minimize ohmic losses. While this geometry minimizes the electron travel path between cells and yields high power densities, it poses challenges for effective sealing at its edges. The tubular design in which the anode, electrolyte and cathode consist of overlaid layers in a tubular geometry is also common. The fuel is fed to the outermost layer (the anode), while air is injected along the tube centerline to the innermost layer (the cathode). The advantage of the tubular geometry is that sealing is simplified, however, long electron paths between cells connected in series increase ohmic losses and reduce the overall power density. A third geometry, known as the monolithic design has been proposed that employs a highly three-dimensional internal architecture to produce very high power densities.

Current Market and Future Prospects

SOFC systems are generally large, are primarily designed for the stationary power sector with a size range approaching several megawatts, and owing to their high reliability and low emissions, they are attractive for critical load applications. Such systems fall into the distributed generation (DG) category. DG avoids transmission losses, reduces grid congestion, peak demand charges, and is compatible with various "smart-grid" concepts. SOFC systems also provide combined heat and power (CHP) with very high efficiencies.

Stationary fuel cell shipments grew by 10 percent between 2009 and 2010 reaching 7,400 units.[38] These shipments encompass micro-CHP and uninterrupted power systems all the way to multi-megawatt prime power installations. Not all of these systems belong to the SOFC category. Some smaller stationary systems belong to the PEM type (for example, ClearEdge Power). Two other fuel cell types (not reviewed in this article), namely molten carbonate fuel cells (MCFC) and phosphoric acid fuel cells (PAFC) are of the larger class and have also contributed to sales in this category. FuelCell Energy in partnership with POSCO Power in South Korea markets MCFCs; it has installed more than 40 MW of capacity in South Korea.[39] POSCO Power has announced plans to build the world's largest fuel cell installation with a size of 60 MW by 2013.[40] UTC Power's 400 kW PureCell belongs to the PAFC type. Several of these systems are providing power to supermarkets, and commercial and residential properties across the United States.

Bloom Energy created a major splash in the media in 2010 with the announcement of their SOFC device, the Bloom Box; current customers include Adobe, AT&T, Bank of America, Coca-Cola, eBay, FedEx, Google, Safeway, Staples, Walmart, and many more. Bloom's device runs on natural gas or biogas. An interesting business model employed by Bloom is that customers can purchase the power produced by the Bloom Boxes without actually purchasing the device itself. Bloom Energy continues to own and operate the units, while the customers sign a contract to buy the energy produced for ten years at a fixed rate. By eliminating the upfront capital expense of the Bloom Box, this financial model eases the financial burden of entry into the market.

Similar to automotive fuel cells, stationary fuel cells must be price-competitive with conventional technologies. However, the acceptable price point of stationary fuel cells is considerably higher than that of automotive fuel cells, primarily because stationary systems are required to exhibit lifetimes (80,000 hours) that are an order of magnitude longer than for transportation applications. Table 3 lists some of the DOE targets for large scale stationary systems.[41] SOFC systems have demonstrated lifetimes in excess of 25,000 hours. The high operating temperatures of SOFC pose problems for long-term durability due to compatibility issues between stack components, and well as repeated thermal cycling. Much of the research effort in SOFCs today is to discover and develop electrolyte and electrode materials that can provide the required fuel cell performance (oxide conductivity) at operating temperatures as low as 600 °C.

The Hydrogen Infrastructure

Although hydrogen is the most abundant element in the universe, it does not exist in its elemental state on earth. Instead it occurs as a compound such as water, or in the form of hydrocarbons. Today, the most cost-effective and common way to make hydrogen is from hydrocarbons, primarily by steam methane

Table 3
Technical Targets for 100 kW–3 MW Combined Heat and Power and Distributed Generation Fuel Cell Systems Operating on Natural Gas[42]

Characteristic	2011 status	2015 targets	2020 targets
Electrical efficiency at rated power	42–47%	45%	>50%
CHP energy efficiency	70–90%	87.5%	90%
Equipment cost, natural gas	$2,500–$4,500/kW	$2,300/kW	$1,000/kW
Installed cost, natural gas	$3,500–$5,500/kW	$3,000/kW	$1,500/kW
Equipment cost, biogas	$4,500–$6,500/kW	$3,200/kW	$1,400/kW
Installed cost, biogas	$6,000–$8,000/kW	$4,100/kW	$2,100/kW
Number of planned/ forced outages over lifetime	50	50	40
Operating lifetime	40,000–80,000 h	50,000 h	80,000 h
System availability	95%	98%	99%

reforming (SMR).[43] Large centralized SMR plants produce most of the hydrogen used today to power automobiles and for other industrial uses. Unfortunately, the use of methane (which is the principal component of natural gas) to make hydrogen is not sustainable; natural gas, as any fossil fuel, is a limited resource. Furthermore, the process of converting any hydrocarbon to hydrogen generates CO_2 as a byproduct, which is simply released to the atmosphere. Hence, both the labels commonly associated with fuel cells—"clean" and "sustainable"—are called into question when the source of hydrogen is a fossil fuel. According to the DOE[44] the distributed generation of hydrogen by reforming natural gas or liquid fuels (including bio-derived fuels such as ethanol and bio-diesel) have good potential for development. Of these, small-scale natural gas reformers are closest to meeting hydrogen production cost targets.

The other option is to extract hydrogen from water. Typically, this is done by electrolysis. Electrolysis is energy intensive, and requires the input of electricity. Electricity supplied by the grid is largely based on coal and natural gas with an average efficiency of only around 33 percent[45] and the concomitant release of CO_2 and other emission products. Hence, again, the labels "efficient", "clean", and "sustainable" are jeopardized when grid electricity is employed to generate hydrogen. One electricity pathway that avoids the use of fossil fuels is nuclear power. It can be argued that converting nuclear power to hydrogen preserves some of the positive labels associated with fuel cells. However, nuclear power continues to be controversial owing to the problematic disposal of its radioactive waste, and the recent disaster in Fukushima has rekindled safety concerns.

Another sustainable pathway for hydrogen is the use of electric power from renewables including wind, solar, hydro, etc., to electrolyze water. The hydrogen produced by such means could be stored locally in pressurized tanks, or distributed off-site by trucks or pipeline to supply fuel cell devices. In addition, the stored hydrogen can also perform a second function that is particularly relevant for renewable sources. One of the inherent difficulties associated with renewable power sources is their intermittency, which leads to large and unpredictable fluctuations in the output power. It has been proposed that such fluctuations could be smoothed out by coupling the renewable source to large-scale electric storage systems such as batteries. As an alternative to batteries, it is also possible to store electricity in the form of hydrogen. During periods of excess power, some of the hydrogen generated by electrolysis could be stored in buffer tanks. During lean periods, the hydrogen could be passed back through a fuel cell to generate electricity and thus provide a load-leveling function to the grid.

It is useful to examine some conversion efficiency numbers for producing hydrogen from renewables. Let us consider the case where the renewable power is being produced by photovoltaic panels. The corresponding efficiency for commercial PV systems is about 15 percent. The efficiency of electrolysis is about 70 percent. Therefore, the combined efficiency for converting sunlight into hydrogen via this pathway is only about 10 percent.

Other renewable pathways also exist that do not involve solar electricity although these are somewhat further away from commercialization. One pathway, known as the thermochemical conversion of sunlight to hydrogen, exploits solar thermal energy.[46,47] Thermochemical reactors are high temperature reactors constructed out of ceramic materials. They contain an aperture through which sunlight concentrated by a factor up to 10,000 is focused into a cavity raising its internal temperature to around 2,000 K. At such high temperatures, it is possible to drive certain reactions that produce hydrogen by thermochemical water splitting. One example of such a process is the zinc oxide thermochemical cycle, which consists of two steps. In step 1, ZnO is dissociated into Zn and O_2 in the presence of concentrated sunlight at temperatures around 2,000 K. The oxygen is typically vented from the reactor and the zinc vapor is rapidly quenched, separated and saved. In step 2, the Zn from step 1 is reacted with steam in a second, lower temperature reactor to produce H_2 and ZnO. The two steps are illustrated below:

Step 1: $ZnO \rightarrow Zn + 1/2O_2$
Step 2: $Zn + H_2O \rightarrow H_2 + ZnO$

As shown, Zn from step 1 is supplied to step 2, and ZnO from step 2 is recycled back to step 1. Hence, the net inputs are sunlight and water, and the net outputs are H_2 and O_2, of which the H_2 is stored for further use. The promise of this method is that the theoretical thermochemical conversion efficiency of sunlight to hydrogen can approach 40 percent.

Thermochemical cycles are currently being researched in laboratories and a 100 kW system is currently being demonstrated at the pilot plant scale. It is possible to conceive of thermochemical plants where a large array of mirrors focuses sunlight into a reactor situated atop a central receiving tower and hydrogen is produced at an industrial scale. However, thermochemical devices have yet to be demonstrated outside of the laboratory, and the commercialization aspects remain to be worked out. A number of engineering issues including the design of high-temperature materials, heat transfer, reaction kinetics, ZnO powder handling, high-temperature gas separations, etc., must be solved prior to commercialization.

Another sustainable pathway for hydrogen production is by photobiological water splitting. Certain microbes such as cyanobacteria already exist in nature that consume water in the presence of sunlight and emit hydrogen and oxygen. Unfortunately, the conversion efficiency is so low that vast areas would have to be brought under cultivation to make reasonable amounts of hydrogen. Furthermore the simultaneous release of oxygen with hydrogen reduces efficiency and complicates hydrogen collection. Therefore, although this pathway is sustainable and promising, it involves land use issues, and the prospects for commercialization are as yet unresolved. Research in this field is currently focused on improving the efficiency of the photobiological conversion, for example, by genetic engineering.[48]

Summary

This article has presented an overview of fuel cell technology. The basic principles of fuel cells have been described along with their three main application areas including transportation, portable power and stationary power. The internal architectures, component materials and process reactions of polymer electrolyte membrane fuel cells, direct methanol fuel cells, and solid oxide fuel cells have also been presented. For each type of fuel cell, the current market status and future prospects are listed. Fuel cells offer the promise of clean, efficient and reliable power for cars, buses, portable electronics, and homes and businesses. In a limited number of application areas, this promise is already being fulfilled and commercialization has been achieved. However, there is tremendous opportunity for growth if challenges arising from cost, durability and the hydrogen infrastructure can be successfully addressed. The recent trends for all of these metrics are very positive, and it is expected that fuel cells will continue to find increasing success in future years.

Notes

1. Bacon, F. T, and Fry, T. M., "The Development and Practical Application of Fuel Cells," *Proceedings of the Royal Society of London. Series A, Mathematical and Physical Sciences* 334 (1973): 427–452.

2. Warshay, M., and Prokopius, P. R., "The Fuel Cell in Space: Yesterday, Today and Tomorrow," NASA Technical Memorandum 102366, 1989.

3. Dismukes. K., "Fuel Cell Power Plants," National Air and Space Administration, April 2002, http://spaceflight.nasa.gov/shuttle/reference/shutref/orbiter/eps/pwrplants.html (accessed July 1, 2012).

4. The 2011 Fuel Cell Patent Review. Fuel Cell Today, (2011) Johnson Matthey PLC.

5. Ibid.

6. Ibid.

7. 2010 Fuel Cell Technologies Market Report. Breakthrough Technologies Institute. (June 2011).

8. Ibid.

9. Ibid.

10. Ibid.

11. Ibid.

12. Bruijn, F. A. de, Dam, V. A. T, and G. J. M. Janssen, "Review: Durability and Degradation Issues of PEM Fuel Cell Components," *Fuel Cells* 8, no. 1 (2008): 3–22.

13. Huang, X., Solasi, R., Zou, Y., Feshler, M., Reifsnider, K., Condit, D., Burlatsky, S, and Madden, T., "Mechanical Endurance of Polymer Electrolyte Membrane and PEM Fuel Cell Durability," *Journal of Polymer Science: Part B: Polymer Physics* 44 (2006): 2346–57.

14. Bazylak, A., "Liquid Water Visualization in PEM Fuel Cells: A review," *International Journal of Hydrogen Energy* 34 (2009): 3845–57.

15. Srouji, A.-K., and M. M. Mench, "Freeze Damage to Polymer Electrolyte Fuel Cells," In *Polymer Electrolyte Fuel Cell Degradation*, ed. M. M. Mench, E. C. Kumbur, and T. N. Veziroğlu, (Elsevier Inc., 2012), 293–333.

16. Asensio, J. A., Sanchez, E. M., and P. Gomez-Romero, "Proton-Conducting Membranes Based on Benzimidazole Polymers for High-Temperature PEM Fuel Cells. A Chemical Quest," *Chemical Society Reviews* 39 (2010): 3210–3239.

17. Bose, S., Kuila, T., Nguyen, T. X. H., Kim, N. H., Lau, K.-T., and Lee, J. H., "Polymer Membranes for High Temperature Proton Exchange Membrane Fuel Cell: Recent Advances and Challenges," *Progress in Polymer Science* 36 (2011): 813–843.

18. Debe, M. K., "Electrocatalyst Approaches and Challenges for Automotive Fuel Cells," *Nature* 486 (2012) : 43–51.

19. Bashyam, R., and Zelenay, P., "A Class of Non-Precious Metal Composite Catalysts for Fuel Cells," *Nature* 443 (2006): 63-66.

20. Xu, Y., Shao, M., Mavrikakis, M, and Adzic, R.R., "Recent Developments in the Electrocatalysis of the O_2 Reduction Reaction," In *Fuel Cell Catalysis*, ed. Marc T. M. Koper (John Wiley and Sons, Inc., 2009), 271–315.

21. Ahluwalia, R. K., Wang, X., Kwon, J., Rousseau, A., Kalinoski, J., James, B, and Marcinkoski, J., "Performance and Cost of Automotive Fuel Cell Systems with Ultra-Low Platinum Loadings," *Journal of Power Sources* 196, no. 10 (2011): 4619–30.

22. Hermann, A., Chaudhuri, T., and Spagnol, P., "Bipolar Plates for PEM Fuel Cells: A Review," *International Journal of Hydrogen Energy* 30 (2005): 1297–302.

23. Tawfik, H., Hung, Y., and Mahajan, D., "Metal Bipolar Plates for PEM Fuel Cell— A Review," *Journal of Power Sources* 163 (2007): 755–67.

24. Yang, J., Sudik, A., Wolverton, C, and Siegel, D. J., "High Capacity Hydrogen Storage Materials: Attributes for Automotive Applications and Techniques for Materials Discovery," *Chemical Society Reviews* 39 (2010): 656–75.

25. Sakintuna, B., Lamari-Darkrim, F., and Hirscher, M., "Metal Hydride Materials for Solid Hydrogen Storage: A review," *International Journal of Hydrogen Energy* 32 (2007): 1121–40.

26. Fuel Cell Vehicles (From Auto Manufacturers). Fuel Cells 2000. Breakthrough Technologies Institute (BTI), January 2011, http://www.fuelcells.org/info/charts/carchart.pdf (accessed 17 May 2012).

27. Department of Energy Targets, California Fuel Cell Partnership, http://cafcp.org/progress/technology/doetargets (accessed July 3, 2012).

28. Ibid.

29. See Note 7.

30. See Note 27.

31. Bubna, P., Brunner, D., Gangloff, J. J., Advani, S.G., and Prasad, A. K., "Analysis, Operation and Maintenance of a Fuel Cell/Battery Series-Hybrid Bus for Urban Transit Applications," *Journal of Power Sources*, 195 (2010): 3939–49.

32. International Fuel Cell Bus Collaborative. http://www.gofuelcellbus.com/, 2012 (accessed July 3, 2012).

33. The Fuel Cell Today Industry Review 2011. Fuel Cells Today. © Johnson Matthey PLC. 2011.

34. Ibid.

35. LaMonica, M., "Apple Fuel Cell Patent Applications Envision 'Weeks Without Refueling'," December 2011, © 2012 CBS Interactive. http://news.cnet.com/8301-11128_3-57347294-54/apple-fuel-cell-patent-applications-envision-weeks-without-refueling/ (accessed July 5, 2012).

36. See Note 33.

37. Technical Plan-Fuel Cells. DOE, http://www1.eere.energy.gov/hydrogenandfuelcells/mypp/pdfs/fuel_cells.pdf (accessed July 3, 2012).

38. See Note 33.

39. See Note 33.
40. See Note 33.
41. See Note 37.
42. See Note 37.
43. Simpson A.P. and Lutz, A.E. "Exergy Analysis of Hydrogen Production via Steam Methane Reforming," *International Journal of Hydrogen Energy* 32 (2007): 4811–20.
44. Technical Plan-Production. DOE. http://www1.eere.energy.gov/hydrogenandfuelcells/mypp/pdfs/production.pdf (accessed July 3, 2012).
45. Amin, M., and Stringer, J., "The Electric Power Grid: Today and Tomorrow," *MRS Bulletin* 33 (2008): 399–407.
46. Steinfeld, A. "Solar Thermochemical Production of Hydrogen-A Review," *Solar Energy* 78, no. 5 (2005): 603–15.
47. Perkins, C., and Weimer, A., "Solar-Thermal Production of Renewable Hydrogen," *AIChE Journal* 55, no. 2 (2009): 286–93.
48. McKinlay, J. B., and Harwood, C.S., "Photobiological Production of Hydrogen Gas as a Biofuel," *Current Opinion in Biotechnology* 21 (2010): 244–51.

13

Nuclear Power: Is It Worth the Risks?

Kathleen M. Saul and John H. Perkins

The major challenge of a green energy economy lies in finding "promising new technologies" that can replace fossil fuels. Thus, a major question arises, "Can we label nuclear power a promising new technology?" or "Is nuclear power 'green' and thus a valuable part of the technology portfolio that should replace fossil fuels?"

Advocates of nuclear power point to its ability to generate electricity with low-carbon emissions, its established role as a base-load power source in numerous countries, its plentiful supplies of fuel, and they argue the existence of new technology and operational practices that render nuclear power clean, green, cost-effective, and safe. Thus, nuclear power deserves a prominent place in the green energy economy. Opponents question claims of low-carbon emissions over the total nuclear fuel cycle and life cycle of the plant, as well as assertions about cost-effectiveness and safety. Nuclear power in this school of thought deserves no, or at most a minimal, role in the green energy economy.

Because of its large scale and capital intensity, questions about the inclusion of nuclear power in the green energy economy demand a "yes" or "no" answer, not "maybe" or "sort of." The middle path is not feasible. As a result, governments considering the nuclear option face Hamlet's pivotal dilemma—"to nuke or not to nuke."

To grasp the arguments surrounding this dilemma requires understanding the history of nuclear power in the United States, the country that shaped and led the evolution of this technology. The most important feature of this history is that decision making on nuclear power has never been simply a matter of science and engineering. While technical matters always occupied a place in debates about nuclear power, value judgments and politico-economic factors have also played important roles. Moreover, even the scientific and engineering communities never reached a consensus on the methods for evaluating the technology. Today's debates are no different. Decision makers and citizens continue to face the same morass of technical and nontechnical issues that have always bedeviled nuclear power.

After reviewing the history of this technology, this chapter provides a critical analysis of examples of policy recommendations that favor or oppose, respectively, nuclear power as an important component of the world's new energy economy. Each of these assessments shows how the respective policy recommendations considered or failed to take into account the lessons from the history of nuclear power.

Finally, the chapter suggests an integrated framework—Political Ecology—that can incorporate qualitative and quantitative data in discussions of nuclear power. Political Ecology will not lead to a scientifically unambiguous answer about nuclear power. However, the framework encompasses the lessons of history, value judgments, concerns about safety, financial investment, and electrical production, allowing political and business leaders and citizens to better understand the dilemmas surrounding energy choices and to more well-informed choices.

Changing Patterns in Decision Making on Nuclear Power, 1939–Present

As described below, decision making about nuclear energy occurred in six phases in the United States, each marked by a shift in policies and politics or in the economic and social contexts surrounding governmental control and promotion of nuclear technologies.

Phase I: The Beginning of the "Atomic" Age, 1939–1954

Seldom has a scientific discovery moved from the laboratory bench to a world-altering technology as fast as the discovery that the uranium atom could fission, releasing huge amounts of energy. That part of the nuclear story began in 1939 and culminated in the dropping of atomic bombs in 1945. Under the Atomic Energy Act of 1946, the Atomic Energy Commission (AEC) launched efforts to vastly expand the nuclear arsenal of the United States and to explore the potential for electrical generation. In 1951, the United States became the first country to generate electricity from this energy source. By 1957 two commercial scale plants had come on line: Santa Susana, CA (20 MWe maximum output) and Shippingport, PA (60 MWe).[1]

This early history of nuclear power in the United States established a number of features about decision making that persist to this day: (a) a high reliance on federal government support in the form of research, subsidies, and regulation; (b) close links between civilian and military issues, due to commonality of technical skills, and because the fission of uranium fuel inevitably produces plutonium, a component of nuclear weapons; (c) an aura of secrecy and security concerns about the technology; and (d) a vision that prowess in nuclear technology confers diplomatic, military, and commercial strength in both the domestic and international political arenas.

Decision making during Phase I focused primarily on the desire for weapons, followed by the pursuit of international leadership in nuclear technology for electric power. Issues of energy security, pollution from competing fuels, or climate change, while present, had no significant voice in the military/political machinery that created the US program in nuclear energy.

Phase II: Origins of Nuclear Power as a Private-Public Partnership, 1954–1962

The Atomic Energy Act of 1954 allowed the US government to share secrets with private companies as a way to promote the development of nuclear power.[2] Despite high expectations for the new law, the eight years following its passage brought only small progress toward the envisioned new era of electricity generated by nuclear fission.

One challenge stemmed from the inability and unwillingness of most private companies to accept the liability for accidents that might release significant amounts of radiation. Industry spokesmen made it clear that neither manufacturers nor utilities would embrace this new technology unless the federal government indemnified them from that liability. As a result, in 1957, the Price-Anderson Nuclear Industries Indemnity Act amended the Atomic Energy Act of 1954 and limited the liability claims from an accident to $560 million.[3] Private insurance held by the reactor owners/operators covered the first $60 million; Congress would pay additional damages up to $500 million (1957 dollars). Compensation for any damages beyond that level would require further action by Congress.

Congress then asked the AEC to estimate the maximum damage that might occur from an accident with a nuclear reactor. In 1957, AEC reported that a major accident could lead to up to 3,400 deaths, 43,000 injuries, or property damages valued at up to $7 billion.[4] AEC scientists felt an accident of such magnitude was extremely unlikely; still, they shied away from estimating the actual chances of such an event.

The legacy Phase II includes the public-private partnership as the major pattern of ownership in the US nuclear power industry. While public entities do operate a few of the nuclear power plants in the United States (Tennessee Valley Authority (TVA), for example), government supported private industry has dominated and looks to continue doing so into the foreseeable future.

Second, the inherent safety problems of the technology became known. Knowledgeable scientists predicted catastrophic nuclear accidents; their projections were confirmed years later by accidents at Three Mile Island, Chernobyl, and Fukushima. Even so, passage of the Price-Anderson Act signaled that a majority in Congress believed that the probabilities of major accidents were low enough to proceed. Private insurers have been less confident. As a result, renewable of the Price-Anderson Act persists as a fundamental condition for private investment in nuclear power.[5]

Third, the light water reactor (using water as the reaction moderator) became the predominant technology for the US industry. Despite continued efforts by the engineers to develop radically different designs, light water reactors continue to be the preferred technology.

Phase III: Nuclear Power Enjoys a Boom and Suffers a Bust, 1962–1978

By early 1962 only a handful of commercial plants had begun operations: Dresden I (867 MW per unit, outside Chicago, 1959), Yankee Rowe (600 MW, in northwestern Massachusetts, 1960) and Indian Point I (275 MW, outside New York City, 1962).[6] The AEC's Congressional overseer, the Joint Committee on Atomic Energy, became impatient with the pace of developments and persuaded President John F. Kennedy to ask the AEC for a forward-looking document. The ensuing report, *Civilian Nuclear Power . . . a Report to the President—1962* dramatically changed the atmosphere surrounding nuclear power and catalyzed the transformation of AEC into a vigorous proponent of nuclear power.[7]

The AEC proposed nuclear power as the answer to soaring national demand for electricity. Between 1960 and 2020, total electricity consumption was expected to increase twelve-fold.[8] Nuclear proponents envisioned 50 percent of total US electricity coming from nuclear power by the year 2000, and about 90 percent by 2020. In addition, a new "breeder reactor" would generate more fuel than it burned, vastly enhancing supplies of uranium fuel, and reducing problems of waste disposal through spent fuel reprocessing.

The pace of construction picked up remarkably. During the "boom" of the US nuclear power industry from 1967 to 1978, the AEC issued ninety-eight construction permits and oversaw the construction of a fleet of 133 power reactors. However, no commercially successful breeder reactor has yet come on line.

Despite the apparent success of the nascent industry, a period that began with high hopes for a new technology ended with bickering, dissolution of the AEC, and civil disobedience at construction sites. The industry never came close to transforming the US electrical industry to predominantly nuclear technology. In fact, the experiences with the construction of nuclear power plants highlighted a number of challenges that collectively devastated the confidence of investors, many political leaders, and substantial numbers of citizens. First, companies and utilities rapidly scaled-up reactor sizes from less than 100 MW in the 1950s, to near or more than 1,000 MW by the mid-1960s. They sought to capture important economies of scale through increased size, but instead outran the skill base of construction workers and the knowledge of regulators.[9] As a result, many projects floundered in re-work, lengthy delays, and budget overruns as contractors tried to meet the AEC's changing requirements.

Environmental issues created problems. Nuclear power had been viewed as cleaner than existing coal-fired operations, which were coming under pressure to reduce emissions by the Clean Air Act of 1970. However, by mid-1971, concerns about thermal pollution of discharges from nuclear plants surfaced. Thermal pollution could have been resolved by the installation of cooling towers for dissipating the waste heat from the reactors. But even today, only 62 of 104 operating reactors rely on cooling towers.

A lack of confidence in engineered safety features also generated complaints. No person knowledgeable about nuclear technology had denied the potential for catastrophic risk, but proponents had argued that the combination of (a) remote siting; (b) redundant engineered safety features; and (c) proper design, construction, operation, and maintenance made the risks "acceptably" low.

The AEC attempted to demonstrate the acceptability of these risks by developing a new method for assessing them: probabilistic risk assessment (PRA). Before PRA, the AEC mandated sites and reactor designs that would not expose any *real person* to radiation above the limits considered tolerable in emergencies. Proponents of PRA suggested that regulations should aim to reduce the probability of serious injury or death to a *statistical person* below a level considered insignificant. "Safe enough" became a calculated probability of harm that political leaders could accept as "low enough."[10]

Other problems lay in the radically changing energy outlook in the 1970s. Electrical industry and utility analysts wrongly assumed that the rate of high growth of the 1950s and 1960s would continue indefinitely. When actual growth rates dropped, companies cancelled or delayed new nuclear construction. Sixty-three nuclear plants fell by the wayside even after having received construction permits.

New frameworks for thinking about energy also emerged in the 1970s and partially eclipsed the framework upon which *Civilian Nuclear Power* had relied. The Ford Foundation study, *A Time to Choose* (1974), identified efficiency as the most important factor in planning for the energy economy.[11] Spurred by the Arab oil embargo of 1973, *A Time to Choose* argued that energy policy should focus first on minimizing demand, not maximizing supply. The virtually unlimited supply of electricity generated by nuclear power was no longer the key consideration in discussions about energy.

Finally, controversies swirling around the AEC as both champion and watchdog of the nuclear industry led Congress to break the agency into two parts in 1975. The Nuclear Regulatory Commission (NRC) would regulate, while the Energy Research and Development Administration (later the Department of Energy) would promote nuclear power.

Significant segments of the public voiced concern over the legacies of Phase III. The security of one's health, family, home, community, and place, all could be threatened by a nuclear accident. Costs and efficiency affected individual pocket-books directly, in utility bills, or indirectly, through government and tax-payer subsidies to the industry. In addition, low investor confidence challenged the premise that nuclear power should be (or could be) a public-private partnership. These issues from Phase III continue to play prominent roles in today's nuclear energy debates.

Phase IV: Accidents and Waste Management Tarnish the Image of Nuclear Power, 1979–1988

Two major nuclear accidents occurred during Phase IV. In 1979, Three Mile Island—2, near Harrisburg, PA, suffered a major loss of coolant accident, partial core meltdown, and release of considerable radioactivity, mostly in the form of gases. In 1986, Chernobyl—4 near Kyiv, Ukraine, suffered a spike in reactivity followed by steam and/or hydrogen explosions, and fire. Massive amounts of radioactive debris contaminated Ukraine, Belarus, Russia, and many countries of central and Western Europe. An Exclusion Zone approximately the size of Rhode Island surrounds the ruined reactor to this day.

The Nuclear Waste Policy Act of 1982 acknowledged a growing problem associated with the disposal of spent fuel from nuclear reactors and the need to "promote public confidence in the safety of disposal of such waste."[12] The Act outlined the responsibility of the US government to select and develop the site(s) for a permanent waste disposal facility. Yucca Mountain, NV ultimately emerged as the first choice site for the spent-fuel repository. As of this writing, however, political opposition from the State of Nevada and the President has left the choice of repository site unresolved.

The major legacies of Phase IV were the preeminence of safety concerns and the intractability of waste management. Earlier knowledge of large accidents stemmed from theoretical studies, modeling, and statistical calculations of risks. Three Mile Island and Chernobyl provided empirical verification of the catastrophic potential of mishaps. Still, proponents of nuclear power in the United States rationalized the outcomes of the accidents by claiming that Three Mile Island demonstrated that safety features, especially containment buildings, worked as planned. United States and Western European proponents dismissed the significance of Chernobyl, because the reactor had no containment building and because the accident stemmed from an ill-advised and ill-timed experiment, not commercial operation. Proponents of nuclear power put their faith in engineered safety features and technical solutions to waste disposal problems. Many members of the general public, however, did not embrace this faith.

Phase V: New Policies Fail to Promote New Plants, 1989–2005

By 1989, proponents of nuclear power again saw steady increases in the overall US demand for electric power as motivation for the revival of the quiescent nuclear industry. First, they needed to remove two barriers to this goal: the lack of standard designs for nuclear power plants, and the cumbersome process under which utilities needed to obtain a construction permit first, and apply for an operating license only once the construction ended. Proponents believed that standardized designs and a one-step licensing process would relieve the uncertainties and risks of building new plants.

The NRC already had begun establishing a process to standardize designs. The agency believed that nuclear engineers would submit a handful of designs for advanced certification. The designs would be for essentially complete, except for necessary site-specific elements, such as cooling water intake structures. Up front safety reviews and public hearings would produce certified designs from which utilities could choose.[13]

In 1992 Congress followed with amendments to the Atomic Energy Act of 1954 allowing early site approval and the issuance of combined construction and operating licenses.[14] Utilities would first get approval regarding the hydrological, geological, seismic, and meteorological features of a proposed site. Subsequent application for the construction-operation license, combined with selection of a standard, pre-approved design, would lead to faster and cheaper completion of safer plants.

Contrary to expectations, by 2005 not a single company had stepped forward to build a new nuclear power plant. Instead, utilities met increased demand by investing in projects to improve efficiency of electricity use and to increase the capacity factor of existing plants. Companies built some new power plants during this period, but they used coal, natural gas and, in a few places, wind, solar, and geothermal resources. Advocates of a "nuclear renaissance" remained frustrated.[15] Resumption of active construction of new nuclear plants required still something else to bring utilities back to nuclear power.

Consumers and citizens mostly forgot about the nuclear industry after the 1990s. Fears inspired by Three Mile Island and Chernobyl faded from memory, and most people outside the nuclear and utility industries had little knowledge or curiosity about the source of their electricity.

Phase VI: More New Policies Stimulate Proposals for New Plants, 2005–Present

The presidential election of 2000 generated more change in nuclear power than the industry had seen since the late 1970s. President George W. Bush appointed Vice President Dick Cheney to head a task force to forge a new national energy policy. Their report, dated May 2001 and entitled *National Energy Policy Report of the National Energy Policy Development Group: Reliable, Affordable, and Environmentally Sound Energy for America's Future*, strongly embraced the project of invigorating the American nuclear industry.[16] Before the end of 2001, the US Department of Energy followed with its report, *A Roadmap to Deploy Nuclear Power Plants in the United States by 2010.*[17] *A Roadmap* recommended financial incentives to motivate design and construction projects. It also put into place a 50–50 cost sharing program to help the first movers demonstrate the NRC's revised site permitting and reactor licensing procedures.

NuStart Energy Development, a company formed in 2004 by ten power companies and two reactor vendors, sought to devise standard methods for preparing applications for the permits needed for constructing new nuclear power plants. As a result, NuStart received $260 million under the Department of Energy's cost sharing program as a "first mover."

In spite of the promotional steps advanced by the Bush-Cheney Administration, additional stimulus had to be added to generate investment commitments in new nuclear power plants. The Energy Policy Act of 2005 (EPAct) contained a number of initiatives that finally prompted applications for construction-operating licenses.[18] One of these was a production tax credit of 1.8 cents per kilowatt-hour for the first 6,000 megawatts of installed capacity, provided the application for construction-operation arrived by the end of 2008 and construction began before 2014.

Next, the EPAct again renewed the Price-Anderson Indemnity Act of 1957 and extended its expiration date to December 31, 2025. Under the new revision, nuclear power plant operators had to obtain $300 million per plant in liability insurance and contribute another $10 million annually to an industry pool. Congress would supplement the funds available by indemnifying all other liability up to approximately $10 billion.

The EPAct also created a "delay risk insurance" policy for the power companies. It authorized the Department of Energy to cover part of the cost of delays due to changing regulations and lawsuits brought by opponents of nuclear power. This protection against delays would equal up to $500 million for the first two new nuclear power plants, and up to $250 million for each of the following four new plants.

The final incentive from EPAct came in 2007: loan guarantees. As initially conceived, the federal government would insure loans to the builders of new plants for up to 80 percent of the total financing for a plant—the Department of Energy later amended this amount to 100 percent of the debt obligation for the plant.[19] This provision reduced the risks to lenders, a strong incentive for private funds to flow into the nuclear enterprise once again. Financial analysts convinced Congress that, without loan guarantees, banks and Wall Street investors would not support construction of new plants.

Retrospectively, loan guarantees almost certainly were the missing link in federal policy from 1989 to 2005.[20] Before the existence of the loan guarantee program, no applications for construction-operation permits arrived at the NRC. After the loan guarantee program came into effect, applications began to arrive. The Department of Energy issued the first loan guarantee for $8.3 billion in February 2010, for Georgia Power's project to build two new AP1000 Westinghouse pressurized water reactors at the Vogtle Nuclear Power Plant near Waynesboro, Georgia.[21]

Guaranteed financing, in the form of loan guarantees or Construction Work in Progress financing (CWIP), also provided the incentive for some utilities to take interest in the construction of new nuclear power plants. For example, Progress Energy Florida was able to garner support for its plans to charge Florida customers in advance of its nuclear plant construction. As of January 2009, Progress Energy has been recovering costs for its planned Levy County facility—costs including those related to the construction of the plant itself, nuclear related transmission expenses, and the annual expensing of pre-construction costs, such as costs related to site selection.[22] Since Progress Energy did not apply for loan guarantees from the DOE, being able to recoup some of the costs up front was crucial for the construction plans to move forward.

The legacies of Phase VI lie in the institutional world of power companies, reactor manufacturers, the finance industry, the US Department of Energy, the US Nuclear Regulatory Commission, and state utility regulatory agencies. Overwhelmingly, the evidence indicates that the private finance industry will not touch investment in new nuclear power plants unless the financial risks are reduced to close to zero. Federal loan guarantees accomplish this by assuring the lenders that, in the event of default, the lenders will still receive their money back from taxpayers. Even without federal loan guarantees, a utility company can finance a plant provided their state regulatory agency allows the company to increase rates to cover the expected costs. In essence, the utility obtains the necessary funds from customers with no requirement to pay any interest.

The legacies left by Phases I–VI still shape the assessment of nuclear power. Four driving features are particularly noteworthy: (a) Nuclear power plants operate under an all-encompassing shroud of public policy; no major investments in nuclear power would have ever been made without strong support from government. (b) The transferal of health and financial risks associated with nuclear accidents from the power plant owners and operators to taxpayers and citizens became a part of the industry standard with passage of Price-Anderson in 1957; the industry has insisted on this transfer ever since. (c) The transfer of the nuclear investor's financial risks to taxpayers and utility customers emerged after 1978 as the only way to obtain new investment capital into this industry. (d) Taxpayers and customers generally have little specific knowledge of the nuclear power industry, and the fears of accidents fade from memory; many remain firmly convinced that nuclear power creates unacceptable dangers, but many others accept the benefits of nuclear power for electrical generation.

Current Assessments of Safety and Costs: The Nuclear-Accepting Position

The legacies of Phases I–VI underlie the nuclear Hamlet's dilemma: should we, or should we not, embrace nuclear power? Two recent studies, both of which accept nuclear power as potentially valuable or even essential, offer valuable insights, yet both suffer important gaps in their analyses of nuclear power.

The first study, from the National Academy of Sciences, assesses all energy issues facing the United States to 2035 and beyond: *America's Energy Future: Technology and Transformation* (2009).[23] The second example, *Beyond Smoke and Mirrors: Climate Change and Energy in the 21st Century* (2010), was written by Burton Richter, the Paul Pigott Professor of Physical Sciences *Emeritus* at Stanford University and recipient of the Nobel Prize in Physics.[24]

First, consider the issue of safety as portrayed in the two studies. Both *America's Energy Future* and the Richter book recognize the safety of nuclear power as a fundamental issue that, if ignored, can scuttle the entire nuclear power industry. In both, the management of safety lies in design, proper operations of plants, careful monitoring of component parts, and a strong regulatory system. *America's Energy Future* briefly mentions the use of PRA, which predicts lower frequencies of core-damaging accidents in newer plants, and which has become part of NRC's "risk-informed" regulatory processes.

America's Energy Future contains a passing mention of the Chernobyl accident and no mention of Three Mile Island. Richter's book, in contrast, discusses both of these accidents. In neither case, however, do the respective authors incorporate accidents, their severe disruptions of life, and the potential for health-related problems as critical components of safety. Instead, both studies implicitly assume that safety can be "managed" satisfactorily. In essence, accidents result from bad handling of a good technology. Past catastrophic accidents and the ever-present potential for future catastrophes tend to be ignored. In both books, engineers learn from past mistakes so that "such a thing will never happen again." In short, accidents are peripheral, not central, to the concept of safety.

We maintain, in contrast, that the potential for calamitous accidents associated with nuclear power must be integral to the concept of its safety and pose the question, "Should the intrinsic potential for catastrophe affect the final decision to adopt or not adopt nuclear power?" We also assert that if consideration of accident potentials evokes only a discussion of good engineering and the regulatory principles to avoid them, and fail to consider the wider impacts of accidents, then the concept of safety remains incomplete and inadequate for decision making.

A similar deficiency attends treatment of the costs of electricity from nuclear power and other sources. These comparisons rely heavily on calculations of levelized costs, which place the costs of nuclear power in or near the range of power generated from advanced coal and biomass.[25]

America's Energy Future defines the levelized cost of electricity as ". . . the average cost of generating a unit of electricity over the generating facility's service life. The levelized cost is computed by dividing the present value of the estimated full life-cycle costs of the generating facility by its estimated lifetime electricity production. The result is usually expressed in terms of cents per kilowatt-hour."[26] Thus, levelized cost of electricity appears to provide an objective, quantitative, scientific method for comparing the economic costs of generating electricity by different technologies.

While levelized costs can be useful for comparing technologies, the limitations of the method pose serious problems for decision makers. A key concern revolves around life-cycle costs and, more importantly, those costs not included.

The cost estimates in *America's Energy Future* include (a) capital construction, (b) financing, (c) operation and maintenance, (d) fuel, and (e) decommissioning. Useful as these costs are, they do not provide all necessary information, because the list excludes two vital issues: insurance costs and those resulting from damages due to accidents involving any significant releases of radioactive material.

As noted previously, nuclear power plant operators must obtain $300 million per plant in liability insurance from a private insurer and contribute $10 million annually to the industry pool. Federal taxpayers absorb liability damages above that level. Are the liability costs covered by private insurance adequate? Cleanup, compensation payments, and other costs associated with the accident at Fukushima in 2011 may reach $250 billion, so the amount not covered by private insurance could be considerable.[27] Levelized cost calculations do not account for these extra costs.

Justifying the omission based on the difficulty of estimating monetary costs of serious accidents will not reassure skeptics of nuclear power. True, damages are not easily monetized, but omitting them, and remaining silent about the omission, undermines the argument that calculated low, levelized costs of nuclear power make it preferable to other sources of electrical generation.

Other problems also plague calculations of levelized costs. The analyst must compute a "present value" of full life-cycle costs, yet calculating present values entails selecting a discount rate. Unfortunately, no objective means exist to pick the proper discount rate. Too high a rate will reduce present value and thus yield low estimates of costs. Too low a rate will inflate present values and thus the estimates of levelized costs. The analyst could use a range of discount values and the associated range of costs, but that approach runs the risk of letting decision makers justify their already-made decision using costs that might not reflect real world conditions.

A more subtle problem of using levelized costs in decision making stems from a bias in the basic approach. The method assumes a lower cost is preferable and asks, "What is the customer/citizen willing to pay for electricity?" Implicitly, the use of levelized costs of electricity assumes that as long as the cost of a technology is lower than competing technologies, the customer/citizen will be happy.

Unfortunately, the method ignores a different question that is equally important: "What is the customer/citizen willing to accept for the damages and risks imposed by the technology?"[28] This question focuses inquiry on the often unacknowledged dark side of all energy technologies and asks what the customer/citizen would voluntarily accept as compensation for the harms and risks. Omission of this question prematurely closes the analysis and leaves decision makers without important information about the public's willingness to accept or reject nuclear power.

As discussed earlier, nuclear power has always been a child of the state. Both *America's Energy Future* and the Richter study acknowledge the importance of public policy, but neither of their respective assessments covers the political tensions stemming from the necessity of heavy governmental involvement as a force for nuclear power's survival.

Similarly, neither of the two assessments discusses the important fact that US policy for the nuclear industry transferred risks to health and to economic well-being from the private sector to citizens and taxpayers. Price-Anderson in 1957 and federal loan guarantees after 2005 socialize risk while leaving potential profits privatized. These transfers, therefore, leave important ethical questions unexamined. Similarly, neither assessment fully embraces the need to assess the impacts of potential catastrophic accidents.

The common failing of each of these two assessments is their lack of historic frameworks for analysis. They ignore lasting legacies of history that are impossible to quantify. As a result, they fail to incorporate a full consideration of both safety and costs associated with nuclear power. Unfortunately, many decisions to proceed with nuclear power rely on such incomplete assessments.

Current Approaches to Costs and Safety:
The Nuclear-Skeptic Positions

Two non-governmental organizations (NGOs) also assessed nuclear power in recent years but drew substantially different conclusions than the studies discussed above. Neither flatly rejects nuclear power, but both remain very skeptical about it. The Natural Resources Defense Council (NRDC), organized in 1970, has long worked on issues of energy.[29] Its methods of analysis incorporate physical science, but NRDC emphasizes economics, law, and litigation. The Union of Concerned Scientists (UCS), also organized in the early 1970s, has focused on nuclear power and on climate change.[30] Its methods include public policy analysis with an emphasis on scientific considerations. Even so, both NGOs have been very attentive to the qualitative factors that formed the history of nuclear power.

Proponents of nuclear power argue that nuclear stands out among existing technologies in the battle against climate change because of its ability to provide base-load power without emitting carbon dioxide or other greenhouse gases. Unfortunately, as the NRDC pointed out in its 2005 position paper and subsequent fact sheets, that statement overlooks an important fact: it will be years, if not decades, before that electrical power could be on line in the United States to help reduce emissions. In addition, despite the flurry of applications to the NRC in 2007 and 2008, only two have received their construction-operating license to build two new reactors.[31]

The NDRC and UCS both fault nuclear power for the lack of a long-term geologic repository for its waste, whether from the mining and milling of ore or from the production of electricity. The US fleet of reactors discharges spent fuel at a rate of about 2,000 metric tons per year.[32] The spent fuel sits in cooling pools for five to ten years before being moved to large steel and concrete casks filled with inert gas. The accident at Fukushima highlighted the vulnerability of the spent fuel pools at each reactor location—cracks in the pools or a lack of adequate cooling water can lead to releases of radioactive material. The radioactive decay and heat generation inside the dry casks and the forces of nature outside may weaken the casks before a permanent storage solution is found.

In contrast to other green energy options, like wind and solar power, which can be added in small increments, current US nuclear power plant designs call for units that produce 1,200 to 1,600 MW each. Electrical capacity added in large, lumpy increments leads to extended periods of excess capacity in the system. Thus, although current capacity factors of nuclear power plants may exceed 90 percent, that number may drop as more units come on line. A decrease in the capacity factor will drive up the levelized cost of nuclear power.

Publications from the Union of Concerned Scientists primarily address the safety of nuclear power plants and the cost of building new ones. For the former, the UCS examined the fourteen special inspections launched by the NRC in 2010 in response to safety equipment problems, security issues, and special events at the plants. It found that even after forty years of operating experience, US nuclear power plants continue to experience problems with safety related equipment and worker errors that increase the risk of core damage.[33] Many of the problems arise from known but misdiagnosed or unresolved issues. This implies the plants either do not have the technical expertise to correctly identify the causes of problems, or that they do not have the proper impetus to fix them in a timely manner. Indeed, many repairs are delayed until scheduled refueling outages, allowing them to worsen before they get fixed.

The UCS also looked closely at the subsidies for existing and new reactors. They argue that subsidies must be taken into account when comparing commercial nuclear power to other options for combating climate change, because of their impact on the cost of electricity produced.[34] For example, all nuclear facilities benefit from artificially low costs of uranium—companies pay no royalties for mines on public lands in the United States, and the industry receives a special depletion allowance equal to 22 percent of the ore's market value.[35] Reactor owners pay little for the massive volumes of water they consume, often receive priority access to water supplies, and in return, alter stream flows and temperatures with their discharges. All reactors also benefit from the Price Anderson cap on accident liability. Moreover, new reactors will benefit from (a) federal loan guarantees, which lower the cost of debt; (b) accelerated depreciation—a tax break of $40–$80 million per year; (c) the ability in some areas to charge ratepayers now for power plants in construction; and (d) production tax credits. All of these subsidies have the effect of transferring the risks of nuclear power from the owners and operators of power plants, to ratepayers and/or taxpayers. The owners/operators do not pay the full cost of nuclear power but do reap the financial rewards.

Additional concerns about safety stem from proposals for the use of new reactor designs that have yet to be tested under production conditions: the Westinghouse AP1000 or AREVA's Evolutionary Power Reactor, for example.[36] Proponents stress the enhanced safety from passive safety features. But will all these engineered safety features really function as expected? Engineered safety features of designs from the 1960s and 1970s were also predicted to be essentially without risk on the basis of models,[37] but Three Mile Island and Fukushima demonstrated that they were not.

Despite the sensitivity the NRDC and UCS have shown to the legacies of nuclear power's history, they and other skeptics of nuclear power rarely provide full analyses of the engineering, ethical, and political-economic challenges of supplying a steady stream of base-load electricity to the US grid with renewable energy. For example, solar and wind power generate electricity where and when the sun is shining or the wind is blowing, generally in places far removed from consumption centers. In addition, solar and wind installations require large land areas. Proponents of nuclear power often cite the low amount of land required for a plant that can generate a steady stream of 1,000 MW or more of electricity: about 245 hectares. In contrast, the land area needed to generate the equivalent of 1,200–1,600 MW from solar or wind would be much larger: over thirty-eight thousand hectares.[38] Large solar or wind installations provoke conflict and litigation based on aesthetic concerns (huge transmission structures or unsightly turbines) and destruction of habitat for endangered species (such as the desert tortoise).

Hydropower faces a similar problem. Many good hydropower sites are already developed. Proposals for new plants endure intense criticisms due to forced relocations of people and destruction of wildlife habitat. For example, the Three Gorges Dam in China will generate about 18,000 MW, but has required the relocation of 1.2 million people.[39] In the Pacific Northwest, conflicts over fish habitat led to dam removal, and it is highly unlikely that any large, new hydroelectric facilities will be built in the United States.

Finally, the manufacture of wind turbines and energy-saving compact fluorescent bulbs and electric vehicles requires rare earth elements, the production of which results in toxic wastes.[40] Moreover, China currently supplies about 93 percent of the rare earth materials used in the United States, so reliance on these materials currently generates concerns about the security of US energy supplies.

The NRDC and UCS have improved upon existing analyses of nuclear power by reflecting on history and including a more comprehensive analysis of safety and costs, the factors that proponents of nuclear power generally shortchange. However, they shortchanged the analysis of the downside of the options to nuclear power. Thus, a major lesson for both sides of the nuclear debate is to of nuclear assess the strengths and the weaknesses of *all* energy technologies. We propose the use of a Political Ecology framework as one way of doing just that.

Political Ecology: A New Framework for Assessing Nuclear Power

"Political Ecology"[41] encompasses both quantitative and qualitative parameters, but does not result in a bottom-line single number, such as a levelized cost or the risk of failure, that can be used in decision making. Instead, the framework produces a narrative that illuminates the critical issues involved in resolving Hamlet's dilemma: should we or should we not embrace nuclear power in a new energy economy?

Historically, purely quantitative methods never have been the basis for decisions about nuclear power. More importantly, purely quantitative measures *should not* be the sole basis for decision making about nuclear power, because such an approach will inevitably omit considerations that lie at the heart of public concerns about the technology. As discussed above, some attributes cannot be fully quantified and/or monetized in any meaningful way.

Political-Ecological assessment incorporates the usual factors of electrical supply and demand, available technologies, and the two major attributes of nuclear power often overlooked: safety and its complete range of costs. It yields a multi-dimensional narrative that introduces the need to make value judgments and not to rely solely on numbers to guide decisions. Political Ecology will not produce an unambiguous conclusion about the wisdom of using or avoiding nuclear power. Rather, it forces decision makers to think about many facets of nuclear power and thus guides them to a more comprehensive assessment.

The Political-Ecological framework (see Figure 1 below) begins with a focus on technology, which links nature and natural resources to people and their material

Figure 1
The Political-Ecological Analytical Framework

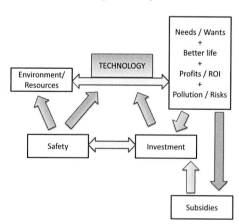

Note: "ROI" is return on investment

wants and needs. As described by Greenberg and Park in 1994, Political Ecology is an exploration of (a) the conflicts between people, their productive activities, and nature, and (b) the influence of culture and political activity on all three.[42]

For energy technologies, "environment/resources" includes the natural resources or supplies of fuels to power electrical generation (natural gas, uranium, coal, wind, solar radiation, water, and land). Technology (power plants, wind turbines, solar panels) allows people to turn natural resources into electricity to satisfy human needs and wants. Access to electricity brings about a better life, and profits and return on investments (ROI), but also pollution and risks. The double-headed arrow under "technology" suggests the mutual interactions between human wants and needs and natural resources: satisfying a want or need using technology makes no sense if the required natural resource is not accessible, does not exist, or if no practical or affordable technology exists. Similarly, no need exists to develop a technology to harvest a natural resource if nobody wants or needs the electricity produced.

Figure 1 draws attention to the idea that both "safety" and "investment" affect choices of technology. Furthermore, safety and investment interact: increased safety usually costs money, and failure to include safety measures puts the investment at greater risk and thus results in a higher cost. "Safety" also affects "environment/resources," because gaining access to any natural resource inevitably disturbs the environment and, in the case of coal or uranium mining, puts people's health at risk.

Finally, Figure 1 draws attention to the interactions among needs and wants, subsidies, and investment. Investors will not supply money unless they perceive low risks and/or adequate returns on their investment. Through various means, government may step in with a subsidy that catalyzes the investment. Subsidies draw money from other activities, through the tax and regulatory systems, into projects that otherwise would die for lack of private investment support. As an alternative, government can become the investor, but again the funds will inevitably come through the tax and regulatory systems. Nuclear power is not unique in its reliance on government subsidies, but many scholars have concluded that subsidies to nuclear power are of overwhelming importance to the existence of the technology.[43] Nuclear power would never have emerged from private investors alone without those subsidies.

For nuclear power, the Political-Ecological framework draws immediate attention to the two factors most prominently incomplete in positive assessments of nuclear power: safety and costs/investment/subsidies. Within the Political-Ecological framework, a calculation of levelized cost of electricity will appear as a necessary—but by no means sufficient—tool for comparing the costs of nuclear power with alternatives. Similarly within this framework, it will be impossible to ignore the important subsidies that underlie nuclear technology.

Assessment of renewable technologies such as biofuels, hydropower, wind, and solar also will benefit from use of the Political-Ecological framework. Issues of subsidies, investments, and safety also attend these technologies, and any valid assessment must develop a narrative to capture these issues. For each renewable energy source the narrative will be different. For example, an examination of wind or solar would highlight the engineering challenges of providing a steady, stable supply of electricity to the grid. The discussion would include the loan guarantees, low interest rates, and tax breaks helping to fund the large solar installations and wind farms.[44] Although neither wind nor solar generates the type of toxic and radioactive spent fuel associated with nuclear power, the narrative might discuss the use and disposal of hazardous chemicals used in equipment manufacture. Understanding the multitude of characteristics for each energy technology will help us all make better decisions about the future of the energy economy.

Conclusions

Finding technologies that can replace fossil fuels remains a challenge for the green energy economy. Devising a quantitative tool that can capture the ethical, economic, and political problems created by these different energy systems presents an additional challenge. The Political Ecology framework, described here for the case of nuclear power, allows for a more complete assessment and comparison of various technologies than do calculations of levelized costs or return on investment. First, the entire energy economy comes into consideration, as questions of which technology to use leads to questions about harvesting natural resources and how best to address needs and wants. Safety and risks also become decision-relevant factors that can lead to rejecting or accepting a candidate technology.

Finally, energy technology appears not as a disembodied physical process but as a component of a socio-technical system in which the *context* of the physical technology is as important as the *technical system* itself.

As for Hamlet's dilemma: should we, or should we not, embrace nuclear power as a necessary green energy technology? Those who promote nuclear power have yet to demonstrate that this choice justifies billions of dollars in investment. Those who oppose nuclear power have yet to find a clean technology capable of providing base-load power. Using a Political Ecology framework may help us all find the mix of technologies that will best meet the social and environmental needs in an ever-changing world.

Notes

1.   "Nuclear Energy in California," *The California Energy Commission*, March 2011, http://www.energy.ca.gov/nuclear/california.html (accessed January 10, 2012); Shirk, Willis L., "'Atoms for Peace' in Pennsylvania," *Pennsylvania Heritage Magazine* XXXV, no. 2 (Spring 2009) http://www.portal.state.pa.us/portal/server.pt/community/history/4569/it_happened_here/471309 (accessed January 23, 2012).

2.   "*Atomic Energy Act of 1954, as Amended. (P.L. 83-703),*" *US Nuclear Regulatory Commission*, http://www.nrc.gov/reading-rm/doc-collections/nuregs/staff/sr0980/rev1/vol-1-sec-1.pdf (accessed December 22, 2011).

3. *"The Price-Anderson Act: Background Information,"* *American Nuclear Society*, November 2005, http://www.ans.org/pi/ps/docs/ps54-bi.pdf (accessed July 2, 2008); *"Price-Anderson Act of 1957, United States,"* http://www.eoearth.org/article/Price-Anderson_Act_of_1957,_United_States (accessed September 29, 2008); *Price-Anderson Nuclear Indemnity Act*, Web. (accessed May 16, 2009).

4. US Atomic Energy Commission, *Theoretical Possibilities and Consequences of Major Accidents in Large Nuclear Power Plants. WASH—740* (Washington, D.C: Government Printing Office, 1957).

5. Saul, K. M., *The Renewed Interest in New Nuclear Construction in the United States: Lessons from History, the Media, and Interviews* (Olympia, WA: The Evergreen State College, 2009).

6. *"1945–1959: Bringing Energy to the World,"* *Bechtel Corporation*, http://www.bechtel.com/BAC-Chapter-3.html (accessed January 24, 2012); *"Decommissioning Nuclear Facilities,"* *World Nuclear Association*, July 2011, http://www.world-nuclear.org/info/inf19.html (accessed January24, 2012); *"Economic Benefits of Indian Point Energy Center,"* (Washington D.C: Nuclear Energy Institute, 2004).

7. US Atomic Energy Commission, *Civilian Nuclear Power: A Report to the President, 1962.* (Oak Ridge, TN: US Atomic Energy Commission, 1962).

8. Based on *Civilian Nuclear Power . . . a Report to the President—1962*, Figure 3, 44; Calculations performed by the authors.

9. See Note 5

10. Ibid.

11. *A Time to Choose: America's Energy Future*, Ford Foundation, Energy Policy Project, (Pensacola, FL: Ballinger Publishing Company, 1974).

12. *"Nuclear Waste Policy Act of 1982,"* http://epw.senate.gov/nwpa82.pdf (accessed December 23, 2011).

13. *"Nuclear Powerplant (sic) Design Standardization"*. Hearings before the Subcommittee on Energy Conservation and Power of the Committee on Energy and Commerce, House of Representatives, July 25 and December 10, 1985, (Washington D.C: US Government Printing Office, 1986), 40.

14. Krauss, C., "Senate Votes to Simplify Nuclear-Plant Licensing," *The New York Times*, February 1992, D2; "House Votes to Speed Licensing of Nuclear Plants," *The New York Times*, May 1992, D2.

15. *Energy Policy Act of 1992, H.R. 776.* Web. (accessed November 12, 2008).

16. *National Energy Policy Report of the National Energy Policy Development Group: Reliable, Affordable, and Environmentally Sound Energy for America's Future* (Washington D.C: US Government Printing Office, 2001).

17. US Department of Energy, Office of Nuclear Energy, Science and Technology, Near Term Deployment Group; Nuclear Energy Research Advisory Committee, Subcommittee on Generation IV Technology Planning. *A Roadmap to Deploy New Nuclear Power Plants in the United States by 2010, Volume 1, Summary Report.* (Washington D.C: US Department of Energy, 2001).

18. *Energy Policy Act of 2005, Public Law 109-58*, August 2005. Web. (accessed January 12, 2012).

19. Schlissel, D., Mullett, M., and Alverez, R. "Nuclear Loan Guarantees: Another Taxpayer Bailout Ahead?" *Union of Concerned Scientists*, March 2009, http://www.ucsusa.org/assets/documents/nuclear_power/nuclear-loan-guarantees.pdf (accessed December 31, 2011).

20. See Note 5.

21. "Loan Guarantees Offered for New Vogtle Units," *World Nuclear News*, February 2010, http://www.world-nuclear-news.org/NN-Loan_guarantees_offered_for_new_Vogtle_units-1702104.html (accessed January 24, 2012).

22. McNulty, B., *Nuclear Power Plant Cost Recovery in Florida*, Louisiana State 2008 Energy Summit, 2008. http://www.enrg.lsu.edu/Conferences/energysummit2008/es2008_mcnulty.pdf (accessed December 31, 2011).

23. National Academy of Sciences, National Academy of Engineering, National Research Council of the National Academies, *Overview and Summary of America's Energy Future: Technology and Transformation* (Washington D.C: National Academies Press, 2010).

24. Richter, Burton, *Beyond Smoke and Mirrors: Climate Change and Energy in the 21st Century* (Cambridge, United Kingdom: Cambridge University Press, 2010).

25. US Energy Information Administration, "Levelized Cost of New Generation Resources in the Annual Energy Outlook 2012," July 2012, http://www.eia.gov/forecasts/aeo/electricity_generation.cfm (accessed December 10, 2012).

26. *"Levelized Cost of New Generation Resources in the Annual Energy Outlook 2011."* America's Energy Future, 56.; US Energy Information Administration, December 2010, http://205.254.135.7/oiaf/aeo/electricity_generation.html (accessed December 22, 2011).

27. "Fukushima Cleanup Could Cost up to $250 Billion," *News on Japan*, June 2011, http://newsonjapan.com/html/newsdesk/article/89987.php (accessed December 23, 2011).

28. John A. Dixon et al., *Economic Analysis of Environmental Impacts* (London: Earthscan Publications, 1994).

29. Cochran, T. B. et al., *Position Paper: Commercial Nuclear Power* (Washington, D.C: Natural Resources Defense Council, 2005); Geoffrey H. Fettus, "Nuclear Facts," (Washington, D.C: Natural Resources Defense Council, 2007).

30. Koplow, D, *Nuclear Power: Still Not Viable Without Subsidies* (Cambridge, MA: Union of Concerned Scientists, 2011); Lochbaum, D, *The NRC and Nuclear Power Plant Safety in 2010: A Brighter Spotlight Needed* (Cambridge, MA: Union of Concerned Scientists, 2011).

31. Wald, M, "Federal Regulators Approve Two Nuclear Reactors in Georgia," *The New York Times*, February 2012.

32. Koplow, *Nuclear Power: Still Not Viable Without Subsidies* (Cambridge, MA: Union of Concerned Scientists, 2011). 96.

33. Lochbaum, *The NRC and Nuclear Power Plant Safety in 2010: A Brighter Spotlight Needed* , *xii* (Cambridge, MA: Union of Concerned Scientists, 2011).

34. See Note 32

35. See Note 32

36. Clements, T., and Olson, M, "US NRC Slams Westinghouse AP1000's Flawed Design," *Nuclear Monitor*, http://www.nirs.org/mononline/nm697.pdf (accessed November 6, 2009); Fairewinds Associates, *Nuclear Containment Failures: Ramifications for the AP1000 Containment Design* (Burlington, VT: Fairewinds Energy Education Corp., 2010); and "Nuclear Expert Warns of Safety Flaws in AREVA's EPR," *Greenpeace*, November, 2009, http://www.greenpeace.org/international/en/press/releases/nuclear-expert-warns-of-safety/ (acccessed February 25, 2012).

37. Bickel, J. H. *Evergreen Safety and Reliability Technologies, LLC*, Evergreen, CO. Telephone Interview, March 16, 2009.

38. *"Backgrounder: A Comparison: Land Use by Energy Source—Nuclear, Wind, and Solar".* Entergy, http://www.entergy-arkansas.com/content/news/docs/AR_Nuclear_One_Land_Use.pdf (accessed January 24, 2012).

39. Kennedy, B, "China's Three Gorges Dam: China's Biggest Construction Project Since the Great Wall Generates Controversy at Home and Abroad," *CNN*, 1999, http://www.cnn.com/SPECIALS/1999/china.50/asian.superpower/three.gorges/ (accessed December 31, 2011).

40. McLendon, R., "What are Rare Earth Metals?," *Mother Nature Network*, June 2011, http://www.mnn.com/earth-matters/translating-uncle-sam/stories/what-are-rare-earth-metals (accessed December 24, 2011).

41. The term Political Ecology first emerged in the 1970s as people became increasingly aware of how highly politicized the natural environment had become. The 1979 anthology *Political Ecology* defined Political Ecology as a "way of describing the intentions of radical movements in the United States, in Western Europe, and in other advanced industrial countries." The definition used in this chapter first appeared in the inaugural edition of the *Journal of Political Ecology* in 1994. In their introduction, Greenberg and Park described Political Ecology as an exploration of the conflicts between people, their productive activities, and nature, and the influence of cultural and political activity on all three. That set the stage for the diversity of subject matter that has become "Political Ecology."

42. Greenberg, J. B., and Park, T. K., "Political Ecology," *Journal of Political Ecology* 1, no. 1 (1994): 1–12. Print.; Perkins, J. H., *Geopolitics and the Green Revolution: Wheat, Genes, and the Cold War* (New York: Oxford University Press, 1997); Saul (2009).

43. Badcock, J., and Lenzen, M., "Subsidies for Electricity Generating Technologies: A Review," *Energy Policy* 38 (2010): 5038–47. Caldicott, H., *Nuclear Power Is Not the Answer to Global Warming Or Anything Else* (Victoria, Australia: Melbourne University Publishing, 2006); Deutch, J et al., *The Future of Nuclear Power: An Interdisciplinary Study* (Cambridge, MA: Massachusetts Institute of Technology, 2003); Levendis, J., Block, W., and Morrel, J., "Nuclear Power," *Journal of Business Ethics* 67, no. 1 (2006): 37–49, Print; Parker, L, and Holt, M., Nuclear Power: Outlook for New US Reactors, *Congressional Report*, March 2007, http://www.fas.org/sgp/crs/misc/RL33442.pdf (accessed January 1, 2012); Saul et al. (2009).

44. Lipton, E., and Krauss, Cl., "A Gold Rush of Subsidies in Clean Energy Search," *The New York Times*, November 2011, http://www.nytimes.com/2011/11/12/business/energy-environment/a-cornucopia-of-help-for-renewable-energy.html?pagewanted=all (accessed January 12, 2011).

14

Why a Green Energy Economy Requires a Smart Grid

Miriam Horn and Elizabeth B. Stein

About 80 percent of global greenhouse gas emissions come from energy generation and use. In the United States alone, electricity generation and use is one of the biggest sources of pollution on the planet, accounting for more than one-fifth of the world's CO_2 emissions. Clearly, the path to climate stability—or instability—runs through energy.

Three recent studies by leading scientists map "The Road to 2050," and California and the European Union (EU), two regions committed to carbon limits, agree on the core elements of a climate stable economy.[1] Far more of the economy must run on electricity, including transport and industry. Electricity must be low carbon, with 55 percent or more coming from renewable resources. Also, electricity must be distributed by a smart grid, making full use of advanced sensing, communication, and control technologies to manage the intermittency of wind and solar power, the new demands created by widespread deployment of electric vehicles, and the challenges and opportunities of a decentralized system.

Over the next several years, policy makers around the world will decide how to spend an estimated $13 trillion on grid investment.[2] The question is whether they will follow the traditional path of building (or retrofitting) large-scale fossil plants and reinforcing utility monopolies or begin building infrastructure required for a low-carbon economy—one that spawns a new era of growth and job creation while serving the world's energy needs.

If the world is to meet the climate challenge, decision makers must choose the latter option. The iron inflexibility of the twentieth century grid—a one-way street from giant power plants to consumers—simply cannot accommodate the changes demanded by a clean energy future: reducing the vast waste in the system, integrating a range of new clean resources, including flexible demand and renewable generation, and enabling a new world of applications for electricity (including electric vehicles) and new energy services. The challenge of enlarging the market share of electricity, while simultaneously making a transition to clean generation, demands a grid and energy markets as open and flexible as the internet.

Eliminating Waste and Losses

Cutting Line Losses and Excess Voltage

The losses, inefficiencies, and unreliability of the current electric grid act as an immense drag on economies around the globe. According to the U.S. Energy Information Administration, electricity transmission and distribution losses in the United States are about 7 percent of all electricity transmitted in the country.[3] A utility named Xcel Energy estimates that the smart grid can reduce these losses by 30 percent, utilizing optimal power factor performance and system balancing[4] to better manage non-working reactive power ("VAR optimization"). Reactive power, measured in volt-ampere reactive (VAR), supports voltage for system reliability.[5] It is an ancillary service,[6] typically supplied by conventional generation, but does not "travel well" over long distances, making precise location key to efficient dispatch. Using capacitor banks to optimize VAR allows for more efficient switching.[7]

In addition to line losses, enormous amounts of energy are squandered through consistent overvoltage, maintained to ensure that adequate voltage reaches the farthest edges of the network at times of peak demand. Off-peak, that excess voltage is not only wasteful but leads to higher energy bills for customers.[8] A study by Current Group of 1,700 substations across the United States found excess voltage on every distribution feeder. Each 1 percent reduction in voltage translates into a reduction in kilowatts of between 0.2 percent and 1.3 percent.[9] Current Group CEO Tom Willie estimates that using conservation voltage reduction (CVR) technologies could save as much as 5 percent of total generation (and avoid an equally significant percentage of carbon emissions, since much of that baseload power comes from carbon-intensive coal).

Integrating voltage and reactive power controls ("volt/VAR optimization" or "VVO") delivers the greatest savings, and the technologies are advancing rapidly. In late 2011, ABB subsidiary Ventyx, for instance, launched a distribution management system that combines advanced sensing and two-way communications, capacitor controllers, geospatial modeling, and mathematical optimization. Oklahoma Gas & Electric, in a 2010 test of VVO on four circuits, enabled peak demand reductions of up to 2.4 percent; it has begun work to deploy the new Ventyx system on 400 circuits[10] in order to achieve a 75-megawatt load reduction by 2020.[11]

Enabling Fuller Use of Existing Transmission Capacity

A second strategy for wresting more value out of the existing electric in-frastructure is "dynamic thermal rating" of transmission lines. The capacity of transmission is not static but varies with ambient temperature and heat-induced sagging. High-tech sensors that measure temperature and sag allow a line to be operated more closely to its limit. More widespread use of dynamic line ratings could provide an additional 10–15 percent capacity for 95 percent of the time, and up to 25 percent more transmission capacity for 85 percent of the time, according to the U.S. Department of Energy (DOE) 2009 "Smart Grid System Report."[12] That unlocked capacity can help bring energy from large-scale renewable resources in remote areas into cities and other regions with high energy demand.

Maintaining Power Quality and Reliability

In most parts of the world, a key economic value will be the smart grid's role in maintaining reliability and power quality. Even in the United States, on an average day, half a million people are hit with blackouts at least two hours long, shutting down productivity. With an aging grid coming under increasing stress and more intense storms, large-scale blackouts are on the rise, at a cost of approximately $100 billion a year.[13] The 2003 blackout across the northeastern United States is estimated to have cost $6 billion.

By providing visibility into the grid—with technologies like phasor measure-ment units ("PMUs"), which provide voltage and current readings thirty times a second at precise locations along power lines, smart meters, and other sensors that can pinpoint outages—together with the capacity to prevent or island problems and automated self-healing, a smart grid will shrink those losses.[14] In Alabama, the deployment of 1.4 million smart meters set the stage for an accelerated re-sponse to the historic tornados that hit the state on April 27, 2011. The meters gave responders a clear sense of where rapid repairs would help the greatest numbers of customers as well as where, tragically, the damage was so extensive that there was nothing left to reconnect to the system.[15] In California, benefits arising from Pacific Gas and Electric Company's (PG&E) smart grid deployment are expected to include system reliability improvements of 10–20 percent. The consulting firm McKinsey projects that grid applications and advanced metering will produce approximately $72 billion in annual benefits in the United States by 2019, primarily as a result of improved grid efficiency and reliability.[16]

Avoiding New Peak Capacity

In advanced economies, the proliferation of electronics and air conditioning and a shift from twenty-four-hour industrial loads is creating an ever "peakier" demand. At times of greatest demand, power becomes extremely expensive—as much as an order of magnitude more costly than normal prices.[17] The sharp spikes in demand also require a vast overbuilding of capacity. A study for the Federal Energy Regulatory Commission found that reductions in peak demand could avoid the need for two thousand peaking plants,[18] each of which costs hundreds of millions of dollars to build. Those plants will be called into service as little as 100 hours a year.[19]

Peak power also exacts high costs to human health. The plants used to supply it are often among the dirtiest, emitting high levels of nitrogen oxides and other health-damaging pollutants. In the United States, during the few hours of the year when air conditioning use forces the dirtiest peaking plants into service, millions are warned to stay indoors owing to poor air quality. A report from the National Academy of Sciences on "Unpriced Consequences of Energy Use and Production" estimates that in 2005 alone, environmental externalities from U.S. electricity production cost $120 billion. The report notes that this figure in fact underestimates the true costs because it does not include the costs of climate change or damage to ecosystems.[20]

While grid operators have some ability now to reduce peak demand, this "demand response" (DR) capability is largely manual, requiring human intervention. Often, this entails an actual phone call from a utility to an enterprise that has agreed in advance to briefly dim its lights, allow its thermostat to vary by a few degrees, or delay an industrial process till a later hour—all for an agreed-upon price. A 2009 report produced for the Department of Energy by Alcoa Power Generating, Inc. and Oak Ridge National Laboratory found that only 5 percent of customers nationwide are enrolled in DR programs, and called DR "the largest underutilized reliability resource in North America."[21] A Federal Energy Regulatory Commission (FERC) study found existing DR programs, focused on large industrial users, delivering 37 GW nationwide. Without new programs, it suggested, that capacity would grow to just 38 GW by 2019.[22] A smart grid will almost quadruple those savings: large-scale deployment of AMI, enabling technologies, and dynamic pricing will foster peak reductions of 150 GW by 2019, equivalent to about two thousand peaking plants.[23] Oklahoma Gas and Electric, which has committed to invest in the smart grid to avoid any new power plant construction for at least a decade, has already canceled plans for two 165 MW peakers scheduled to be built in 2015.

Enabling New Clean Demand-Side Resources

With its ability to enable communication across the entire network and support transmission of real-time price signals, a smart grid will greatly expand the number of demand-side resources that can participate in the market, including potentially billions of devices in residences, commercial enterprises, and industry. Software will allow consumers to decide at what price their home or business will buy electricity to run machinery or appliances, or sell electricity generated or stored on site back into that market. Whirlpool—the world's largest appliance maker—estimates that shifting defrost cycles in refrigerators could alone make available 5 GW of flexible demand.

Over the next twenty years, tapping the full potential of a smart grid to automate DR could profoundly reshape investments in electricity infrastructure. According to a 2008 report by the Brattle Group for the Edison Foundation, an aggressive program of energy efficiency and DR could decrease the amount of new generation needed in the United States from 214 to 111 GW.[24] The total investment needed will still exceed $1.5 trillion, but far more of that money will flow into the distribution system—that is, into communities, businesses, and homes—vastly expanding economic opportunities.

Beyond halving the need for new capacity, DR can improve the environmental characteristics of the energy mix, as well as the business case for developing renewable energy. Although renewable energy is not always available, its near-zero marginal cost makes it among the first resources to clear the market whenever it is available. Wind, in particular, which peaks at night in some regions, and hydroelectric and geothermal resources, which are productive at all times, comprise a disproportionate share of the generation mix during off-peak periods.[25]

Thanks to reforms that have opened up some U.S. wholesale markets to demand resources and investments in new technologies by aggregators such as EnerNOC and Comverge, DR has already begun to transform the mix of electric-supply resources in some regions. For example, within the PJM Interconnection, a regional transmission organization (RTO) that covers thirteen states and the District of Columbia, enough low-cost demand response bid into the capacity market for 2014–2015 such that 7 GW of coal capacity failed to clear the market.[26]

Creating New Revenue Streams for the Manufacturing Sector

Some large industrial electric customers have a particularly valuable and profitable role to play in electricity markets. Those who can respond within seconds to a signal from the grid, making small adjustments in their power draw to regulate frequency, can sell that "ancillary service" at a premium price. Traditionally, such a frequency regulation service has been provided by fossil-fuel generators at a high economic and environmental cost, compromising the efficiency of these power plants.

A pioneer in this kind of industrial demand response is Alcoa, Inc.—whose ten aluminum smelters and associated fabricating facilities in the United States represent a combined average load exceeding 2,600 MW.[27] A 2009 DOE study found that the electrolysis process used for smelting could make Alcoa an ideal supplier of frequency regulation.[28] It could not only provide this regulation service without compromising its own efficiency but also ensure *superior* service, because it could adjust its demand to follow the correction signal far more quickly than a fossil generator could ramp up or down. A new federal policy (FERC Order 755) requiring higher compensation for "faster ramping" resources will open the door for still greater participation by storage and demand resources.[29]

Turning Buildings into Resources

In the commercial building sector, where building automation and variable energy pricing (at a minimum, hourly) is increasingly the norm, engineers and manufacturers are designing technologies to integrate intelligent buildings into energy markets. For instance, they are developing switches that toggle between passive cooling and mechanical air conditioners based on price signals from the smart grid.[30]

The U.S. Green Building Council (USGBC) has taken note of the smart-grid-enabled opportunities for building owners and operators to reach beyond the building envelope by creating a new set of credits for commercial buildings that participate in demand–response markets through aggregators. Environmental Defense Fund (EDF), an environmental non-governmental organization headquartered in New York City, and the Lawrence Berkeley National Laboratory (LBNL) will analyze the data from USGBC's initial pilots to determine real-time emission outcomes from this participation, and refine the credits to ensure they deliver significant pollution reductions.

Enabling New Clean Supply Resources

Integrating Intermittent Renewable Generation

The intermittent nature of wind and sun poses a major challenge to electric grid management. Instructive here are findings from the National Renewable Energy Lab's Wind and Solar Integration Studies for the Eastern[31] and Western[32] interconnects, and the International Energy Agency's (IEA) 2011 Smart Grid Technology Roadmap.[33] Both found that maintaining grid stability when renewables supply 20 percent or more of the total demand will require real-time system information, enhanced forecasting and monitoring,[34] smart transformers, and other advanced distribution management technologies to respond to supply fluctuations. A KEMA study similarly finds that accommodating the 33 percent renewable portfolio standard in California will require "major alterations to system operations" to balance volatility.[35]

Consequent to such technical considerations, China and Europe have matched their leadership in renewables deployment with leadership in deployment of smart grid technologies. Ireland's transmission system operator, EirGrid, is deploying smart grid technologies—including high-temperature, low-sag conductors and dynamic line rating special protection schemes—to manage the high proportion of wind energy on its system. Operation of the system is being improved through state-of-the-art modeling and decision support tools that provide real-time system stability analysis, wind farm dispatch capability, improved wind forecasting, and contingency analysis. Such smart grid approaches are expected to facilitate real-time penetrations of wind to up to 75 percent by 2020 (EirGrid, 2010). In Spain, Red Eléctrica has established a Control Centre of Renewable Energies (CECRE), the first in the world for managing an entire nation's wind farms larger than 10 MW. Smart technologies also will be crucial to wean renewables off the need for gas-fired back-up plants. The California Council on Science and Technology study referenced at the beginning of this chapter found that hitting climate stabilization targets will require "zero-carbon balancing" of renewables, adjusting demand to match intermittent supply, or storage to shift the time of supply.[36]

That load shifting and storage will also greatly improve the economics of wind, which is most productive at night when demand and prices are at their lowest. Increasing night-time demand will avoid curtailment and boost prices for wind energy. Networked energy storage additionally will help boost the capacity factors and financial viability of wind turbines. In regions with constrained transmission resources, storage will serve as a kind of queue, where (rather than being curtailed) wind can "wait in line" until space opens up on the power line.[37]

Managing Distributed Generation

Distributed generation adds another set of challenges that a smart grid will be critical to solving, such as how to manage multidirectional energy flows and keep voltage stable with millions of small sources of supply, most of them owned and controlled by someone other than the utility. A recent survey from Accenture[38] found that 72 percent of utilities are concerned that their grids will face challenges or require upgrades as solar photovoltaic (PV) penetration approaches 24 percent. The challenges are especially great when these distributed renewable resources constitute a higher portion of total supply, as in the springtime, when the sun is shining intensely but air-conditioning load is low and the whole system is running below 60 percent capacity.

To get ahead of those challenges, San Diego Gas & Electric (SDG&E) is deploying smart sensors across its network, gaining unprecedented real-time visibility into what is happening on its wires, at substations and on transformers to pinpoint where fortification is most needed. Its initial findings point to battery storage, rather than larger transformers, as the most effective means to smooth output.[39]

In the EU, Energias De Portugal is using smart technologies to predict and isolate problems, limiting the need to disconnect distributed generators from the grid. With additional smart technologies, it anticipates being able to go further, harnessing those "edge" resources to support voltage, rather than having to push out excess voltage from central station power plants to reach far-flung, rural customers.[40]

Enabling New Applications for Electricity

The electrification of industry and transport will create vast new business opportunities but will also place immense new demands on the grid. The charging demand of a plug-in electric vehicle (PEV) can be comparable to the load of an entire household: 3.3 kW.[41] Early adopters of electric vehicles, like Prius buyers, are likely to be clustered in particular neighborhoods.[42] Depending on the load of individual homes before PEV adoption, the effect could be a sudden doubling of load in those localized areas, challenging parts of the system, even before widespread adoption.[43] If neighbors arrive home after work and all plug in to charge their cars, the risk of brown-outs will increase, as will power plant pollution, as fossil generators kick on to meet the spike in load.

Smart grid and smart market rules can mitigate those challenges. Electric vehicles, for instance, can be incentivized with dynamic pricing to "smart charge" when electricity is abundant, clean, and cheap. In wind-rich areas, programming EVs to charge at night will bring additional value. Like load-shifting and storage, it will increase the market for power at times when wind is most productive and thus boost the economic value of those generators, further accelerating wind development. As an exceptionally flexible load, capable of rapidly adjusting the rate of charging (for frequency or voltage regulation) and of kicking on or deferring charging to capitalize on clean supply, EVs are an ideal partner for renewable generation. In San Diego, which has one of the world's highest densities of electric vehicles, SDG&E has patented a smart transformer to stagger charging, and is testing price incentives.

"Smart charging" also will improve the emissions profile of electric vehicles. Converting the light duty fleet to electric by 2030 would reduce CO_2 emissions by 39 percent compared to gasoline "even if electricity is generated using the current coal fuel mix."[44] But if the power for electric vehicles is generated from renewable sources, both power plant and vehicle emissions of CO_2 as well as NO_x, volatile organic compounds (VOCs), SO_2, and mercury are eliminated.[45] If the technology challenges can be overcome, electric vehicles might even serve as distributed storage for the grid, with the vehicle's battery power available to be drawn upon under certain conditions (so-called V2G, or "vehicle to grid"). A recent analysis by Pacific Northwest National Laboratory (PNNL) found that smart-grid-enabled EVs could provide all of the backing resources needed for an additional 14.4 GW of wind energy in the Pacific Northwest.[46]

Principles for Design

Given the enormous potential economic and environmental benefits of smart grid development, the following question arises: how should regions, or nations, go about actually designing and building a smart grid? Guiding principles can support the transition from our legacy system of fossil generation to utilization of smart-grid-enabled resources. In regions where there is no legacy fossil-fuel system, building a clean distributed system from the start could enable a leapfrog past the developed world.

Leading-edge projects around the world point to several key principles for success:

(1) Openness. The opening up of the telecom world to competition, together with the advent of the internet, with its open architecture, revolutionized communications. The world went from black rotary phones to global connectivity and the iPhone; from phone books to Facebook. Communications, once a stodgy, mature market sector, exploded into a vast new arena of wealth and job creation. Opening the energy sector to entrepreneurs and innovation will usher in a similar explosion in clean energy resources and applications, together with dynamic economic growth. That openness must extend to the regulatory process, the design phase, the architecture of the technology platform, and the energy marketplace.[47]

(2) Regional specificity. One of the greatest potential benefits of the smart grid is its ability to empower local communities, enabling them to rely to an unprecedented extent on local resources, like small-scale generation matched to flexible demand. Smart grid designers will need to dive deep into local needs, constraints, resources, and competitive strengths in order to make this potential a reality.

(3) A systems view. To maximize returns on infrastructure investment, communities will need to quantify their economic, environmental, and social goals and weigh conventional solutions against the decentralized promise of a smart grid. Only by evaluating their electricity system as a whole will they be able to recognize, and realize, the immense added value that comes from integrating multiple technologies to achieve a desired end.

(4) An iterative process. Given the pace of innovation and range of global experiments underway, building smart grids by phases will enable everyone to learn from the mistakes and successes of others, and to analyze the rich data stream generated by smart grid technologies with a view toward making continuous improvements.

A powerful example of these key principles in action is now in its sixth year in Austin, Texas. The Pecan Street Project (Pecan Street) was launched in 2008 by a group that included the University of Texas (UT), the Austin Technology Incubator, the city and the utility it owns (Austin Energy [AE]), EDF, and the Greater Austin Chamber of Commerce. Today, the project combines a world-leading "energy Internet" demonstration project in the Mueller neighborhood with a consortium working to make Austin a center of smart grid research and economic development. The idea of an energy internet had been advanced by people like Robert Metcalfe, co-inventor of Ethernet and founder of 3Com, who in 2009 suggested that if we recognize energy as "mostly a networking problem, and take the Internet as our guide, then the power grid will get smart, switched, asynchronous, symmetrical, redundant, storage-intensive at various levels of aggregation, and standardized in layers. Will energy be solved with some sort of smart grid Manhattan Project? More likely, like the Internet, the 'Enernet' will be built by fiercely competing teams of professors, venture capitalists and scaling entrepreneurs."[48]

Open Process/Open Platform

Pecan Street began as, and remains, a broadly inclusive, nonproprietary collaboration, with the city utility just one partner among many. This is in keeping with Austin Energy's rich history as a green energy innovator. Under the leadership of former General Manager Roger Duncan, Austin Energy developed the nation's first green building program, which grew into the U.S. Green Building Council's Leadership in Energy and Environmental Design (LEED) rankings.[49]

At Duncan's urging, Phase 1 of Pecan Street cast the broadest possible net—for both participants and ideas. Industry partners included many of the companies working at the leading edge of grid innovation, including IBM, Intel, Microsoft, Oracle, Applied Materials, Cisco, and GE Energy. Additional participants came from clean-tech startups (many of them Texas-based), global energy analysts, utility customers, and members of the Austin community. In all, nearly 200 people were involved.

Pecan Street spent its first year in an almost unbounded brainstorming. The participants organized themselves into a dozen teams, each focused on an issue: distributed generation, efficiency and demand response, storage, transport, and of course, the smart system needed to integrate and manage it all. The teams considered not only technology innovation but also new utility business models, behavioral economics and workforce training. They looked beyond electricity to opportunities for smart management of the gas and water systems, and beyond the utilities to every possible lever for change, including city planning and building codes.

The distributed generation team—led by Miriam Horn—laid out in its vision statement the range of possibilities it had been encouraged to explore: "An 'open-source' pilot will allow inventors and young companies to test and demonstrate emerging technologies, positioning Austin to attract those innovative start-ups to locate their design or manufacturing facilities within the city. Almost anything will be possible here, aimed at the ultimate goal of harvesting every excited electron . . . ." "Austin will get a first look at breakthroughs in organic thin films, solar cells with roughed-up surfaces or photonic crystals that bend and capture more light; nanoantennae that harvest infrared light, including heat radiated from the earth at night; kite generators that catch high-altitude winds; roads plumbed to serve as giant solar thermal collectors; closed loop industrial ecologies, where (for instance) waste nutrients from landfill biogas projects are fed to algae and the algae converted to fuel. Microbial electrolysis, artificial photosynthesis, viral batteries—the whole clean energy future will be explored and envisioned here."

Regional Specificity

That sky's-the-limit vision was tethered to reality by a simultaneous deep dive into local constraints and opportunities. The team considered both electrical and thermal loads and assessed all possible supply options within the utility service area: ground-mount and rooftop solar, microwind, microhydro, cogeneration (with district heating and cooling), geothermal electricity, and combustion of landfill gas. Critical supports included a rooftop study funded by the U.S. Department of Energy's Solar America Cities Initiative and a model (developed by UT Professor Mack Grady) that forecasts the harvest and cost of electricity from any solar technology at any location and orientation. With these tools, the team determined that photovoltaic solar power within the city limits could supply Austin's full peak load (2,000 MW). It also completed a block-by-block mapping of all cost-effective distributed generation capacity deployable by 2020.

One prime goal of Pecan Street was to understand how, if solar PV were to be deployed at that pace, utility infrastructure and operations would need to change to manage tens of thousands (or in larger cities, millions) of variable electricity sources owned by many different players. With power flowing in many directions, and unpredictable voltage swings as clouds passed overhead and solar panels' power output sagged or surged, integration would be crucial, so that (for instance) community-based batteries, or individual batteries in electric cars, could absorb the fluctuations in solar power to buffer the grid.

Taking advantage of the Texas-size laboratory, Pecan Street also would explore the opportunities presented by the larger electricity network. Could smart home energy management systems, for instance, solve some of the problems with large-scale renewable resources? Could they adjust the demand to follow the ebbs and flows of wind supply, or use night-time wind (frequently overabundant on the Texas grid) to freeze ice as a coolant for the following day?

Building the Idea

Out of this brainstorming came Phase 2, a $25 million American Recovery and Reinvestment Act (ARRA)-funded demonstration of a maximally open, flexible system enabling a world of new energy "apps." The idea was to integrate the full range of clean energy resources, including those just barely invented, in a real Austin neighborhood of 1000 homes.

In the request for proposals (RFP) for the project, the commitment to "open-source" innovation was redoubled. Vendors were advised that their technology solutions would be considered only if they provided an open, interoperable platform that enabled entrepreneurs to "plug-and-play" their innovations. They were advised that all data would need to be shared. Bidders were strongly encouraged to form teams and submit integrated proposals that fully exploited the synergies among solutions.

Drafting the RFP required tackling some of the most critical questions now confronting utilities and regulators. Could utilities play a dominant role without retarding innovation? Telecommunication was transformed by new, disruptive entrants, not incumbents. And electric utilities are among the most conservative companies in the world. Their obligation to provide universal access to secure, reliable power drives them to minimize technology risk. So too, does the immense scale of their investment; with billions of dollars of public or ratepayer funds at stake, technologies are required to be robust and long-lasting, slowing the innovation cycle to twenty or thirty years.

More pressingly, could the utility meter—even the most advanced meter—provide a fast and flexible enough gateway to the customer? Could it supply the data flows needed to support rapid innovation without compromising its two critical utility roles: as tollbooth, keeping track of how much electricity travels between the grid and a particular customer, and as sensing and control node, for monitoring and stabilizing the network? Would it be better to communicate with customers via high-bandwidth, low-latency networks, like cellular, Wi-Fi or cable DSL? Would that, in fact, be the only way that the Facebooks and Twitters of energy could emerge?

Pecan Street technology director Bert Haskell spent nearly a decade at Austin's Microelectronics and Computer Consortium (MCC), in a world where a new generation of technologies was evolved every two years. According to Haskell,[50] "We're not ruling out the use of the utility meter and backhaul as part of the solution, but it has clear limitations. The utility industry knows how to engineer for scale, security, reliability and cost, but it lacks insight into the attributes networks need to support the full sweep of innovation. Their meters are providing data in 15 minute intervals, with a 24 hour delay. If a breakthrough new energy service needs five second data in real-time, we're out of luck for about 15 years."

Carter Williams, former head of innovation at Boeing and founder of Gridlogix (a building management company recently purchased by Johnson Controls), argues that in this earliest phase of innovation, it is critical to explore multiple technology solutions without locking too early into standards that strangle innovation in the cradle.

Sharing this commitment to sustain experimentation, Pecan Street is exploring a range of possible gateways and communications solutions. In one set of homes, a Zigbee system will extract data directly from Landis and Gyr meters without waiting for the data to travel through the utility system and back again. Other experiments will bypass the meter and extract data through the utility backhaul, publishing it securely on the internet so that customers or their designated third parties can access it within a minute's time. Still, others are focused on technologies that can provide granular usage data without any utility input. Intel, for instance, is developing sensing technologies that can read the signature voltage pattern every device displays as it switches on and ramps, a kind of electrical melody as distinctive as a penguin's mating song. Through pattern recognition, Intel will identify the device down to its model number, and then map it against a database that specifies the power draw for every device and appliance in the world.[51]

Long Term Goals

Beyond establishing experimental protocols, moving from concept to deployment required agreeing upon quantified goals for the project and a strategy for working across the entire system to deliver on them. Environmental goals remain foremost: the Pecan Street team asked EDF to join its governance group because environmental protection was a core objective, not something to be realized incidentally. The project participants committed to provide big, measureable improvements in water use and water quality, air quality and—most importantly—carbon emissions. This aligned with the City of Austin's own goals, which included a future in which all new homes would be net zero emitters of greenhouse gases and citywide carbon reductions kept pace with those set by the Kyoto Protocol. In the end, Pecan Street distinguished itself by setting the most ambitious emissions reduction goal of any smart grid project in the United States (and perhaps the world). The integrated systems deployed in the Mueller neighborhood will achieve 65 percent reductions (compared to an average Austin neighborhood) in greenhouse gases and health-damaging pollutants.

Market innovation is an equally important goal. The project is testing a range of alternate business models for utilities, alongside tariffs designed to accurately reflect costs and incentivize behaviors that will keep this energy system of the future operating smoothly and minimize its footprint on water resources, land, air quality, and climate. This is not just an environmental necessity; the current utility business model—based on selling a commodity (kilowatt hours) at a constant price—becomes increasingly less viable as consumers also become energy producers and managers, and as entrepreneurs begin to capitalize on the new flows of data from the smart grid to develop new technologies and services.

The conventional model further fails to internalize many of the costs of a centralized, fossil-based system, or to value the benefits derived from distributed energy resources. The benefits include the following: reduced line losses and avoidance of costs for distribution system upgrades (because the power is being made at the point of use); lowered capital risk (because supply can be deployed in increments more closely matched to growth in demand); reduced vulnerability to volatile fossil fuel prices; high productivity (for solar) at peak, when electricity is most valuable; and environmental values (least-cost path to compliance with clean air requirements, water savings, reduced land impacts, and a hedge against a future carbon price). Distributed generation, meanwhile, also imposes costs on the utility for maintaining the wires and system reliability. Therefore, Pecan

Street has been experimenting with unbundled and real-time pricing models to internalize all those benefits and costs.[52]

Finally, the project committed to creating a replicable process, adaptable by other communities wishing to meet their own economic and environmental goals.

System Solutions

One of the clearest early lessons learned by Pecan Street is that piecemeal investments undertaken to solve isolated problems often wind up as underperforming or stranded assets. As one of the earliest movers in the industry, AE invested in an early generation of "advanced" meters: these meters could send a one-way signal to the utility (avoiding the need for meter readers) but lacked the reverse capability for AE to send usage information and price signals to customers. Until these meters are retrofit, or an alternate gateway is put in, customers remain unable to see and manage their energy use, respond to changing prices, or benefit from new services and technologies as they emerge.

Utilities now are much better equipped to look across their whole system and design integrated solutions. SDG&E, for instance, which, as of the end of 2012, had 3–4 percent of all the electric vehicles in the United States in its service area, and a 2.5 percent per month growth rate for distributed PV,[53] focused its early investments on sensors across its system: at their substations, on their distribution wires, and—with a full rollout of smart meters with two-way communications—at every customer endpoint. Through analysis of the data delivered by these sensors, they can now see exactly which transformers are laboring close to their limit and where the network needs to be fortified, allowing them to target investments to achieve maximum returns.

This whole-system analysis has enabled the utility to find the low-cost path to managing its emerging challenges. Rather than building bigger transformers and wires, it will deploy storage to smooth the output from the high-density PV. This system analysis has illuminated critical technology gaps, some of which its own engineers have set out to fill; it has, for instance, patented a smart transformer that can talk to the cars to stagger charging.

System analysis also has made it possible for the utility to capitalize on the multiple functionalities of each technology and capture the synergies from integration. The charging upgrades are being synchronized with experimental rates for EV charging that are five times lower at "super-off-peak times" (between midnight and 6 a.m.); early evidence suggests that these rates are moving 80 percent of customers to super-off-peak charging.

In Pecan Street, energy storage will serve multiple roles: substituting for spinning reserve, purchased power, or new peaking plants; transforming intermittent renewable generation into firm, dispatchable power, increasing its value to both producer and purchaser; enabling the utility (or its customers) to buy low-cost off-peak energy for use at peak, high-cost times of the day. When combined with demand reductions and demand response, it will become even more powerful and more affordable, since smaller systems are required.

The integration of the electric system can (and should) ultimately link distributed resources all the way with bulk power markets. Recently, EDF succeeded in revising the rules governing Electric Reliability Council of Texas (ERCOT) to enable demand response to bid directly into the wholesale market. The supply and demand resources being created within the Pecan Street demonstration will provide an opportunity to pilot and refine these transactions. Figure 1 gives a sense of the multiple integrations required.

Pecan Street's demonstration project is close to the Texas State Capitol, in the Mueller neighborhood, which is a mixed-use redevelopment of the former Austin municipal airport. All the homes are built to the LEED platinum standards and are less than ten years old. All three of the major utilities—electric, gas, and water—are engaged, providing meter integration and research support to project researchers and participating companies.

Figure 1
Pecan Street and ERCOT

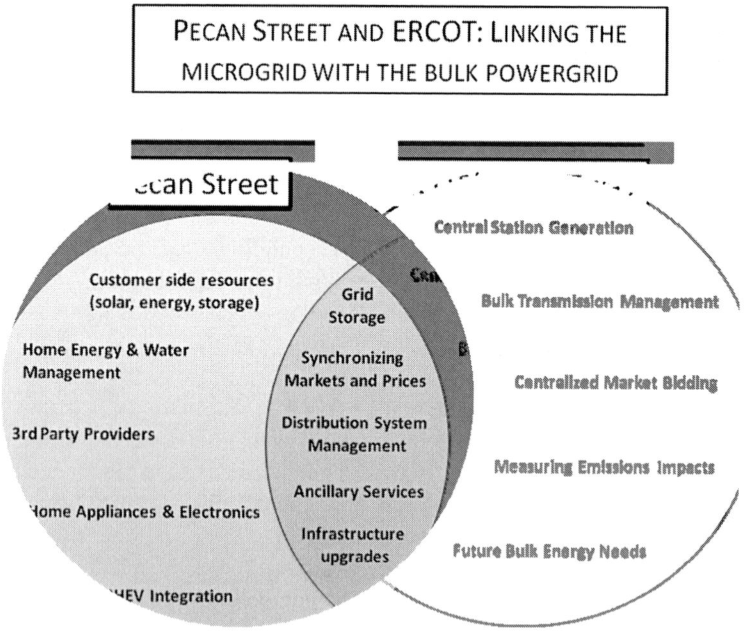

PECAN STREET AND ERCOT: LINKING THE
MICROGRID WITH THE BULK POWERGRID

Pecan Street

Customer side resources
(solar, energy, storage)

Home Energy & Water
Management

3rd Party Providers

Home Appliances & Electronics

HEV Integration

Grid
Storage

Synchronizing
Markets and Prices

Distribution System
Management

Ancillary Services

Infrastructure
upgrades

Central Station Generation

Bulk Transmission Management

Centralized Market Bidding

Measuring Emissions Impacts

Future Bulk Energy Needs

Source: Environmental Defense Fund, 2012. An iterative process.

Home Energy Management Systems

These operating platforms for consumer smart grid products and services will provide energy management as well as home security, health care monitoring, entertainment, and other services. In many of the homes, these products and services will ultimately measure and report real-time usage of water and natural gas, in addition to the measurement and reporting of electricity, which is already ongoing.

Intel will install home automation and management systems in up to fifty homes. Researchers from Intel Labs, including experts in sensor networks, system automation, and consumer preferences, will work directly with residents to understand the customer value proposition and to develop unobtrusive means of identifying energy use patterns by each type of device in the home.

Sony will test and deploy an internet-ready home energy management system (HEMS) in 500 homes, streaming energy-use data through participants' television sets and providing control options tailored to each customer's preferences. As the company's first venture into home energy management, the project will tap Sony Group's products, including lithium-ion home batteries and entertainment services.

Whirlpool will provide connected smart appliances with advanced information technology and energy management capabilities. In some homes, it will build end-to-end smart appliance solutions including its own home energy management system and networking components. It will also work with other vendors to quantify and expand the benefits gained by integrating smart appliances with electric vehicles, solar panels, and home services systems. Again, Best Buy's Geek Squad will work with Whirlpool to install and maintain the smart appliances.

Landis+Gyr is deploying several hundred E350 FOCUS AX smart meters and a networking platform to enable, Intel, Sony, and Whirlpool to integrate pricing and demand management information from utility distribution systems into their home services systems.

Home Solar Charging of Electric Vehicles

More than 150 Pecan Street residents have rooftop solar PV. Two-thirds of those families also have Chevy Volt plug-in hybrid electric vehicles, one of the highest-density deployments of plug-in vehicles in the United States. A SunEdison system will allow those vehicles to be charged directly from solar PV. In some homes, the addition of batteries will allow solar energy to be stored during the day to charge the vehicle at night. All systems will integrate into the Intel, Sony, and Whirlpool home services systems.

In December 2011, Greentech Media called the Pecan Street smart grid test in Austin "the most ambitious EV-solar-smart-grid integration project in the

United States. The goal is to get EVs, rooftop solar, smart appliances and even household batteries all communicating to shape peak loads, ease strain on the grid, and hopefully save everyone a bit of cash."[54]

Data Analysis

The most important part of Pecan Street is its rigorous analysis of the rich streams of data at an unprecedented level of detail. EDF and a UT team with access to a supercomputer at the Texas Advanced Computing Center are analyzing the data from the Mueller houses to ensure the project meets its ambitious emissions goals. The research team has installed energy measurement equipment, developed by Incenergy, an Austin-based start-up, in hundreds of residences to capture energy usage information in fifteen-second increments for the whole home and major appliances.

The initial baseline data has already led to significant innovations, including determining the precise southwest orientation for solar PV that better matches the panels' productivity with Austin's late-peaking load (late peaking because it remains hot well into the night). The researchers expect that their data will, over time, yield important findings on how technologies interact, how customers respond, and how changes in one part of the system affect other parts of the system—and total emissions. That will allow continuous refinement: identifying, for instance, the most effective way to optimize solar for the competing needs of car battery charging, battery storage, and grid supply.

The Future of Pecan Street

Data analysis, more fundamentally, will allow rigorous tracking of progress toward notable environmental goals: mapping changes on the demand side against power plant dispatch to track real-time emissions impacts, and revising the deployment as necessary to stay on course. Pushing beyond the Mueller pilot, Pecan Street has launched a research consortium composed of the Industry Advisory Council, UT researchers and experts from EDF, the Galvin Electricity Initiative, and Underwriters Laboratory to expand field testing of consumer smart grid architectures and interoperability standards.

Pecan Street technology director Bert Haskell and Dr. Robert Hebner, the director of UT's Center for Electromechanics and former acting director of the National Institute of Standards and Technology (NIST), will lead the research agenda. They will focus on determining what data is needed to drive innovation and track performance and discovering which new technologies and services customers value most.

Pecan Street and the National Renewable Energy Laboratory (NREL) are further building a smart-grid-interoperability research lab to enable utilities, vendors, and others to test the performance and integration of multiple companies' home smart grid systems with electricity, gas and water distribution, and back office systems.

The Pecan Street commitment to an open platform, innovation, transparency, and system optimization has at times tested the limits of the utility's willingness to share critical data—without which it is impossible to, for instance, map the most congested parts of the distribution system, which are the most valuable locations for distributed generation. Resolving those issues around data access will be critical to driving the kind of innovation that remade computing and telecom.

Though Austin Energy is owned by the city, nearly 75 percent of power in the United States is delivered by investor-owned utilities, regulated by public utility commissions (PUCs). For these utilities, state policymakers play a crucial role. It is up to them to push for open design and an open market that enables entrepreneurs and prices resources accurately. It is the job of policymakers to set economic and environmental goals, to require utilities to model and plan across their whole system to achieve these goals, and to set rates of return for utility investments linked to how well they perform on achieving these goals.

Pecan Street has designed and built the Pike Powers Laboratory and Center for Commercialization, which will promote research, commercialization, and educational opportunities for University of Texas students, faculty, and start-ups from UT's Austin Technology Incubator (ATI). Established in 2013, this is the nation's first non-profit smart grid research lab.[55]

Policy Options to Maximize Returns on Smart Grid Investments: Lessons from California

In developing good policy to drive smart grid investments that support the new energy economy, the first step is to come to agreement on ultimate goals. Does the city, state, or nation hope to achieve, for example, economic development, provision of reliable, low-cost power, or the protection of natural resources?

California's role as the first mover on Smart Grid laws and regulation builds on its long history as a leader in developing energy policies to simultaneously advance an innovation-centered economy and achieve ambitious goals for clean air, greenhouse gas reduction, and protection of water and habitat—policies that are often adopted by other states, or catalyze advances in national policy.

California also led the nation on building and appliance standards. The California Energy Commission (CEC) first adopted appliance efficiency standards in 1976[56] and followed with building efficiency standards ("Title 24") in 1978.[57] Both sets of rules are updated periodically to keep pace with technological changes, and as of 2013, the CEC's standards were credited with $74 billion of cumulative energy savings.[58] With its early-mover status and scale (a large enough market that manufacturers cannot afford to ignore it, even when California gets ahead of the pack), the state's appliance standards have influenced other states and federal efficiency standards.[59]

Beyond product standards, California additionally has led the way in developing transformative business models for utilities. In 1982, it became the first state to decouple utility revenues from energy sales in order to drive large-scale utility investments in reducing electricity demand.[60] Though decoupling was suspended in 1996, during electric restructuring, decoupling resumed in 2004 after the California Electric Crisis of 2000–2001.[61] The limitations of decoupling have become apparent: relying on the utility as the primary actor in advancing efficiency has constrained both financing sources and innovation, and resulted in relatively low energy savings per dollar invested. But this effort laid the ground for more radical rethinking of utility business models, and still more importantly, illuminated the need to open energy markets to third parties and the new technologies and services they can provide. Altogether, California's rocky path through the energy crisis set the stage for today's audacious, far-reaching regulatory reforms.[62] A 2002 state law established a requirement that California's major state energy agencies produce an Integrated Energy Policy Report every two years, assessing the industry and markets to keep policy apace with market realities and to "develop energy policies that conserve resources, protect the environment, ensure reliability, enhance the state's economy, and protect public health and safety."[63]

The California Public Utilities Commission (CPUC) also established a requirement that the investor-owned utilities (IOUs) comprehensively plan for energy procurement, independent of rate cases.[64] This requirement simultaneously sets the stage for planning smart grid deployments, in that utilities have become accustomed to having to plan both energy requirements and infrastructure needs well in advance, and is apt to be affected by the smart grid rollouts, as smart technology brings greater levels of renewables and EVs online and makes load more responsive to grid conditions.

In its *2003 Energy Action Plan*, California added another innovation: instituting a "loading order" giving preference to low-carbon resources; it calls for policymakers to "decrease electricity demand by increasing energy efficiency and demand response, and meet new generation needs first with renewable and distributed generation resources, and second with clean fossil-fueled generation."[65] The same year, California established one of the most aggressive renewable portfolio standards in the United States. More recently, a 2011 amendment raised the target to 33 percent of retail sales by 2020,[66] with the expressed aims of displacing fossil fuels, reducing air and climate pollution, stimulating investment in green technology in California, and promoting energy independence.[67]

Taxing an "Antiquated Distribution Infrastructure"

The commitment to shift priority to demand reductions, intermittent renewables and distributed resources focused attention on what a 2005 staff report by the California Energy Commission (CEC) called an "[a]ntiquated distribution infrastructure . . . not compatible with the advanced loading order technologies."[68] Full realization of the loading order, as per the CEC, would require advanced technology on the distribution system, including specialized metering and other sensors, communications, and controls, in order to facilitate expanding demand response and dynamic pricing, track and meter distributed generation, and bring Renewable Portfolio Standard (RPS)-mandated levels of renewable generation into the system.[69]

California's commitment to lead on greenhouse gas reduction further raised the performance requirements for the electric grid. The California Global Warming Solutions Act of 2006 (Assembly Bill 32) directed the California Air Resources Board (CARB) to cut carbon emissions to 1990 levels by 2020,[70] and to adopt "rules and regulations . . . to achieve the maximum technologically feasible and cost-effective greenhouse gas emissions reductions . . ."[71] The recent "road-maps to 2050" for California, developed by the California Council on Science and Technology[72] and researchers at Lawrence Berkeley National Labs,[73] have converged on a consensus that achieving these carbon targets will require the following: broad electrification of the economy (including transport and industry); scale deployment of low-carbon renewable supply; and a smart grid to manage the challenges of this new clean electric economy, including intermittency of wind and solar power, multidirectional electron flows from new distributed resources, and new load.

Recent rates of adoption and new targets for distributed solar PV and electric vehicles add yet another layer of demands on California's energy infrastructure. As of March 2013, California led the nation in solar projects, with an installed capacity of 1,515 MW.[74] In some communities, one in ten residences is now selling power back to the grid.[75] Policy and financing have both driven adoption: the PUC attributes 60 percent of the recent growth to funding through its California Social Initiative program.[76] In mid-2011, California Governor Jerry Brown announced a distributed clean energy goal (separate from the RPS) of 12 GW by 2020.[77] Meanwhile, in January 2012, the California Air Resources Board approved a package of vehicle-emission regulations, including a regulation that should result in 1.4 million "zero emission vehicles" on California roads by 2025 (15.4 percent of new vehicle sales in that year).[78]

Recognizing the Need for a Smart Grid

All of these commitments—to scale deployments of renewables, distributed resources, and electric vehicles—led to recognition by Senator Alex Padilla (Senate District 20) and others that the grid would need to be modernized to support that transformed system. The result was Senate Bill 17, which became a law in October 2009. The law requires the California Public Utilities Commission (CPUC), in consultation with stakeholders, to determine requirements and deadlines for utilities to submit smart grid deployment plans (and to report annually to the Governor and the Legislature on the status of plans and deployment, costs, and benefits to ratepayers and Commission recommendations). The law specifically identifies dynamic optimization and the integration of smart technology, renewables, demand response, and plug-in vehicles as key priorities for the smart grid.[79]

Implementation of the Smart Grid Law began with a 2010 order by the CPUC directing the state's three IOUs—Pacific Gas and Electric, Southern California Edison, and San Diego Gas & Electric—to plan smart grid deployments to meet state goals, including market access and environmental requirements. Specifically, the Smart Grid Deployment Order required that in their deployment plans, IOUs should identify how their vision of the smart grid will address three points proposed by EDF and adopted by the CPUC:

- "Enable maximum access by third parties to the grid, creating a welcoming platform for deployment of a wide range of energy technologies and management services;
- Have the infrastructure and policies necessary to enable and support the sale of demand response, energy efficiency, distributed generation, and storage into energy markets as a resource on equal footing with traditional generation resources; and
- Significantly reduce the total environmental footprint of the current electric generation and delivery system in California."[80]

The goal of opening market participation well beyond the utility was particularly underscored in the decision:

"The vision must address how the plans will enable consumers to capture the benefits of a wide range of energy technologies and energy management products and services that may, or may not, be provided by the utility, while protecting consumers' privacy . . . ."[81]

The decision set a deadline of July 2011 for the plans, which were to include a vision statement, deployment baseline, strategies for grid security and cyber security, a roadmap, cost and benefit estimates, and metrics to measure progress toward the goals of the smart grid law. The order provided for these metrics to be developed through a subsequent process, and EDF has been working with the CPUC and the IOUs to develop those metrics. Drawing on functional goals suggested by the legislation as well as recommendations of EDF, the CPUC specified that the utility smart grids must do the following:

(a) Be self-healing and resilient;
(b) Empower consumers to actively participate in grid operations;
(c) Resist attack;
(d) Provide higher quality power and avoid outages;
(e) Accommodate all generation and energy storage options;
(f) Enable electricity markets to flourish;

(g) Run the grid more efficiently;

(h) Enable penetration of intermittent power generation sources;

(i) Allow a wide range of energy technologies and management services;

(j) Enable and support the sale of demand response, energy efficiency, distributed generation, and storage into wholesale energy markets as a resource, on equal footing with traditional generation resources; and

(k) Significantly reduce the total environmental footprint.[82]

Simultaneous with the planning process, all three utilities have been deploying smart meters and other sensors across their systems. The data they are gathering from these sensors are providing unprecedented insight into their networks: where the weaknesses are, the most cost-effective options for fortifying these sections, and the capability for advanced modeling using data from ubiquitous sensors to understand whole-system ramifications of any particular technology deployment and target investment. An important next step for policymakers in California and elsewhere will be to require that kind of data gathering, analytics, and modeling to ensure investments achieve the desired outcomes in enabling the green energy economy.

California is just one of many places in the world leading the way. The European Union (EU) has committed to reduce carbon emissions to 80–95 percent below 1990 levels by 2050; similar to California, the EU has put forward a roadmap to meet these carbon targets, which calls for greater electrification and renewable generation and a modernized grid to support both.[83] South Korea plans to invest $27.5 billion in smart grid technologies and another $37.5 billion in new and renewable energy by 2030, to create 1.2 million jobs, add $74 billion to GDP, and avoid approximately $7.9 billion in costs for energy imports and new power plant construction. China plans to invest $625 billion in the smart grid and some $300–$450 billion in renewable energy by 2030, creating 11.1 million jobs.[84]

The economic value of smart grid investments is already being felt in California and across the United States. The Silicon Valley Leadership Group found a 129 percent growth in local employment in smart grid sectors from 1995 to 2009, with half of the jobs in manufacturing.[85] A report from Duke University's Center on Globalization, Governance and Competitiveness[86] found 17,000 new smart grid jobs in 334 locations in thirty-nine states, with the potential for nearly 300,000 jobs by the end of 2012. Looking to 2020, SDG&E's smart grid plan forecasts that its investment of $3.5 billion will return $3.8–7.1 billion with benefits in reliability, energy savings, and carbon reductions.[87] PG&E's investment of $1.2 billion is expected to return $2 billion in economic benefits.

These numbers likely understate the case to a very significant degree. A March 2011 white paper[88] argues that system transformations, such as the one now beginning in energy, achieve their most profound effects by transforming the broader economy and ways of life. The railways, electrification, and the rise of information technology created enormous economic growth "not merely in the rail or energy or IT sectors, but because of the opportunities created by innovation in those sectors for the economy as a whole. Indeed, in each case the growth in a given sector was far outstripped by the growth in the broader economy. That growth came about because the innovations themselves changed what was possible for economic production, and those changes drove massive and repeated investment in new business models, products, and modes of production."[89]

New technologies, especially networks, transform our way of life and our very perceptions of what is real and what is possible. The network revolutions in the late nineteenth and early twentieth centuries—beginning with railways, but continuing with telegraphs and electrification—transformed our perception of space and time. These new networks did not just give us new modes of transportation, communication, and working; their consequences, including globally standardized time and the blurring of day and night through electric lighting, transformed our sense of time in ways that artistic and literary luminaries of the period—from Paul Cézanne and Marcel Proust to Franz Kafka, James Joyce, and Salvador Dalí—grappled with in their works.[90] These new ways of connecting helped bring us modernity. Similarly, to fully appreciate the ramifications of today's new networks, including the next-generation electric system, it is critical to recognize the network itself as the innovation platform and to create a sufficiently open structure for "the energy networks of tomorrow, and the markets that govern how energy is produced, distributed and used . . . to allow the entire economy to discover the growth potential."[91] That is the promise charted by the open platforms and markets now being developed in Texas, California, Europe, China, Korea, and other leading edges of the global smart grid revolution.

Notes

1. "Report Maps California's Energy Future to 2050," California Council on Science and Technology, May 2011, http://ccst.us/publications/2011/2011energy.php

    James H. Williams, Andrew DeBenedictis, Rebecca Ghanadan, Amber Mahone, Jack Moore, William R. Morrow III, Snuller Price, Margaret S. Torn, "The Technology Path to Deep Greenhouse Gas Emissions Cuts by 2050: The Pivotal Role of Electricity," November 2011, <http://www.sciencemag.org/content/early/2011/11/22/science.1208365.abstract?sid=ba72f8c7-16a6-4236-8e3e-6f5773788788>.

    "Communication From The Commission To The European Parliament, The Council, The European Economic And Social Committee And The Committee Of The Regions," Energy Roadmap 2050. 2011, <http://ec.europa.eu/energy/energy2020/roadmap/doc/com_2011_8852_en.pdf>.

2. "Accelerating Successful Smart Grid Pilots," *World Economic Forum & Accenture*, 2010, 12. <http://www3.weforum.org/docs/WEF_EN_SmartGrids_Pilots_Report_2010.pdf>.

3. "How much electricity is lost in transmission and distribution in the United States?," *United States Energy Information Administration*, US Department of Energy. Web. March 2012. <http://www.eia.gov/tools/faqs/faq.cfm?id=105&t=3>.

4. "Smart Grid: A White Paper," Xcel Energy, February 2008, <http://www.e-renewables.com/documents/Smart%20Grid/Xcel%20Energy%20Smart%20Grid.pdf>, 5.

5. "Principles for Efficient and Reliable Reactive Power Supply and Consumption" Federal Energy Regulatory Commission, February 2005, http://www.ferc.gov/eventcalendar/files/20050310144430-02-04-05-reactive-power.pdf.

6. Ibid.
7. Miriam Horn teleconference with Jessica Harrison (KEMA, Inc.), Tom Willie (President, Current Group), Mani Vadari (President, Modern Grid Solutions) and consultant Roger Duncan, November 22, 2011.
8. Ibid.
9. Mike Prevallet, Tom Johnson. "Integrated Volt/VAR Control," Cooper Power Systems <http://www.smartgridnews.com/artman/uploads/1/Optimizing_Voltage.pdf>.
10. "Ventyx Launches Network Manager (TM) DMS v5.3 With Model-Based Volt/VAR Optimization," IT News Online, December 2011, <http://www.prnewswire.com/news-releases/ventyx-launches-network-manager-dms-v53-with-model-based-voltvar-optimization-135018028.html>.
11. "Ventyx Customer Oklahoma Gas & Electric Recognized as One of Nation's Top 10 Smart Grid Deployments," *Ventyx Press Release* June 6, 2012. <http://www.ventyx.com/en/company/news/press/20120606-oklahoma-ge>
12. "Smart Grid System Report," U.S. Department of Energy, July 2009, <http://energy.gov/sites/prod/files/oeprod/DocumentsandMedia/SGSRMain_090707_lowres.pdf>.
13. "The Pecan Street Project: Building Tomorrow's Electrical Grid in Austin, TX," *Environmental Defense Fund*, (Spring 2011), 4. <http://www.edf.org/sites/default/files/11821_pecan-street-project-report-spring-2011.pdf>.
14. Minkel, J. R., "The 2003 Northeast Blackout Five Years Later," *Scientific American*, August 2008, <http://www.scientificamerican.com/article.cfm?id=2003-blackout-five-years-later>.
15. Redell, Charles, "Alabama Power Finds a New Benefit for Smart Meters: Disaster Recovery," September 2011, <http://www.greenbiz.com/news/2011/09/15/alabama-power-finds-new-benefit-smart-meters-disaster-recovery>.
16. Booth, Adrian, Michael Green, and Humayun Tai, "U.S. Smart Grid Values at Stake: The $130 Billion Question," *McKinsey on Smart Grid* 1 (Summer 2010), 5. <http://www.mckinsey.com/Client_Service/Electric_Power_and_Natural_Gas/Latest_thinking/McKinsey_on_Smart_Grid>.
17. Smith, Rebecca, "Texas Power Grid Falls Short," *Wall Street Journal*, August 2011, <http://online.wsj.com/article/SB10001424053111904823804576502592393033486.html>.
18. "A National Assessment of Demand Response Potential," *Federal Energy Regulatory Commission*, June 2009, xi to xii, <http://www.ferc.gov/legal/staff-reports/06-09-demand-response.pdf>.
19. "The Pecan Street Project: Building Tomorrow's Electrical Grid in Austin, TX," *Environmental Defense Fund* (Spring 2011) 4, <http://www.edf.org/sites/default/files/11821_pecan-street-project-report-spring-2011.pdf>.
20. "The Hidden Costs of Energy: Unpriced Consequences of Energy Production and Use," National Academies of Sciences Committee on Health, Environmental, and Other External Costs and Benefits of Energy Production and Consumption; National Research Council, et al., <http://www.nap.edu/catalog.php?record_id=12794>.
21. Todd, DeWayne, and Brian Helms, Mike Caufield, Michael Stark, Brendan Kirby and John Kueck, "Evaluation of the Demand Response Capabilities of Alcoa Inc," January 2009, 1, <http://certs.lbl.gov/pdf/dr-alcoa.pdf>.
22. "A National Assessment of Demand Response Potential," *Federal Energy Regulatory Commission*, June 2009, Pages xi to xii, <http://www.ferc.gov/legal/staff-reports/06-09-demand-response.pdf>.
23. "A National Assessment of Demand Response Potential," *Federal Energy Regulatory Commission*, June 2009, xii, <http://www.ferc.gov/legal/staff-reports/06-09-demand-response.pdf>.

24. "Transforming America's Power Industry: The Investment Challenge 2010–2030," *The Brattle roup*, November 2008, 2, <http://www.brattle.com/_documents/upload library/upload725.pdf>.

25. Herig, Christy, "Using Photovoltaics to Preserve California's Electricity Capacity Reserves," *National Renewable Energy Laboratory*, September 2011, 2 <http://www.nrel.gov/docs/fy01osti/31179.pdf>.

26. Dosunmu, Ade, "Up in Smoke: Seven Gigawatts of Coal Retrenches from PJM,"*Energy Pulse*, <http://www.energypulse.net/centers/article/article_display.cfm?a_id=2451>.

27. Todd, DeWayne, Brian Helms, Mike Caufield, Michael Stark, Brendan Kirby, and John Kueck, "Providing Reliability Services Through Demand Response: A Preliminary Evaluation of the Demand Response Capabilities of Alcoa Inc.," January 2009, 27, <http://certs.lbl.gov/pdf/dr-alcoa.pdf>.

28. DeWayne Todd, Brian Helms, Mike Caufield, Michael Stark, Brendan Kirby, and John Kueck. "Providing Reliability Services Through Demand Response: A Preliminary Evaluation of the Demand Response Capabilities of Alcoa Inc.," January 2009, 5, <http://certs.lbl.gov/pdf/dr-alcoa.pdf>.

29. FERC Order 755. Paragraph 72.

30. The Ivanovich Group, "Industry Research & Report: Smart Grid," *Danfoss*, Oak Park, IL, June 2011, <http://www.smartgridinformation.info/pdf/4528_doc_1.pdf>.

31. "Eastern Wind Integration and Transmission Study," *The National Renewable Energy Laboratory*, February 2011, <http://www.nrel.gov/docs/fy11osti/47078.pdf>.

32. "Western Wind and Solar Integration Study," *The National Renewable Energy Laboratory*, May 2010, <http://www.nrel.gov/docs/fy10osti/47434.pdf>.

33. "Technology Roadmap: Smart Grids," *International Energy Agency*, 2011, <http://www.iea.org/publications/freepublications/publication/smartgrids_roadmap.pdf>.

34. Mani Vadari, Mike Davis, "Investigating Smart Grid Solutions to Integrate Renewable Sources of Energy into the Electric Transmission Grid," Presented at UHV 2009 conference in Beijing, China.
    Pratt, R. G, et, al., "The Smart Grid, an Estimation of Energy and CO2 Benefits," January 2010, 2.4–3.4, <http://www.pnl.gov/main/publications/external/technical_reports/PNNL-19112.pdf>.

35. "Press Release: KEMA Sees Opportunities for Storage Applications in High-Penetration Renewables," *KEMA*, June 2010, <http://renewablesbiz.com/article/10/06/kema-sees-opportunities-storage-applications-high-penetration-renewables>.

36. "Research Evaluation of Wind Generation, Solar Generation, and Storage Impact on the California Grid," *KEMA*, June 2010, <http://www.energy.ca.gov/2010publications/CEC-500-2010-010/CEC-500-2010-010.PDF>.

37. "Market Analysis of Emerging Electric Storage Systems,"*DOE/NETL* , July 2008, 6–7.

38. "Achieving High Performance with Solar Photovoltaic (PV) Integration," Accenture, 2011, <http://www.accenture.com/SiteCollectionDocuments/PDF/Accenture-Achieving-High-Performance-Solar-Photovoltaic-Integration.PDF>.

39. Infanzon, Armando, "Author interview," *SDG&E*, November 2011.

40. Stilwell Andrade, Miguel, "Author interview," Inovgrid, EDP Distribuicao. EDP (Energias de Portugal), October 2011.

41. Faruqui, Ahmad, Ryan Hledik, Armando Levy, and Alan Madian. "Will Smart Prices Induce Smart Charging of Electric Vehicles?," *The Brattle Group*, July 2011, 4.

42. Faruqui, Ahmad, Matthew E. Kahn, and Ryan K. Vaughn, "Green Market Geography: The Spatial Clustering of Hybrid Vehicles and LEED Registered Buildings," *The B.E. Journal of Economic Analysis & Policy*, 2009, 20.

43. Faruqui, Ahmad, Matthew E. Kahn, and Ryan K. Vaughn, "Green Market Geography: The Spatial Clustering of Hybrid Vehicles and LEED Registered Buildings," *The B.E. Journal of Economic Analysis & Policy* (2009): 2.

44. Yuhnke, Robert E., and Michael Salisbury, "Ozone Precursor and GHG Emissions from Light Duty Vehicles—Comparing Electricity, Natural Gas and Biofuels and Transportation Fuels," 2010, 1.

45. Ibid.

46. Tuffner, F., and M. Kintner-Meyer, "Using Electric Vehicles to Meet Balancing Requirements Associated with Wind Power," <http://energyenvironment.pnnl.gov/pdf/PNNL-20501_Renewables_Integration_Report_Final_7_8_2011.pdf>.

47. Chesbrough, Henry, "Reinvention," *Technology Review*, MIT, November–December 2011, <http://www.technologyreview.com/computing/38872/>.

48. Metcalfe, Bob, "Enernet: Internet Lessons for Solving Energy," Polaris Venture Partners, March 2009, http://www.polarisventures.com/documents/Enernet.pdf.

49. At the 2011 Gridweek, Roger Duncan was awarded a Leadership Role for his "visionary" impact on the utility industry.

50. Conversation with Miriam Horn, New York, January 2012.

51. The USNAP Alliance is working to solve this problem for global manufacturers struggling to make products compatible with the various communications protocols being adopted by different utilities and regions of the world. Their solution is a card that can be "snapped" into a device to enable it to talk to any smart grid, like the USB port on a PC that allows a myriad of applications. Similarly flexible solutions will need to be provided for the gateway itself. <http://www.usnap.org/>.

52. "The Value of Distributed Photovoltaics to Austin Energy and the City of Austin," Clean Power Research, L.L.C. Austin Energy, 2006, <http://www.ilsr.org/wp-content/uploads/2013/03/Value-of-PV-to-Austin-Energy.pdf>; Lovins, Amory and Karl Rabago, et al. "Small is Profitable," 2003, <http://www.smallisprofitable.org/207Benefits.html>.

53. Email correspondence with Josh Gerber of Sempra Utilities, dated March 21, 2013 (on file with authors).

54. "Top Five Electric Vehicle Initiatives of the Year," GreenTech Media, December 2011, <http://www.greentechmedia.com/articles/read/top-five-ev-initiatives-of-the-year/>.

55. "Pike Powers Laboratory and Center for Commercialization," Pecan Street Inc., March 2013, <http://www.pecanstreet.org/projects/lab/>.

56. *See* the California Energy Commission's website describing *Historical Regulations & Rulemakings*, http://www.energy.ca.gov/appliances/previous_regulations.html (accessed April 8, 2013).

57. "California's Energy Efficiency Standards for Residential and Nonresidential Buildings," Title 24, Part 6, of the California Code of Regulations, <http://www.energy.ca.gov/title24/>.

58. Ibid.

59. Carlson, Ann E., "Commentary, Energy Efficiency and Federalism," *Mich. L. Rev. First Impressions* 63 (2008): 107, <http://www.michiganlawreview.org/assets/fi/107/carlson.pdf>.

60. Mufson, Steven, "In Energy Conservation, Calif. Sees Light; Progressive Policy Makes It a Model in Global Warming Fight," *Washington Post* February 2007, <http://www.washingtonpost.com/wp-dyn/content/article/2007/02/16/AR2007021602274.html.

61. American Council for an Energy-Efficient Economy. "State Energy Efficiency Policy Database," 2011, <http://www.aceee.org/sector/state-policy/california>.

62. Ibid.

63. Pub. Res. Code § 25301(a).
64. "Long-Term Procurement Plan History," *California Public Utilities Commission*, 2011, <http://www.cpuc.ca.gov/PUC/energy/Procurement/LTPP/ltpp_history.htm>.
65. "Implementing California's Loading Order for Electricity Resources," *California Energy Commission*, July 2005, E-1.
66. "Order Instituting Rulemaking Regarding Implementation and Administration of the Renewables Portfolio Standard Program," Filed: Public Utilities Commission. San Francisco Office, Rulemaking 11-05-005, May 2011, <http://docs.cpuc.ca.gov/word_pdf/FINAL_DECISION/134980.pdf>.
67. Ibid.
68. Sylvia Bender, et, al., "Implementing California's Loading Order for Electricity Resources," California Energy Commission, July 2005, E-4, <www.energy.ca.gov/2005publications/.../CEC-400-2005-043.PDF>.
69. Id., at E-13–E-14.
70. AB 32, Part 3, Section 38550.
71. AB 32, Part 4, Section 38560.
72. "California's Energy Future: The View to 2050," California Council on Science and Technology, May 2011, <http://www.ccst.us/publications/2011/2011energy.pdf>.
73. Williams, James H., Andrew DeBenedictis, Rebecca Ghanadan, Amber Mahone, Jack Moore, William R. Morrow III, Snuller Price, Margaret S. Torn. "The Technology Path to Deep Greenhouse Gas Emissions Cuts by 2050: The Pivotal Role of Electricity," *ScienceExpress*, November 2011. Page 2.
74. PV installation statistics are available at the "Go Solar California," *State of California, California Energy Commission & California Public Utilities Commission*, <http://www.gosolarcalifornia.ca.gov/ (accessed March 21, 2013).
75. Lee, Morgan, "San Diego Leads State in Rooftop Solar," *UT San Diego*, <http://www.utsandiego.com/news/2012/jan/23/san-diego-leads-state-in-rooftop-solar-power-gener/>.
76. Lee, Morgan, "Electricity from rooftop solar hits milestone in California," *UT San Diego*, November 2011, <http://www.utsandiego.com/news/2011/nov/10/rooftop-solar-hits-1-gigawatt-mark-in-california/>.
77. Hsu, Tiffany, "Gov. Brown pushes 12-gigawatt clean-power goal," July 2011, <http://articles.latimes.com/2011/jul/26/business/la-fi-small-renewables-20110726>.
78. "California Air Resources Board Approves Advanced Clean Car Rules," News Release, California EPA Air Resources Board, January 2012, <http://www.arb.ca.gov/newsrel/newsrelease.php?id=282>.
79. *See* California Senate Bill 17 (Padilla), Electricity: Smart Grid systems. (Approved by the Governor and filed with the Secretary of State, October 11, 2009).
80. California Public Utility Commission, "Decision Adopting Requirements for Smart Grid Deployment Plans Pursuant to Senate Bill 17 (Padilla), Chapter 327, Statutes of 2009," (Decision 10-06-047 June 24, 2010), at 34 and 139, <http://docs.cpuc.ca.gov/word_pdf/FINAL_DECISION/119902.pdf> *Id.* at 34.
81. Id. at 4.
82. Id. at 133–34.
83. "Communication from the Commission to the European Parliament, the Council, the European Economic & Social Committee & the Committee of the Regions," Energy Roadmap 2050, 2011, <http://ec.europa.eu/energy/energy2020/roadmap/doc/com_2011_8852_en.pdf>.
84. Byrne, John. Gridweek Presentation. Center for Energy and Environmental Policy, University of Delaware, September 2011, <http://www.pointview.com/data/2011/09/52/pdf/John-Byrne-CSWZQKEA-13354.pdf>.

85. "Smart Grid Deployment and the Impact on Silicon Valley," Silicon Valley Smart Grid Task Force, 2011, <http://svlg.org/docs/smartgrid_sv_2011.pdf>.
86. Lowe, Marcy, Hua Fan, and Gary Gereffi, *U.S. Smart Grid: Finding new ways to cut carbon and create jobs* (Durham: Duke University, 2011), <http://cggc.duke. edu/pdfs/Lowe_US_Smart_Grid_CGGC_04-19-2011.pdf>.
87. SDG&E, *Smart Grid Deployment Plan* (San Diego: SD&G, 2011), 284, <http:// sdge.com/sites/default/files/regulatory/deploymentplan.pdf>.
88. Huberty, Mark, Huan Gao, Juliana Mandell with John Zysman, *Shaping the Green Growth Economy: A review of the public debate and the prospects for green growth* (Berkeley: The Berkeley Roundtable on the International Economy, 2011). <http:// brie.berkeley.edu/publications/Shaping-the-Green-Growth-Economy-BRIEreport. pdf>.
89. Huberty, Mark, Huan Gao, Juliana Mandell with John Zysman, *Shaping the Green Growth Economy: A review of the public debate and the prospects for green growth* (Berkeley: The Berkeley Roundtable on the International Economy, 2011), 27–28, <http://brie.berkeley.edu/publications/Shaping-the-Green-Growth-Economy-BRI-Ereport.pdf>.
90. Kern, Stephen, "The Culture of Time and Space 1880–1918," Harvard University Press, 1983. Chapter 1.
91. Australian Academy of Technological Sciences and Engineering (ATSE), "Green Growth—Energy: Industry Opportunities for Australia." Australian Research Council, 2013. Document available at: https://www.atse.org.au/Documents/Publications/ Reports/Energy/Green%20Growth%20Energy%20Report/ATSE%20Green%20 Growth%20Report.pdf.

15

Integrated Green Energy Approaches

Peter D. Lund

Introduction to Green Energy on an Urban Scale—Understanding the Broader Picture

As the main source of major challenges in energy and climate, cities or urban areas in general form an interesting case for analyzing the green energy economy. Interestingly, by employing green technologies that address energy and climate issues, cities may simultaneously offer solutions to these grave problems, in addition to creating new business opportunities and economic growth. A city has a dual character—it is the problem, and at the same time, it can be the solution.

Most of the global energy use is concentrated in urban areas. Depending on the source used, 50–70 percent of our total energy demand comes from cities. With growing urbanization, this share is expected to increase in the future so that by 2040, close to 80 percent of all energy demands may stem from urban energy use.[1,2] Presently, around half of all people live in cities, but by the middle of this century, this number may increase to 70 percent. As more than 80 percent of global energy is derived from fossil fuels, which causes release of carbon dioxide emissions, the source of energy and climate problems can now easily be tracked back to urban lifestyles and how we handle energy on a city scale. It is important to notice that two energy sources dominate carbon emissions above any other perceivable factors. Coal and oil account for 80 percent of all global CO_2 emissions.[3] The key to mitigating climate change truly lies in replacing oil and coal by sustainable substitutes.

Coal represents one quarter of all the fuel needed in power plants to produce electricity. Because of its high specific emissions (1 MWh of electricity produced by a coal power plant releases 1 ton of CO_2), coal-based electricity represents

40 percent of all the CO_2 released into the atmosphere. An additional factor that affects the emissions of coal-based electricity is the overall conversion efficiency from fuel to electricity, which, in traditional thermal power plants, is around 40 percent. Making use of the waste heat—which represents more than half of the energy content of the fuel—from power plants could be one of the most import-ant measures to improve fuel efficiency, which, in turn, could reduce emissions significantly without any change in the fuel mix. Cogeneration or combined heat and power (CHP) plants offer an effective way of pushing the conversion efficiency from fuel to final energy to as far as 90 percent. This chapter focuses on cogeneration, and the other different options available will be discussed later.

The importance of finding effective and sustainable ways to deal with power production and electricity use cannot be overemphasized for the simple reason that demand for electricity is increasing faster than energy supply. The relative weight of electricity in our final energy use is set to rise in the decades to come. Electricity now represents 20–25 percent of the final energy use of cities in gen-eral, but by 2050, it could be over 50 percent. The driving factor for this increase is the usefulness of electricity and the ease with which it provides almost any kind of service, and specifically its role in fueling the whole digital economy.

The other half of the climate problem is related to oil, as gasoline is the basis for almost all modes of transport, particularly personal vehicles. A simple calcu-lation shows that if all Chinese adopted the American way of moving around by car, there would be a 150 percent increase in the world oil demand. Clearly, this would no longer be a climate issue alone but a resource issue, as oil reserves are limited. At the present levels of consumption, oil reserves will last around forty years, but serious shadows have also been cast on the adequacy of the present economically recoverable reserves, which may run out much faster. Needless to say, a multitude of different options—ranging from structural changes in trans-port modalities to improved energy efficiency in the transport sector, alongside introduction of biofuels—will be necessary in the years to come.

The second important dimension to understanding urban areas with regard to green energy is their relation to innovations. It is well-known that social inter-actions and communication are central to new innovations and their penetration into the market. A city is both a source and sink of know-how, that is, it can generate both a technology push and a market pull to bring new innovations to the market. Cities represent a huge energy market and the public sector alone may represent around 10 percent of all urban energy demands (e.g., public buildings). Local policymakers may easily influence the demand for green energy and be frontrunners in opening new markets. Indeed, an increasing number of cities worldwide are demonstrating success in pursuing such green energy policies, not only from the climate or energy perspective but also by creating economic growth through green energy. Different types of public–private partnerships, which may involve the development of local energy sources and technologies, pave the way for new business and jobs. A successful green energy policy would simultaneously bring solutions to society's two big challenges, namely jobs and climate. New innovations, such as green energy, generate new jobs, whereas old industries such as paper and pulp, car manufacturing, or chemicals are losing jobs.

This is the main framework in which green energy approaches should preferably be analyzed to understand their full potential, though a stronger technological perspective will be considered here when analyzing integrated green energy approaches. Nevertheless, it is important to understand that technology has many more dimensions than just the environmental, technical, or cost-related ones, which have traditionally been emphasized as the primary drivers in energy change. Consequently, our decisions and choices in energy use have been dominated by factors such as energy security, environment, and economics. These "three Es" are still important and valid but are not adequate to combat climate change and the major societal issues of unemployment, energy poverty, negative health impacts, etc. The old thinking in energy leads to a too-narrow, silo-like thinking, and ultimately to less effective (financial) resource utilization. The more modern framework explained earlier, which is perhaps complementary to the old framework, combines "three coupled loops," namely innovations and know-how, job creation and economic growth, and energy and climate solutions, into one package.

Such a green triangle (see Figure 1 below) forms the basis of the European Union's (EU) future strategy until the year 2020.[4] Europe hopes to create close to one million new jobs in ten years from its renewable energy and energy-efficiency targets only. Meanwhile, through its twelfth-five-year state program, China aims to create an "innovation China" with forty million new skilled

Figure 1
Green Energy Triangle

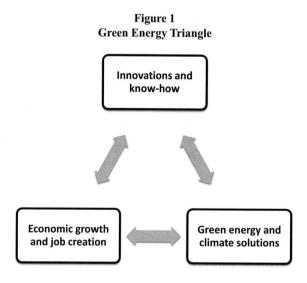

jobs and is pouring hundreds of billions of dollars into investments in clean energy and energy efficiency. These examples support the new thinking that green growth and green energy go hand in hand. One may argue that increased economic activities would require more energy anyway, irrespective of whether they are green or gray, that is, energy and growth are coupled, which has been quite obvious in some emerging economies such as China. In most highly in-dustrialized Organisation for Economic Co-operation and Development (OECD) countries, however, decoupling has taken place, and in Sweden, for example, the electricity demand has not increased notably (about 0.1–0.3 percent per year) since the early 1990s despite economic growth. Therefore, the major economic component associated with green energy must be derived from new technologies and innovations through which green energy is produced, rather than only from non-carbon-energy production, which is essentially an energy commodity. We could even counter-argue that investing in traditional energy with the goal of creating new jobs, for example, by building more traditional power plants, would be highly inefficient when compared to green energy investments with a strong new technology dimension.

A change toward a sustainable future cannot take place without investing in technology and green energy solutions. We may well place sustainable energy technology at the center of the Green Energy Triangle. A number of green energy options are available, ranging from energy-efficient technology to renewable energy production. Some of these, such as solar photovoltaics (PV) or wind pow-er, are now in the process of being scaled up to a mass market and are growing annually in two-digit percentages.[5] With growing volumes, prices are expected to come down further, thus accelerating market take-up. Some technologies, such as marine and wave power, are still in an early stage of development and no winning design has yet emerged, which would be necessary for commercial consolidation and scale-up. Bioenergy in its different forms is used for various purposes (fuel, heat, and electricity) and has an established position within several markets worldwide. The role of new energy technology over the coming decades will be crucial in assisting the shift from the conventional polluting energy sources to more sustainable types. New technologies, broadly defined, could account for as much as a 70 percent reduction in carbon emissions by 2050.[6] For comparison, climate change stabilization would require global greenhouse gas emissions to be reduced by 50–60 percent by the middle of the century.[7]

Some countries have chosen the more radical path of basing most of their future energy supply on renewable energy only. In October 2011, Denmark an-nounced a plan to work toward a 100 percent renewable-based energy system by 2050, and not surprisingly, highlighted the importance of the green-collar jobs that should be created along this path. The Danes make a strong case for increas-ing the use of wind power, but see an even more important role for bioenergy, particularly in bio-cogeneration systems. Many cities worldwide are embracing similar sustainability ideas and a variety of eco-city concepts have emerged.[8,9]

Germany, similarly, has ambitious goals for green energy. Germany plans to rely more heavily on energy efficiency and renewable energy, which should account for 80 percent of all energy by 2050. Before this decision, however, the Germans had already internalized the green energy economy with their feed-in tariff (FIT) system. Germany's efforts have resulted in more than 300,000 new jobs in various renewable energy industries with positive economic benefits.[10]

Cogeneration—a technology that produces both thermal and electrical energy—may play an increasingly important role when all the aspects of green energy discussed above are put together. Commonly used in northern countries with notable heating demands and industries that require high levels of processed heat, the technology allows for a diverse range of fuels, both fossil and renewable, to be used with high fuel use efficiency. Very high wind or solar electricity targets could motivate the conversion of some of the surplus electricity into thermal energy. In such a context, cogeneration could provide an important support function and increase energy system flexibility.

The strong local character of cogeneration utilizing local fuels and expertise, accompanied by its high fuel efficiency, also makes it a strong case for the green energy economy. As shown later in this chapter through a few cases from Finland, bio-based cogeneration originating from local interest and demand could be counted among the most successful examples of the green energy economy.

Cogeneration as a Green Energy Option on an Urban Scale

Cogeneration or CHP is by no means a disruptive new energy innovation like solar or wind energy, but rather a well-established technology. The world's first combined heat and power plant was put into use more than one hundred and twenty years ago in New York by Thomas Edison. The New York City steam system still supplies steam to tens of thousands of customers in Manhattan. But the main scale-up of this energy innovation, which originated in the United States in the late nineteenth century, has actually taken place in Europe.

About 10 percent of all power production worldwide is from CHP. Most of the capacity is found on the European continent. In Denmark, Russia, Finland, Latvia, and the Netherlands, cogeneration accounts for more than 25 percent of electricity, but notable shares are found in many other European countries as well (e.g., Hungary 19 percent, Poland 17 percent, Austria 15 percent, and Germany 12 percent). Cogeneration—often integrated with district heating—was preferred in many former Communist countries, explaining its high share in Eastern Europe. In the United States, the share is around 8 percent, and in China, 12 percent. Outside these regions, cogeneration is marginal. Its potential is, however, large, and by 2030, it could represent up to 20 percent of the total global power production.[11]

Denmark is the leading nation in the use of combined heat and power; more than half of all the country's power is derived from this technology. The use of CHP in Denmark is concentrated in urban and municipal areas. Roughly two-thirds of all buildings are served through district heating systems fed by CHP. There are close to 1,000 CHP plants of varying sizes in Denmark. The largest ones provide over 95 percent of the heat needed in the country's ten largest cities, but there are many examples of small community and institutional building installations as well. The main fuels used in cogeneration are natural gas and biomass.[12,13]

The Danish authorities estimate that such an extensive use of CHP yields energy savings of the order of 30 percent and reduces the country's emissions by 15–20 percent a year, in comparison with a more traditional baseline (i.e., separate power and heat production). Denmark represents a success story of the potential of cogeneration on an urban scale. Two factors played a major role in this. First, without the persistent Danish energy policy favoring cogeneration since the days of the oil crisis in the 1970s, the country would never have been able to build up the extensive infrastructure needed for deploying cogeneration in urban areas. Second, the Danish legislation has ensured that district heat (which originates from CHP plants) needs to be priced at a non-profit or modest profit level. This is highly important because cogeneration in cities may offer an almost monopolistic heat supply setting to private CHP or municipal energy companies, which could then easily misuse their position unless proper regulations are in place. As a result, in Northern Europe, including Denmark, district heating is almost always the cheapest option for consumers. The socioeconomic impact of CHP is good and it is viewed among the majority of consumers as the preferable option.

In another country leading in the use of cogeneration, namely Finland, where cogeneration accounts for 38 percent of electricity, CHP was originally introduced through industrial use. The country had (and still has) an extensive forestry industry. The production of pulp and paper not only requires huge amounts of steam and heat but also generates large flows of biomass waste, which is an excellent fuel for CHP. The bio-CHP plants—created to capture this dual dynamic—also started to supply heat to the surrounding communities.

Through the extensive experience gained from these activities, a strong technology industry was created around CHP, particularly in connection with biomass and developing the necessary technology to convert different solid fuels into heat. Leading boiler or gasification (e.g., fluidized bed technology) manufacturers are today found in the country. The Finnish success story in CHP comes largely from the kind of technology procurement environment that was created, that is, industries not only tendered for better and cheaper technology, but also rewarded the winners sumptuously with orders and long-term business relations.[14]

Contrary to many renewable energy sources that represent a single technology solution, cogeneration is rather a family of technologies utilizing different fuels in gaseous, liquid, or solid form. A common feature of all CHP alternatives is high fuel efficiency. Biomass and biogas would be the main renewable sources for cogeneration. It is common to employ different waste streams in CHP plants, for example, residuals from forestry industries, organic municipal waste, and landfill gas. Using such biowaste offers a double economic benefit, as waste is eliminated and low-cost fuel is provided for the power plant.

The flow diagram of a cogeneration plant is shown in Figure 2. The conversion from fuel to final energy can be achieved through different thermodynamic cycles, depending on the type of fuel used. A solid biomass (or waste) CHP plant would be quite similar to an ordinary thermal power plant except for a few additional components (e.g., heat exchangers, different turbine construction) that are needed to draw heat from the process. For biogas or gasified biomass, two different options can be used, either a gas engine or a gas turbine (often a combined cycle gas turbine [CCGT]), both of which need some modifications to recover heat. The heat withdrawal or heat recovery may lead to a minor decrease in the electrical output compared to an ordinary power plant.

Existing technological and engineering solutions give rise to five basic types of CHP plants: steam back-pressure turbines; combined cycle plants; steam condensing turbines; gas turbines; and engines with heat recovery.[15] Normally, CHP plants are on the megawatt power scale, even up to 1,000 MW, but small units or so-called micro- and mini-CHPs can also be found. Micro-CHP provides a few kilowatts of output power and would typically be employed in residential buildings, while mini-CHP provides a power output of more than

Figure 2
Process Flow of Separate Heat and Power Production (Left)
and CoGeneration (Right)

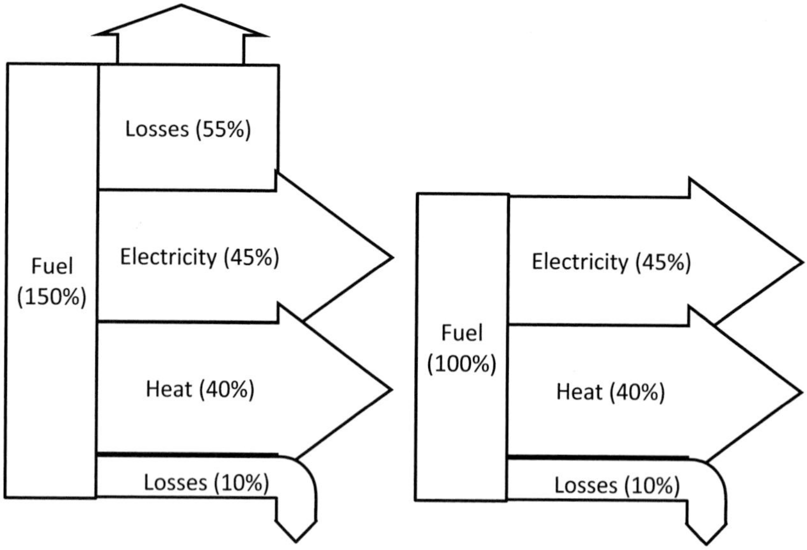

100 kW. Advanced energy technologies such as fuel cells, Stirling engines, and micro-turbines would fit well in both micro- and mini-CHP applications. Though still marginal in market size, micro- and mini-CHPs considerably extend the utilization range of cogeneration, and CHP projects do not necessarily need to be large to be feasible, nor is major infrastructure needed; a house unit may be enough. This could play an important role not only in serving the various sectoral energy needs of many emerging economies but also when opting for larger distributed energy generation schemes as a strategy for meeting broader demand.

Cogeneration systems find use both among public and private utilities and among industrial clients. Important industrial sectors where CHP plants are found include paper and chemicals, food, and refineries, where much process heat and steam is required. Industrial CHP plants would therefore often produce relatively less electricity than, for example, municipal CHP plants. If located close to a municipality, an industrial CHP could even deliver heat to private clients through a district heating network. In Europe, natural gas and coal account for two-thirds of the fuels used in cogeneration. The share of renewable energy sources is around 10 percent but is on the rise.

Table 1 (below) summarizes the characteristics of typical small-scale CHP plants below the 1-MW power level. The investments are around €1,000 per kW (electricity) but may vary considerably depending on the technology employed, size, and application. An important design parameter of cogeneration systems is the ratio of power to heat in the plant. Depending on the technology used, this ratio may range from 0.1 to beyond 2. Solid fuel plants have a lower ratio than that of gas plants. Owing to a decrease in the space heating demand resulting from

Table 1
Typical Technical and Cost Data of Small-Scale CHP Systems (<1 MW)

	Electrical efficiency (%)	Thermal efficiency (%)	Investment cost €/ kW(el)	Power-to-heat ratio
Gas Engine	>35	>45	700–1000	0.6–1.1
Gas Turbine	30	40	1500	0.5–1.1
Steam turbine	10	70	800–1000	0.1–0.15
Future options:				
Stirling engine	20	70	>2500	0.3
Fuel cell	40	40	5000	1
Microturbine	25	45	2000	0.6–0.8

Source: www.eere.energy.gov/industry; www.cogeneurope.eu.

better building standards and an increase in internal heat gains from different new appliances, together with an increase in the electricity demand, high power-to-heat ratios are desired. A large solid fuel CHP with back-pressure turbines could achieve a power-to-heat ratio of 0.3–0.5, but for combined cycle gas turbines a ratio of 0.7–1.2 is feasible. For example, the natural gas CCGT CHP plant in Helsinki has an electric power output of 630 MW and produces 580 MW of district heat, while the overall fuel efficiency is over 90 percent.

A modern gas engine designed and optimized specifically for cogeneration would yield a ratio of 0.7 to 1. The electrical efficiency of an engine CHP starts from 35 percent in smaller units to 45–50 percent in large engines (>1 MW). A special feature is that they can operate on gases with extremely low to very high calorific values. Possible gas sources are, for example, landfill gas, sewage gas, natural gas, wood gas, pyrolysis gas, or different gases from chemical industries. An engine CHP also could easily utilize different types of liquid biofuels that could be produced from practically any oil-rich crops, or non-vegetable oils or fats, all abundantly available worldwide. This technology could prove quite useful in agricultural areas where a lot of different residuals are readily available.

Cogeneration has traditionally been viewed as an option for conditions with a high demand for heat, which is the case in cold northern climates or in process industries. The next step from a cogeneration system is the so-called tri-generation or poly-generation system, which produces power, hot water, and chilled water at the same time, that is, a system that could efficiently meet all final energy needs found in buildings. The cooling option is achieved by connecting an absorption or compressor chiller to the system, thus converting some of the heat or electricity into chilled water for air-conditioning. In practice, a tri-generation system could flexibly interchange between heating and cooling, thus matching well with the seasonally varying thermal energy demand in cities. If a CHP plant could easily achieve a fuel efficiency of greater than 85 percent when operating in the heating mode, the efficiency may drop to less than 70 percent in the cooling mode. This is a result of the lower coefficient of performance (COP) of chiller units, particularly if absorption cycles with a COP of around one or less are used. Nevertheless, the savings when compared to separate power, heating, and cooling energy systems are still significant.

Helsinki, the capital of Finland, on the northern shore of the Baltic Sea, is a good example of a mid-sized Northern European city (500,000 inhabitants) with a long-term commitment to providing local energy in an efficient way. The one hundred-year-old Helsinki Municipal Utility or Helsinki Energy has more than fifty years of experience in using CHP. Three large centralized CHP plants produce 85 percent of all heat and 98 percent of electricity in Helsinki, but all the fuel is still fossil-based (50 percent natural gas, 50 percent coal). The old plants based on coal are close to the downtown area, whereas the newest gas-CHP is on the periphery of the city. The fuel efficiency exceeds 90 percent. The utility operates an extensive 1,200-km-long district heating network, which covers 93 percent of the heating requirement in the city, or some 7,000 GWh per year. The total electricity production is also close to 7,000 GWh but some 40 percent of this is a surplus sold to the Nordic electricity market to generate revenues for the city.[16]

Helsinki also has one of Europe's largest and fastest-growing district cooling systems. While dominated by heating, the demand for cooling is growing. Helsinki is building a district cooling network to connect larger thermal loads like shopping centers or office areas. It already has a 30-km-long network providing 100 MW of cooling power, but by 2020, the capacity is expected to grow to 250 MW. The source for cooling is cold sea water and chilled water from absorption chillers that run on heat obtained during summer from the CHP plants.

For a customer, district heating and cooling provided by CHP has been a preferred option from economic, technical, and environmental points of view. Though 100 percent city-owned, Helsinki Energy is a market-economy-based municipal utility and actually generates, particularly through its electricity business, a profit close to €300 million ($500 million) that reduces the local income tax by around 10 percent. The unit CO_2 emissions in Helsinki have dropped since 1990 from 0.4 to 0.26 tCO_2/MWh in 2010.[17] Helsinki Energy has been internationally awarded for its pioneering work, for example, with the International Energy Agency (IEA) Certificate of Merit for outstanding work in climate change mitigation. A typical customer in Helsinki would pay €30/MWh (including taxes) for district heating in the summer and €50–55/MWh during the winter (1€ = 1.35$, as of October 2011). The upfront investment cost of connecting to the district heating line would be around $10,000 for a block of five apartments. The electricity price (including taxes) is €135/MWh, with some day-night variation. From a technical point of view, the district heating and cooling technologies are in practice service-free options for the customer, who pays mainly for the service delivered.

Many northern cities have already reached a high fuel efficiency level through cogeneration. For historical reasons, several of the CHP plants in larger cities and those close to seaports such as Helsinki use fossil fuels. To reduce CO_2 emissions further, more sustainable green energy options need to be introduced. In Finland, which has the largest per capita forestry base in Europe, adding a biomass component would naturally be the next incremental step. By 2020, Helsinki plans to co-fire its coal CHP plant with 20 percent biomass. It is also investigating the application of the natural gas network to blend natural gas with some biogas or gasified solid biomass with methane synthesis for use in the CCGT cogeneration plant. But going for a 100 percent bioenergy-based energy system would probably not be feasible, as Helsinki's bioresource base is much more limited than that in small inland cities.

Therefore, in the long term, cities may need to consider making greater use of their inherent local renewable energy sources, coupled with increasing energy efficiency. In the case of Helsinki, this means focusing more on building energy use and local renewable energy use in buildings, in addition to maintaining the high fuel efficiency offered by the present CHP schemes. To investigate these future avenues, two major eco-suburbs are being constructed in Helsinki. Additionally, as part of the need to diversify energy supply to be more sustainable, municipal agencies that use high levels of energy in their operations have also increased the use of sustainable energy. One example is the Helsinki Region Environmental Service Authority (HSY), which operates waste-water treatment and purification plants, landfills, and some of the waste transport in the metropolitan region. With one of the largest landfill gas power plants in Europe, the HSY is able to recover 90 percent of the harmful methane and has grown into one of the principal producers of renewable energy in the region. The biogas serves the in-house heat and power demands, and surplus electricity is sold to the market. Almost 50 percent of the 185 GWh that the HSY needs annually comes from its own renewable energy production. Expected to grow in the coming years, the HSY should be self-sufficient in energy around 2014–2016 and close to reaching full carbon neutrality. The above case demonstrates how single municipal agencies can play an active and important role in enhancing the green energy economy in their region.

Considering electricity and heat demands in parallel may open up interesting possibilities of combining CHP and variable renewable resources such as wind power and solar energy. Taking a closer look at the final energy use in buildings reveals that most of the energy needed is in thermal form. Figure 3 shows typical energy use patterns in households in different regions of the globe.[18]

Figure 3
Final Energy Use in Households by Use

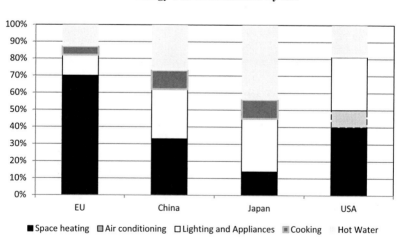

Source: World Business Council for Sustainable Development (2009).

In the EU, as much as 80 percent of the energy consumption is used for heating purposes, and in most industrial countries, thermal energy use would make up around 60 percent of the final energy. Storing thermal energy, either as hot water for heating or chilled water for cooling, on a daily or even longer basis is both technologically and economically feasible. A thermal storage system could add considerable flexibility to an urban energy system and allow the parallel use of CHP and other local renewable sources. Through a smart grid type of control[19] and by converting surplus electricity into thermal energy, use of renewable energy can be increased.

Instead of limiting the use of renewable power or paying nothing for it during periods of high renewable electricity supply (>100 percent of a city's self-use capacity), an alternative option would be to utilize renewable power to create hot or chilled water in order to satisfy immediate thermal energy demands, or to store it for future demands. A CHP system with a district heating (or cooling) network would fit such a strategy ideally without the need to build extensive new energy infrastructure. Next, such an electricity-to-thermal conversion strategy is demonstrated for the case of Helsinki in connection with offshore wind power, which tends to peak during winter, when there is a high demand for heat (for solar this would be the opposite). Figure 4 (left) illustrates the average heat and power demand over the year and the hourly values around late February, when the highest wind speeds are encountered (see Figure 4, right). The self-use of electricity (100 percent of the electricity demand) at the time of the peaking of wind power would put a limit of 490 MW on wind power utilization (Case A in Figure 4, where electricity demand = wind power output).

On a yearly scale, this would lead to a total wind electricity contribution of 1.3 TWh, or 27 percent of the city's electricity consumption. In this case, no electricity would flow outside the city boundaries. If an electricity-to-thermal conversion option (Case B) were added and wind power could satisfy both electricity and heat demands at the wind peak, an almost threefold amount of wind capacity could be employed and more than 70 percent of the yearly electricity demand could originate from wind power. On a yearly basis, the heat component would not be affected much, as only 2 percent of the heat comes from wind in this case. Does this mean that the CHP system needs to be discarded when a large amount of wind power is used (or solar power, for example, in a southern climate)? Not necessarily, as wind power could also be converted flexibly into heat to enable the CHP plant to operate smoothly. Adding a thermal storage system to the energy system would make this option even easier and lead to a more flexible heat-to-power ratio. One might come up with a smaller CHP plant corresponding to the renewable electricity share. However, though it is not shown here, some additional fast-ramping electricity capacity may be necessary to match the minute-scale wind variability. Nevertheless, this example demonstrates that it may be worthwhile to investigate the thermal energy (heating, cooling) integration of renewable electricity, while striving for a high renewable energy

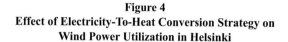

Figure 4
Effect of Electricity-To-Heat Conversion Strategy on
Wind Power Utilization in Helsinki

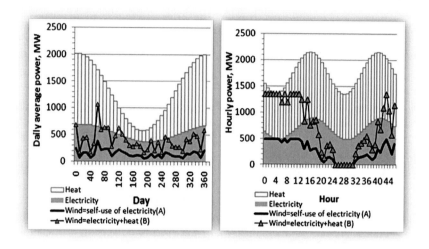

share, particularly in cities with a thermal energy distribution capacity and CHP similar to those of Helsinki.

Green Economy Aspects of CHP

In large cities or metropolises, the prime motivations for cogeneration may be cost-effectiveness and reduction of emissions. In smaller cities or towns, the driving factor more often may be utilizing local renewable energy sources and possibly local technology, and importantly, creating new jobs, that is, wealth creation, though reducing carbon emissions is still a valid argument.

In 2008, the Lahti City Council created a strong vision of cutting its carbon emissions by half by the year 2025 and reducing the energy demand by 15-25 percent by as early as 2015. The main means chosen to realize these goals were use of local energy sources, enforcement of eco-efficient urban planning, and enhancement of public transport schemes. The municipal utility, Lahti Energy, operates a cogeneration plant and a district heating network delivering some 2,500 GWh of energy yearly. The present energy mix is 80 percent fossil and 20 percent renewable.[20]

The city has seen the opportunity of specializing in cogeneration based on waste-to-energy services and has reached a cutting-edge position worldwide. Waste has an important bioenergy fraction; it is readily available and is normally dumped into a landfill. Lahti has been able to manage the whole waste-to-energy value chain in a cost-effective manner, that is, to create the necessary logistics to collect and sort waste and to turn it effectively into electricity and heat. Lahti

Energy has recently invested €160 million (~$200 million) in the world's first full waste-to-energy CHP plant (160 MW), based on Finnish fluidized bed gasification technology. This technology enables them to convert close to 40 percent of the waste into electricity and feed the rest into the district heating network.[21]

Several contractors specializing in the waste business collect a quarter of a million tons of waste annually from a radius of 100-200 km for the new waste CHP plant. The waste collection is so effective that 96 percent of the urban waste produced in Lahti is collected and recycled (see Figure 5). Normally, when waste is dumped into a landfill, a gate fee must be paid, but in Lahti, it is the other way around: the company pays for the waste if it is pre-sorted. This business model is quite radical as traditional waste incineration plants charge for burning the waste, that is, turning it to heat that could potentially be used for district heating.

The city of Lahti has an important catalyzing role in supporting local energy and clean tech companies. It supports technology development and piloting by creating markets for advanced energy technologies. As a result, the main development activities have been located in Lahti, including the national clean tech cluster coordination, which has achieved the creation of 500 green-collar jobs. In line with its proactive green energy innovation policy, the city of Lahti and Lahti Energy have decided to make its unique full-scale gasification plant open for international research groups to pursue cutting-edge R&D at the facility when it opens in the spring of 2012. A further step could be to offer the know-how that has been created more widely, as waste is a major problem in most parts

Figure 5
Municipal Waste Handling in the City of Lahti[22]

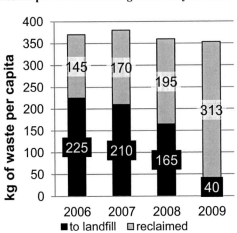

of the world, but could also be a solution as demonstrated by the city of Lahti. In Finland alone, with some five million people, the construction business and other industries produce more than thirty million tons of waste that contain a major share of biomass (>60 percent).

In another case, in the city of Kotka, which has 40,000 inhabitants, the visionary municipal green energy policy has reduced carbon emissions by 40 percent in ten years (see Figure 6). In accordance with this green vision, the business strategy of the local municipal utility, Kotka Energy, was changed from a traditional fossil-fuel-based district heating sales company into a more innovative public-private partnership investing in cogeneration by using local renewable energy sources and waste. The utility, with heat sales of 1,000 GWh a year, has invested both in a bioenergy CHP (130 MW) and a waste-to-energy plant (34 MW). It has planned new investments in a multi-fuel bio-CHP plant (45 MW) and is scaling up its 4-MW wind power demonstration into one on a commercial scale (33 MW) on the city's seafront.[23]

The green energy vision in Kotka is strongly linked to the economy and jobs, as this is a declining industrial region. The city stimulates both investments in green energy and development work. Similar to Lahti, Kotka has identified major opportunities in the waste business, where local companies provide waste-to-energy services to more than half a million people in the region (ten times the population of Kotka). Through proactive green energy investments, emissions have dropped by one million tons of CO_2; this has generated extra revenues from the European Emission Trading Market through selling carbon

Figure 6
The Shift from Fossil Fuels into Renewable Energy in the City of Kotka (Southern Finland)[24]

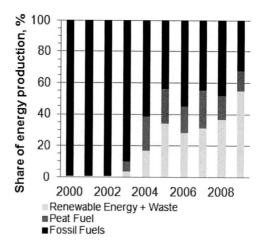

emission rights. On the industrial and energy technology side, a major cluster of wind energy companies (blades, logistics, towers, components) has settled in the region, and a network of innovative local companies is collaborating with universities to develop a unique biorefinery concept that produces not only liquid biofuels but also biogas and biofertilizers from algae. Altogether, more than two hundred new jobs have been created through this, but the true green business opportunities lie in the neighboring Russian metropolis of St. Petersburg, just 100 miles from Kotka, which has five million potential green consumers and grave environmental problems. The International Energy Agency has estimated that Russia could save 30 percent of its primary energy demand-comparable to annual energy use in Great Britain-if the efficiency of the Russian economy approached that of the OECD countries.[25]

Europe offers many green energy success stories like Kotka or Lahti in Finland. One of the first local regions to realize the potential of the green energy economy was the town of Guessing, in the eastern part of Austria.[26] Twenty years ago, Guessing, with 4,000 local inhabitants and some 30,000 in the surrounding region, was among the poorest regions in the whole country. The main source of livelihood came from agriculture, and the migration rate from the city was high. In the early 1990s, the city council decided to move away from fossil fuels and turn Guessing into a green city, producing all of its energy through renewable energy. A step-wise strategy, starting with energy efficiency measures in public buildings, gradually increasing the use of local bioenergy resources, cogeneration, and district heating, and later diversifying to other renewables such as solar energy, has led Guessing in practice to become autonomous in energy and fossil-free. Through its visionary policy of creating demand for green energy and turning the city into a huge living laboratory, Guessing also succeeded in drawing fifty new enterprises into the city. These firms work with green energy technology or offer related services and have generated 1,000 new jobs. In particular, many have developed cutting-edge technology in solar energy and biomass. Guessing is also a major attraction for several tens of thousands of eco-tourists in a year.

This intelligent use of market forces and procurement in utilizing local renewable energy sources could lead to major technology and innovation activities as well. Turning back to Finland, its utilization for over a century of domestic biomass resources in the pulp and paper industries, and its salvaging of the biowaste streams for cogeneration, have led to a strong build-up of know-how in the bioenergy field. Though large industrial investments in the traditional forestry business have vanished from Finland and shifted to Latin America or Asia owing to their cheaper resources and labor, the strong domestic technology base, nourished generously by the Finnish government, still attracts major global technology providers to establish technology centers in Finland around biofuels, bio-refineries, and combustion technology. Similarly, Denmark developed from a dominant force in the global wind turbine market in the 1980s and 1990s into a know-how hub and technology development serving a global industry base.

Denmark's case suggests that though market shares may decrease, increasing global demand for green energy technology may lead to positive and increasing economic results.

The above examples from Europe demonstrate that the green energy economy, when viewed from a technological perspective, may evolve through distinctive phases, from a small start to a major success as depicted in Figure 7.[27] During the start-up phase, which is often launched by a national or local vision of employing local energy resources or energy efficiency, that is, creating a market for green energy, the core technology may even come from abroad, accompanied by local input and know-how. Creating a market may be the single most important element in the commercialization of green energy as demand attracts business and innovations and generates competition, which leads to cheaper technologies and more effective resource use. Once the local market expands and reaches a critical level and if there are local technology industries and manufacturers present in that market, further growth would be sought from foreign markets. If there were no major local manufacturers and if the domestic market were large enough, subsidiaries of foreign companies would be established. The latter has taken place, for example, in Spain and Portugal with wind power, and later, through the acquisitions of such subsidiaries, Spain grew into one of the leading countries in wind power technology. Both paths would thus result in green jobs. The last phase of this evolutionary development would be full expansion to global manufacturing and global markets, while, for example, core technology development may remain in the original location.

Germany has gone through all three phases and is now a leading country both in solar PV and wind power technology worldwide. In 1990, the German wind power industry manufactured for the domestic market only, but twenty years later, the export share is 75 percent, corresponding to almost 20 percent of

Figure 7 Steps in Green Energy Business Expansion[28]

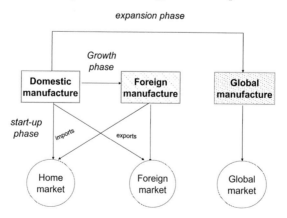

the world market in wind power.[29,30] The industry has a turnover of €6.4 billion and employs 102,000 people in Germany. At the same time, the share of wind power in the German power mix has jumped from zero to more than 6 percent of gross electricity consumption in the country. A similar success story can be told for PV, where Germany represents 75 percent of all installed PV capacity in Europe; in 2010 alone, some 7,400 MW_p of PV was installed.[31] The turnover of the German PV industry was €12.2 billion in 2010, with exports contributing to 50 percent of the turnover. This industry employs 108,000 people in several hundred companies all over the country. There are close to one million PV systems in Germany, with a total capacity of 17,000 MW_p, most of which are on the rooftops of buildings. It is expected that solar PV could reach a capacity of 28,000 MW_p by the end of 2012, meaning that occasionally on sunny summer days all the electricity in Germany could be solar-and in theory, solar and wind together could account for all electricity, even today, during a sunny and windy weekend when the electricity demand drops.

The German government has sharply reduced feed-in tariffs for PV as a result of decreasing PV prices, in order to dampen the sky-rocketing demand for PV and stabilize the turbulent market. The same has been witnessed in 2011 throughout Europe, for example, in Spain, Italy, and in the fall of 2011, in the United Kingdom, which announced that it would halve the support level. Some experts estimate that PV could reach grid parity in a few years' time, that is, PV-produced electricity would cost the same as electricity produced from more traditional energy sources. For an investor in wind or solar, the conditions created by the feed-in tariff have been generous, eliminating most of the financial risks attending new energy technologies and guaranteeing a profitable investment. Similar developments are now foreseen in other countries as well.

The role of local or national policies in promoting a green energy economy must be emphasized, even though policies alone are seldom sufficient to guarantee success. It is almost certain that if policy and support mechanisms are missing when green energy initiatives are launched, little success can be expected. Likewise, the amount of financial resources available is not a prerequisite or guarantee of success, although it can be very helpful. The success stories described here, just a few of the many examples that could also be cited, came about in towns or cities struggling with unemployment and disappearing industries, that is, places that at the first glance would not appear to be the most obvious test sites for the green energy economy. When the economy grows fast and budgets show a surplus, it is not necessary to seek new avenues of prosperity, which options such as green energy offer, when the economy is in decline. On the other hand, a wealthy region accompanied by strong will, wise policy, and adequate support for green energy could leap forward in the sector's development and be in the vanguard of emerging efforts.

A set of other factors can be identified from the examples that seem to play a role when local initiatives achieve success. Such factors include the following:

- As a starting point, avoid "silo" thinking by viewing green energy as an energy commodity only, but understand and exploit its connection to economic growth and job creation as well;
- Focus on strong utilization of local renewable energy sources accompanied by energy efficiency measures, for example, bioenergy in cogeneration schemes;
- Strive for multiple revenue generation from investments in green energy, for example, savings in waste handling gate fees, carbon emission savings trading, and selling surplus green electricity outside;
- Link local know-how and technology to investments in green energy whenever possible;
- Employ intelligent and cost-effective policy tools that catalyze a change and scale-up in green energy investment, e.g., using public procurement or public purchasing power, public sector pioneering in using green energy, and technology procurement-type measures;
- Create strong public-private partnerships in realizing different green energy schemes;
- Support (public) investments in long-term (energy) infrastructure enabling large-scale green energy utilization to happen.

At a more general level, a lack of financial resources often needs to be substituted for by intelligent measures, as demonstrated above. In these cases, any public support would be focused on the process of bringing green energy to the market, that is, on the commercialization process, rather than on subsidizing the product or capacity increases, which would lead to a situation where capacity growth would directly correlate with the subsidies available, and once these ceased, the market growth would stop. The energy impact per public dollar spent on such catalytic (process) measures could be as high as tenfold compared to the traditional (product) measures.[32] To ensure long-term and enduring success instead of sudden hype, which may quickly fade away, local and national green energy policies should also aim to bring the price of green energy down. Once green energy is cheaper, or the ratio of benefits to costs is larger than that of traditional "brown" energy, a major paradigm shift to a green energy economy would take place. The role of creating local markets for green energy should once again be stressed here, as this would attract businesses that commercialize energy technology and would increase competition, which leads to better cost-effectiveness and advances innovation, cutting-edge technology, external flow of ideas, etc.

Finally, technology procurement-a typical catalytic policy measure used in connection with biomass cogeneration-as a process, includes several steps but aims at getting a cheaper and better technology to the market by exploiting two key forms of market forces, namely purchasing power and competition. First, a buyer group that needs the solution sets aside challenging technological and economic goals for the new technology. These goals should naturally be realistic compared to what is achievable, but still challenging enough to draw the latest developments to the market. In the next step, an open call for tenders is arranged with these requirements. On the reward side, the buyers commit themselves to purchasing a certain amount of the product meeting the specifications. The market thus created may be substantial and attract competitors to develop and bid for the tender. The winner(s), that is, the best bid, takes all, for example, in the case of cogeneration, the delivery of the plant(s). However, even the losers may profit from the resulting market surge. The pulp and paper industries used to develop along such lines in the past, which led to world-class technology for the combustion of biomass sources; this practice is associated with cogeneration. In Sweden, technology procurement was successfully applied to energy efficiency in the 1990s, for example, to commercialize heat pumps or energy-efficient lighting systems, which were major successes.[33] Another field that has frequently applied this process is defense.

European Policies in the Green Energy Economy

Green energy efforts at a local level are substantially easier when the overall energy-climate policy framework is in place and favors such initiatives. The EU has been quite proactive in this respect, which has resulted in the favorable growth of green energy. Combined with business, these measures together have established Europe as a leading region worldwide in several areas of green energy technology.

The EU, with its twenty-eight member states, is the world's largest economy. Though each of the member states has far-reaching autonomy in deciding on the realization of practical measures in energy, there are common goals for the EU as a whole and also important common directives to follow. For example, in 2007, Europe set a common target on energy for 2020: 20 percent improved energy efficiency; 20 percent of final energy use should be renewable; carbon emissions should be reduced by 20 percent; and biofuels should represent a 10 percent share of all fuel. Each member state has been given its own targets derived from these common EU goals, on the basis of factors such as a country's current progress in these areas and its potential. Basically, all twenty-eight countries of the EU have implemented the common policy and climate goals in their national policies or are well on their way in doing so. Moreover, Europe has directives for a common internal market in electricity and gas, which is important for large-scale renewable electricity schemes. Europe also has implemented a system of caps and trading to track carbon emissions from its largest emitters. This so-called European Emission Trading System (ETS) covers almost half of all CO_2 emissions from energy in Europe and serves as the main policy instrument toward industries and the power sector. Such goals contribute positively to green economic growth. The "20 percent" directives mentioned above may create close to 800,000 new jobs in Europe.[34]

The EU's legislation in the energy field is vast and comprises some 160 directives or regulations. It also covers the main areas for enhancing a green energy economy, specifically promotion of renewable energy and energy efficiency.

The directives on the eco-design of energy-consuming products cover a range of household appliances and lighting devices, office equipment, and motors. An extensive energy labeling scheme for domestic appliances, which is similar to the U.S. Energy Star program, is also in use. The promotion of cogeneration is further included in the legislation, with some indicative targets for CHP use in Europe, though this is not binding.

As explained earlier, the EU intends to increase the share of renewable energy to 20 percent by 2020 (from 8.5 percent in 2005). The allocations per member state vary considerably. In countries that already have a high share of renewable energy, such as Latvia, Sweden, and Finland, the increase is less than 10 percent, whereas the largest member states, Germany, France, and the UK, are clearly above this threshold. Such binding targets in renewable energy create markets for green energy products. For example, when interpreted in terms of markets, the 20 percent renewable energy goal for 2020 means an investment of €500-600 billion. The six largest European states will account for up to 70 percent of this, meaning that their national policies will steer investments in green technology investments at the macro level. Subsequently, more than half of the investment in renewable energy will most probably go to renewable electricity, most of the remainder to wind power and solar, and some to biomass-based electricity. Looking at the renewable energy targets the other way around means huge future opportunities for local initiatives, as investments are made locally, but governments need to provide supportive measures within their commitments to the Union.

The EU recently revised its energy policy and in 2011 prepared the Energy 2020 strategy for competitive, sustainable, and secure energy, which, in addition to the 2007 "20-20-20-10" goals, puts forward five strategic specific priorities to further enforce a green energy economy. These include the following: achieving an energy-efficient Europe; building a pan-European integrated energy market; empowering consumers and achieving the highest level of safety and security; extending Europe's leadership in energy technology and innovation; and strengthening the external dimension of the EU energy market.[35] Energy-efficiency measures and energy-technology innovations have been given more emphasis than earlier. The EU now requires member states to take the necessary steps to realize the infrastructure for exploiting district heating and cooling, in order to enable the increased use of cogeneration, waste heat, and different renewable energy sources. The European Commission has proposed that all new electric power plants in Europe above 20 MW should be equipped with heat recovery equipment enabling cogeneration to be performed, and should be located closer to the points of the heat demand, for example, urban areas. As 80 percent of final energy use in Europe is heat, strengthening the position of cogeneration would have a major impact on energy and climate once these directives are in place.

From a climate mitigation point of view, most of the world's greenhouse gas emissions need to be eliminated by 2050 to avoid irreversible damage to global ecosystems. More stringent energy and climate goals are under discussion in the

EU and its governing bodies. The 2011 communication "A Roadmap for Moving to a Competitive Low-Carbon Economy in 2050" states that an 80-95 percent reduction of CO_2 emissions will be necessary, meaning a yearly emission level of 1 tCO_2 per capita.[36] The primary energy per capita would be around 2 kW. To achieve such challenging goals, it is envisioned that the whole European power sector needs to be made carbon-free within the next four decades (93-99 percent CO_2 reduction), which will mainly need to be met by renewable electricity capacity. By 2035, a third of all European electricity could originate from new renewable sources. Europe's urban areas foresee carbon dioxide reductions as high as 90 percent. Perhaps needless to say, such a revolutionary European vision would open up tremendous markets for green energy and trigger massive technology development and innovation efforts, which collectively would contribute positively to economic growth but in a more sustainable way.

The EU creates market pull for green energy through the common European policy measures and targets. In addition, stronger technology push efforts will be necessary to bring the price of clean energy to a competitive level. For this reason, the European Commission has launched a €50 billion effort called the Strategic Energy Technology Plan (SET Plan), which aims at technology breakthroughs in key areas.[37] The SET Plan covers a wide range of technologies such as renewables, nuclear, and carbon capture and sequestration (CCS), the emphasis being on green energy. The main part of the realization occurs through public-private partnership programs called European Industrial Initiatives (EIIs), which pull together the best know-how and leading market players in Europe toward a common goal in each technology area. Such EIIs-joint large-scale technology development projects of up to the multi-billion euro scale-have been started in, for instance, solar energy, wind power, smart grids, and CCS. In 2011, the "Smart Cities" innovation partnership was launched to provide cities and urban and rural areas with green energy solutions. This effort integrates renewable energies, energy efficiency, smart electricity grids, clean urban transport such as electromobility, and smart heating and cooling grids, combined with highly innovative intelligence and information and communication technology (ICT) tools. In addition, the European Regional Policy is envisaged to play an important role in unlocking local potentials for such integrated approaches. In several of the renewable energy EIIs, issues related to the large-scale deployment of these sources are addressed, meaning that now there is also a greater focus on energy system issues such as the integration and interfacing of the new technologies with the existing energy infrastructure. Construction of enabling energy infrastructure such as local smart electricity grids or pan-European grids, or even extending these to surrounding countries in the Mediterranean region, would be important in this context.

Summary

Urban environments contain both causes of our energy and climate problems and solutions to these problems. Most of our energy is used in cities and buildings. Cities are also often the source of new ideas and innovations. Combining these factors can offer a unique and enormous opportunity for the green energy economy. Approaching urban energy challenges without coupling green energy and green jobs to these would indeed be a lost opportunity. This conclusion is evident from the case studies and examples presented in other chapters of this book.

There is always a strong technology dimension to the green energy economy, as ultimately it is about investments in clean and efficient technology while reducing fossil fuel dependency and cutting carbon emissions. Here we chose cogeneration as an example technology for our case studies: it is highly fuel-efficient and enables the use of different local bioenergy and waste resources. It has traditionally been considered as the energy technology of cold or northern climates, but could be used more extensively worldwide for simultaneous heating, cooling, and power production. Cogeneration could also be utilized in connection with other variable renewable energy sources. The Finnish case studies support the conclusion that cogeneration can create economic growth and jobs locally.

The importance of the policy dimension in creating successes in green energy cannot be overstated, but this would by no means undermine the central role of the private sector in realizing green energy in practice. Public-private partnerships or different networks provide a better chance of finding the correct solutions to the challenges of local conditions. But green energy successes are not born by themselves. These initiatives need to be backed up with political will and vision. Financial support is necessary to get the wheels rolling. Finally, a good policy would also pay attention to the following factors: creating or expanding markets, lowering costs of green energy, and finally striving for a favorable cost-to-benefit ratio.

Notes

1. Shell International. Future Cities in a Resource Constrained World. www.shell.com/scenarios, 2010.
2. World Business Council for Sustainable Development. Energy Efficiency in Buildings. Transforming the Market, Switzerland, 2009.
3. International Energy Agency, World Energy Outlook, Paris, 2010.
4. Barroso J.M. (President of the European Commission), Europe 2020. Presented at the European Council, February 11, 2010, http://ec.europa.eu/commission_2010-2014/president/news/speeches-statements/pdf/20102010_2_en.pdf.
5. Martinot E., Sawin J., REN21 Renewable Global Status Report 2010. http://www.ren21.net, September, 2010.
6. International Energy Agency, *Energy Technology Perspectives. Scenarios and Strategies to 2050* (Paris, France: OECD/IEA, 2008b).
7. IPCC, "Climate Change 2007," Synthesis Report.
8. Dastur A., Suzuki H., *Eco² Cities. Ecological Cities as Economic Cities* (Washington, D.C: World Bank, 2009).
9. International Energy Agency, *Cities, Towns & Renewable Energy, Yes In My Front Yard* (Paris, 2009).

10. Massive introduction of renewable electricity, which now accounts for close to 20% of all electricity in Germany, reduced the average marginal price of electricity by more than the extra price paid for wind electricity from the feed-in tariff (FIT) scheme some years ago, meaning that an average German household saved money through FIT. In the electricity market, different forms of electricity are used in increasing order of marginal costs. This means that fuel-less sources such as wind and solar with almost zero running costs would always be used first, making the fuel-based sources with higher running costs less attractive, which leads to a lower average consumer price.

11. International Energy Agency, *Combined Heat and Power. Evaluating the Benefits of Greater Global Investment* (Paris, France, IEA, 2008a).

12. Danish Energy Regulatory Authority, *Results and Challenges 2010* (Copenhagen 2011), http://www.energitilsynet.dk/.

13. International Energy Agency, *CHP/DHC Country Scorecard: Denmark. The International CHP/DHC Initiative. Advancing Near-Term Low Carbon Technologies*, 2008c. http://www.iea.org/g8/chp/profiles/denmark.pdf

14. International Energy Agency, *Creating Markets for Energy Technologies* (Paris, France: IEA), 2003.

15. International Energy Agency, *Combined Heat and Power. Evaluating the Benefits of Greater Global Investment* (Paris, France: IEA, 2008a).

16. Helsinki Energia, Annual report 2010. www.helen.fi

17. Helsinki Energia, Environmental review 2010. www.helen.fi

18. World Business Council for Sustainable Development, 2009.

19. U.S. Department of Energy, Smart Grid: Enabler of New Energy Economy, DOE, 2008, http://www.oe.energy.gov/DocumentsandMedia/final-smart-grid-report.pdf.

20. Lahti Energia, Annual report 2010, www.lahtienergia.fi

21. Lahti Energia, Annual report 2010, www.lahtienergia.fi

22. Päijät-Häme Waste Management Ltd. Kujala Waste Treatment Centre. http://www.phj.fi, 2011

23. Kotka Energia, Annual report 2010, www.kotkanenergia.fi.

24. Ibid.

25. International Energy Agency, World Energy Outlook, Paris, 2011.

26. Guessing Municipality General information, 2011, http://www.guessing.co.at/

27. Lund P.D., *Effects of Energy Policies on Industry Expansion in Renewable Energy. Renewable Energy* 34 (2009): 53-64.

28. Ibid.

29. German Wind Energy Association (BWE), Wind Energy Markets. Wind Energy in Germany, 2011. http://www.wind-energy-market.com

30. German Trade and Invest, Wind Energy Industry in Germany, Berlin, 2011a. http://www.gtai.com

31. German Trade and Invest, The Photovoltaic Industry in Germany, Berlin, 2011b. http://www.gtai.com

32. Lund P.D., "Effectiveness of Policy Measures in Transforming the Energy System," *Energy Policy* 35 (2007): 627-639.

33. International Energy Agency, Creating Markets for Energy Technologies (Paris, France, IEA), 2003.

34. Barroso J.M. (President of the European Commission), Europe 2020, Presented at the European Council, 11 February 2010, http://ec.europa.eu/commission_2010-2014/president/news/speeches-statements/pdf/20102010_2_en.pdf

35. European Commission, Communication from the Commission to the European Parliament, the Council, the European Economic and Social Committee and the

Committee of the Regions; Energy 2020. A strategy for competitive, sustainable and secure energy. COM, 639, Brussels, 2010.

36. European Commission, Communication from the Commission to the European Parliament, the Council, the European Economic and Social Committee and the Committee of the Regions; A Roadmap for moving to a competitive low carbon economy in 2050. SEC, 287-89.Brussels, 2011.

37. European Commission, Communication from the Commission to the European Parliament, the Council, the European Economic and Social Committee and the Committee of the Regions; Investing in the Development of Low Carbon Technologies (SET Plan) - A Technology Roadmap. COM, 519. Brussels, 2009.

16

Drivers for Change

Job Taminiau, Young-Doo Wang, and John Byrne

Introduction

The range of challenges reflected upon in the preceding chapters shows how "green energy" is not only a movement for technological change, but it also encapsulates social well-being and progress along a much broader set of dimensions. Conventional challenges such as national security concerns, economic instability, critical infrastructure development, and environmental risk reduction are combined with more fundamental and philosophical narratives of community revitalization, livelihood creation, and a paradigmatic shift in the nature–society relationship toward sustainability. The book, therefore, not only provides an overview of the various challenges presented by a transition to a green energy economy but also highlights its substantial promise.

The contemporary configuration of the modern energy economy maintains a *modus operandi* that ineffectively addresses negative consequences of its operation or inadequately incorporates other perspectives that allow for alternative-energy development. While working toward an alternative-energy future, the practical and philosophical narratives offered by the movement for a green energy economy tackle several of the critical issues that modernity faces. While currently critical components of the modern energy economy, these issues, when properly addressed, can form a driving force toward sustainability. Conceptually, these issues can be divided into five key categories, namely ecology, society, economy, technology, and policy. Several of the main moving parts within these five key categories are considered in this book, and together, they shape a picture that presents a fundamental choice: we face a policy choice between a twentieth-century energy infrastructure (reliant on conventional energy sources and associated with a particular way of life) and a new, "green" energy infrastructure with which to advance into the twenty-first century.

Key drivers for change, which constitute a collective pushing force that provides pressure and elevates the urgency of the policy choice, can be identified in the five key categories of ecology, society, technology, policy, and economy. As such, the choice will be determined by the extent we are able to recognize and act upon these five key drivers of change. Within their associated category, these drivers can be conceptually outlined as follows: (a) the extent we recognize—and, more importantly, maintain action *within*—the boundaries that our ecological surrounding presents; (b) an economic driver, as our increasing dependence on abundant flows of primary energy destabilizes the modern energy economy; (c) a social driver that elucidates the need to expand benefits of change to the global population rather than the developed nations alone; (d) our dependence on technological advancement to explore new frontiers that pose daunting and potentially insurmountable challenges forms a technological driver; and (e) the policy landscape increasingly incorporates long-term environmental and social considerations, thus introducing wholly new, alternative energy development pathways. The convergence of these drivers presents a potential framework of change that opens the door toward a green energy economy.

The green energy economy, however, is still a relatively new concept with much work yet to be done in areas such as research and development (R&D), policy refinement, and market maturation. In the following pages, by considering the concept of the five key drivers, this chapter explores and outlines some of the work that is yet to be done. It will become clear that when acted upon, the window of opportunity these drivers offer can substantially accelerate the transition to a green energy economy and fundamentally alter the way we live our lives.

Ecological Promises and Threats

The twentieth century witnessed a rapid expansion in all facets of life. Often termed the "Great Acceleration,"[1] society saw enormous increases in aspects such as population, gross domestic product (GDP), and automobile use. This strong growth realized substantial benefits and continues to shape the world as we know it today. However, society increasingly grapples with the negative consequences of this growth as issues such as increase in carbon dioxide (CO_2) emissions, pollution, energy use, and water use aggravate concomitantly. As the world faces this spectacle of breakneck-speed development, we are beginning to understand that such a "business-as-usual" development pathway will quickly breach the ecological limits that constitute a safe operating space for human activity.[2,3] In fact, modern society has lost all sense of staying within a space of operation that allows for long-term sustainable interaction with our ecological surroundings. Instead, driven by an expansionist mindset of continuous growth, modern society expands into all natural frontiers in the quest for new resources.

Fueled by copious amounts of energy, modern society not only affects the biological fabric of the natural environment—deforestation, desertification, ocean acidification, and declining biodiversity are just some of the environmental issues modernity faces—but is also actively appropriating the natural environment itself.[4,5] Modern society seeks to transform inherently natural functions, such as genetic reproduction and makeup as well as climate and weather stability, into dimensions available for human decision making.

This understanding—that human activity now wields a force equal to or greater than even geological or climatic processes—led Crutzen and Stoermer[6] to argue that our age is the time of the "Anthropocene." In this, physical earth systems and their functions have been taken over by human decision making, allowing society to choose to actively modify the natural order at every level of nature. For example, Vitousek et al.[7] show that 30–50 percent of the world's land surface has been transformed to meet human needs, and Daly[8] calculates that humans appropriate a significant amount of the photosynthetic capacity of the planet. This notion of the "Anthropocene" reflects an important philosophical transformation. While social organization was previously centered around the natural order, and more importantly, restrained by it, the separation of society away from the natural order created a nature–society relationship in which the value of our ecological surrounding itself is reduced to a reservoir available for extraction and exploitation. As Mumford[9] notes, within this new relationship, "the realities are money, prices, capital and shares: the environment itself, like most human existence, was treated as an abstraction. Air and sunlight, because of their deplorable lack of value in exchange, had no reality at all." Without a "language" and "currency" (Kammen, this volume, Chapter 3) to conceptualize the natural world, to understand the consequences of its degradation, and to value mitigative action as beneficial, the current configuration of modern society lacks the notion of staying within ecological boundaries. This leads to the realization that a notion of sustainability needs to be injected into the nature–society relationship.

The sketched situation forms a driver for change: it is becoming clear that ecological damage forms a threshold barrier to the modern energy project that we are fast approaching, if not already transgressing. The challenge that ecological limits introduce in the policy choice is finding a way to halt the negative consequences of the "Great Acceleration" and establish a basis for interaction that can be termed "sustainable." While the struggle among the different sources of energy (fossil fuels, nuclear, renewables) and their associated technologies allows for a significant opportunity to substantially change the energy landscape (Klare, this volume, Chapter 2), our choices need to be informed of this more fundamental challenge. Energy decision making itself needs to be conducted differently to allow for an acceleration in the transition to a green energy economy and to establish a new relationship with energy (Byrne et al., this volume, Chapter 1).

Realizing Economic Security

The ongoing global economic downturn leads nations around the world to debate means to revitalize economic progress. To address contemporary economic ailments, nations consider options ranging from Keynesian economic stimulus to draconic austerity measures. However, a major case can be made for a transition to a green energy economy as the pathway away from current economic hardship.

The modern energy project establishes a path-dependence that makes society vulnerable to economic shock. The guiding principles of the twentieth-century energy infrastructure, as summarized by Sovacool,[10] established a culture of abundance through large, cheap, highly technical, and short-term solutions to be decided upon by experts and bureaucrats. The dependence on large-scale, "abundant energy machines"[11] and abundant flows of primary energy commodities creates a structure that is difficult to change. It is this path-dependent nature that reduces resiliency and adaptability even though the "old" guiding principles are now contested.[12]

The adverse ecological, social, and economic consequences of the dependence on primary energy commodities (oil, coal, natural gas, nuclear power), however, reveal the crucial need for resiliency and adaptability. Economic security, for example, is degraded due to the price volatility of these energy resources. The price of uranium oxide (U_3O_8), for instance, increased by 403 percent ($11.04–55.64 per million pounds) over 2000–2011.[13] Oil, natural gas, and coal demonstrate similar price volatility.[14,15] This volatility comes with a substantial social cost,[16] as price volatility can spill over into non-energy commodity markets,[17] which can degrade essential livelihoods support markets such as the agricultural market.[18]

In the context of volatile and rising conventional energy prices, the concept of the green energy economy offers solace. A green energy strategy utilizes domestic energy resources, shortens the length of the supply chain, lowers the risk of political dependence on energy and security conflicts, and reduces ecological damage. These economic and environmental benefits motivate energy efficiency spending and savings[19] and form an economic driver for change toward a twenty-first-century energy infrastructure. A comprehensive strategy that accelerates implementation of renewable energy and energy efficiency technologies can further advance such economic benefits.[20,21,22]

Considering the imminent investment cycle in power generation, a green energy strategy can capitalize on a window of opportunity for transformational change.[23] While still in its early stages, the green energy economy has its champions. Germany, for instance, initiated a policy portfolio that, for over twenty-five years, has supported its domestic photovoltaic (PV) energy market.[24] This market now provides substantial economic benefits and forms a strong pillar of the German economic and manufacturing base. The United States already has a substantial renewable energy and energy efficiency sector that provides for millions of skilled, well-paying jobs (Wendling and Bezdek, this volume, Chapter 4).

Similarly, the untapped potential of energy efficiency can, when addressed, substantially increase economic performance (Laitner, this volume, Chapter 5).

Both developing and industrialized nations face a lingering and growing challenge: exacerbated by the global economic downturn, creating employment opportunities that are sufficient in number for a burgeoning population while also offering adequate wages and livelihoods remains a challenging objective. Success along this dimension can contribute immensely to the public's well-being while augmenting social interest for an even greater commitment to sustainability at the highest levels. While a full-fledged green energy economy will require a more fundamental commitment, a green energy manufacturing strategy can provide vital economic benefits as it anchors the middle class and provides millions of jobs (Rynn, this volume, Chapter 6). An example of this line of reasoning can be found in China's PV manufacturing strategy. In 2006, China redirected its attention in a significant way to the PV market by utilizing its strong manufacturing base as a means to enhance competitiveness. In 2010, together with Taiwan, their market share was about 60 percent of the global PV production market (or 14 GWp of the global 24 GWp) (Figure 1). While this manufacturing strategy targets the European and American markets, as China's actual implementation of the technology domestically is still very low, it offers illustration into the advancement of the twenty-first-century technologies and the effect of a policy choice toward such technologies.

Figure 1
Overview of the PV Production Capacity of China and Taiwan. Left Figure: Annual Solar PV Production in MW. Right Figure: Market Share of Global PV Production

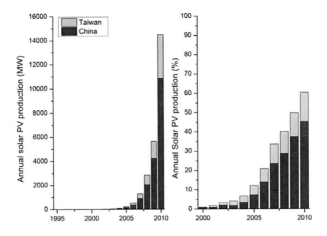

Source: Authors. Data as compiled by Earth Policy Institute (EPI). Data retrieved from http://www.earth-policy.org/data_center/C23.

Reshaping the Way We Live: The Need for New Communities

The challenges of lagging development—such as poverty, disease, and social conflict—stress a social driver to extend the benefits of energy development to the global population. Poverty, for instance, has declined in both relative percentages as well as absolute numbers over 1981–2005,[25,26] but progress has been unevenly distributed, with the majority of progress taking place within China. Energy poverty (the lack of access to energy), similarly, is a key issue within this context as it forms a considerable restraint on socioeconomic development.[27] With about 1.4 billion people who lack access to electricity and 2.7 billion who rely on traditional biomass for their energy needs,[28] expanding energy access is a major twenty-first-century issue. However, energy development along the same trajectory of the modern energy project is likely to further strain ecological limits and further increase economic insecurity and energy dependence. The priority of advancing economic well-being and socioeconomic development, thus, needs to be synergistically intertwined with socioecological progress. These concerns suggest a need for energy development globally to be linked more clearly and deliberately to the green energy economy concept.

In order to expand the benefits of a green energy economy to wider shares of the global population, a re-arrangement of development toward sustainability is required. With more than half of the world's population living in urban areas and the expectation that half of Asia's population and half of Africa's population will live in urban areas by 2020 and 2035, respectively,[29] the social organization of city life will be a key aspect in realizing such a re-arrangement. The relation to energy in urban settings is abstract as cities rely on external sources of energy and the organization of life in cities (e.g., transportation options and the built environment) limits individual agency to pursue alternative energy development. The concept of "eco-cities" can comprehensively address such challenges, as it reshapes what constitutes the urban life (Roaf, this volume, Chapter 7). Keeping social and environmental considerations in mind from the outset, such as emphasizing walkable communities, improves living conditions and urban welfare, as it reduces economic stress, increases resiliency, and enhances environmental performance.

The built environment undergoes substantial and rapid change, especially in the developing world. China, for instance, annually adds about 1.7 billion square meters (18.3 billion square feet) of new floor space.[30] Without a new strategy to construct such new urban settings, energy use and consumption will increase concomitantly. Such a strategy can capitalize on the evolution in buildings toward "regenerative" capacities and improved environmental performance (Syrett, this volume, Chapter 8). Innovative and smart urban design and planning utilizing such state-of-the-art buildings can thus support sustainability across the board.

The organization of social space is, next to the built environment, tremendously dependent on the transportation modalities that are available in day-to-day life.[31]

Shaping urban communities around the automobile as the primary means of transportation can pressure urban design with a high demand for physical space, high costs for oil imports, and substantial pollution and health consequences. To transition to more sustainable transportation options, urban design needs to not only reconsider the organization of transportation modalities (e.g., the layout of roads, subways, and rail systems) but also consider innovative options to fundamentally change the design plan of cities (Newman and Schipper, this volume, Chapter 9).

Changing the Technology Base

The twentieth-century energy infrastructure depends on a momentum of technological advancement to contain and reduce negative environmental and social consequences. Constructed along the guiding principles mentioned earlier, the pursuit for energy, however, engages ever more challenging environmental circumstances with increased technological complexity. The assumption that these new natural and technological frontiers can be understood, and more importantly, controlled, to minimize environmental and social harm reflects a "strategy of resilience"[32]—an expectation that gained experience and knowledge from previous adverse consequences allows for the aversion of such consequences in the future.

An important consideration within the concept of sustainability, however, is the notion of stability. Ever more demanding natural and technological frontiers to exploit previously unattainable resources—whether drilling for deep offshore oil, splitting or fusing the atom, or extracting shale gas and oil—reveal a flaw in this strategy in that the complexity and scope of the new setting makes previously acquired experience insufficient. The assumption that we can design technological complexity with the required level of precision to address "known unknowns" as well as "unknown unknowns" is brought back into reconsideration by the breakdown of the nuclear reactor in Fukushima, Japan, and the Deepwater Horizon oil spill in the Gulf of Mexico. These events highlight the existing instability within technologically complex centralized systems under ever-challenging circumstances.

The increased complexity and sophistication of such energy systems have created a situation in which human authority can no longer be trusted to appropriately deal with unexpected situations, thus elevating the need for additional technological fail-safes. Technological redundancy measures and "passive" measures, for instance, are put in place to deal with adverse consequences. The trained professionals' inability to fully comprehend the dynamics of the technological systems might, in the event of disturbance within the system (as in Fukushima with a combined natural disaster in the form of an earthquake and flood wave and technological failure), lead to a situation in which the operators can't stop the technological cascade or unwittingly worsen the situation. Simple cause-and-effect cascades can thus become catastrophic accidents further limiting the stability of the system. Beck's concept of the "risk society" illustrates this nicely.[33]

Systems that revolve around "authoritarian technics"[34]—that is, systems of operation that draw on inventions of a high order and create complex machines but allow a lower degree of human control—face such a problem of instability. These types of considerations need to be included within the decision-making structure when choosing among technologies (Saul and Perkins, this volume, Chapter 13). When such considerations are firmly incorporated within energy decision making, other alternative technologies become more attractive. The rapid development of these alternative technologies forms a technology driver that opens up a window of opportunity for the implementation of the green energy economy concept. High-efficiency PV (Barnett amp; Wang, this volume, Chapter 10), for instance, allows for an alternative pathway of provision of energy when incorporated within decision making.[35] Similarly, fuel cell technology developments (Prasad, this volume, Chapter 11) allows for a resiliency in the energy system when used as a storage device. When applied in a comprehensive, smart, and innovative strategy, the twenty-first-century energy technologies can synergistically augment each other's functioning. Intermittency issues within a green energy system, for example, can be addressed with energy storage technology capable of storing excess energy for times of high demand (Yonemoto, Hutchings, and Jiao, this volume, Chapter 11).

Choosing a Green Energy Future

Informed by the previous four drivers, a final driver for change is shaped by the changing developments in the policy agenda. Policy agendas around the world are increasingly recognizing the causes for environmental and social distress and many attempts have been made to formulate response strategies. Such "mainstreaming" of environmental and social considerations into the overall policy narrative[36,37] has the potential to minimize contradictions between policies, to introduce specific new considerations into an overall evaluation of policy, and to review the potential alternative energy development strategies in this light.

A main consideration within this context is that these new policy responses are focusing on the long-term, on the order of several decades. The European Union's (EU) *Energy Roadmap 2050* outlines several potential scenarios for energy development in the European Union until the year 2050.[38] Similarly, the EU's *Roadmap for Moving to a Competitive Low-Carbon Economy in 2050*[39] demonstrates how long-term policy response strategies can support environmental and social objectives. Many other countries around the world are also busily working to articulate their own strategies for the future. South Korea, for instance, has sought Green Growth as an official strategy for economic and environmental progress. Within the United States, a diversity of initiatives at the state and local level are complemented by a number of policies and programs at the federal level to promote renewable energy.

Such strategies and scenarios inform long-term development challenges and opportunities and allow for the conceptualization of alternative energy futures.

The complications, challenges, and opportunities that various energy futures offer, moreover, reveals both the difficulties associated with the modern energy system and elucidates potential practical pathways toward a green energy transition (Horn, this volume, Chapter 14). To achieve the goals set forth in such long-term strategies, the appropriate use of sound policy is crucial. Policy can substantially accelerate the implementation of renewable energy technologies, for instance.[40] China's strategy outlined earlier is an example of how policy support can considerably accelerate the advancement of a twenty-first-century technology. Similarly, Spain's aggressive PV policy support strategy realized enormous growth in the sector at extremely rapid rates.[41]

The varying ambitions of different countries toward achieving a clean energy future can be augmented by recognizing the potential of integrated green energy approaches (Lund, this volume, Chapter 15) and of collaboration between and across countries. Through this type of negotiated approach to tackling emerging challenges and forging solutions and innovations, communities can build on unique talents, capture complementarities, and help avoid expensive and time-consuming duplications of effort. A policy strategy for the future, therefore, supports the wider development of human resources and social capital and directs it as a creative and innovative force for economic and technical shifts. A policy portfolio that pursues a green energy future thus suggests the possibility for larger social transformation as a function of long-term strategies.

The Rise of the New Energy Economy

The evolution of the green energy concept to date has already motivated industries and governments to shift their focus and has altered lifestyles in substantial ways. While the final form of this change is uncertain and the stakes remain high, momentum is offered by five key drivers of change. The plethora of efforts currently underway suggest that the green energy economy is one of promise: the promise of new resources, new relationships, and finally new opportunities for shared progress and prosperity.

As to its fruition, however, the promise is hindered by many barriers. From the preceding chapters, it has become abundantly clear that thinking in terms of alternative technologies alone is not enough. The journey toward a New Energy Economy is fundamentally a social one in which options need to be considered from a range of different perspectives that complement the technical viewpoint with considerations of ecology, society, economy, and policy. Effectively, the combined implementation of these various vantage points changes the societal relationship to energy and fundamentally alters the energy discourse.

Such a new line of action emphasizes the local and human aspects to energy development. Day-to-day experience within a green energy economy is thus fundamentally reshaped as it will take place in new communities, revolving around new ways of interaction (e.g., walk-able communities, shared energy generation) and allows for a sense of agency in the articulation of future energy development pathways. During such a transition, the twentieth-century technology base finds that their functional niche is continually diminishing as new technologies arise that engage in the competition for the future.

The combined picture that arises from the preceding chapters is one of a choice between a twentieth-century energy infrastructure and a twenty-first-century energy infrastructure. This book offers an account of the potential of the green energy economy concept as a viable alternative energy strategy that can be positioned as a more appropriate energy infrastructure for the twenty-first century and beyond. To capitalize on this potential, society will need to recognize the fundamental drivers of change detailed in this book and action will need to be formulated that advances a more secure and resilient future energy economy. Such a New Energy Economy offers a promise for the future that emphasizes sustainability as we advance into the twenty-first century.

Notes

1. Steffen, Will, Crutzen, Paul J. and McNeill, John R., "The Anthropocene: Are Humans Now Overwhelming the Great Forces of Nature," Royal Swedish Academy of Sciences, *AMBIO: A Journal of the Human Environment* 36 (2007): 614–21.
2. Rockström, J., et al., "A Safe Operating Space for Humanity," *Nature* 461 (2009): 472–75.
3. IPCC, *Climate Change 2007: Synthesis Report; Contribution of Working Groups I, II and III to the Fourth Assessment Report of the Intergovernmental Panel on Climate Change Core Writing Team Pachauri R K amp; Reisinger A* (Geneva, Switzerland: IPCC, 2007).
4. Shiva, Vandana, "The Seed and the Earth: Biotechnology and the Colonization of Regeneration," In *Close to Home: Women Reconnect Ecology* (Philadelphia, PA: New Society Publishers, 1994), 128–43.
5. Byrne, John, Glover, Leigh, and Martinez, Cecilia, "The Production of Unequal Nature," In *Environmental Justice—Discourses in International Political Economy—Energy and Environmental Policy Vol. 8.* (New Brunswick and London: Transaction Publishers, 2002), 261–91.
6. Crutzen, Paul, and Stoermer, Eugene, "The 'Anthropocene'," *Global Change Newsletter* 41 (May, 2000): 17–18.
7. Vitousek, P.M., et al., "Human Domination of Earth's Ecosystems," *American Association for the Advancement of Science, Science, New Series* 277 (1997): 494–99.
8. Daly, Herman, *Sustainable Growth: an Impossibility Theorem*, In *Debating the Earth: the Environmental Politics Reader*, ed. J.S. Dryzek, and D. Schlosberg (Oxford: Oxford University Press, 1998), 1–3.
9. Mumford, Lewis. *Technics and Civilization* (New York: Harcourt, Brace & Co., 1934). 151–211.
10. Sovacool, Benjamin K., *National Energy Governance in the United States* (Oxford University Press, 2011), Journal of World Energy Law and Business, Vol. 4.
11. Byrne, John, and Rich, Daniel. "In Search of the Abundant Energy Machine," *The Politics of Energy Research and Development* (New Brunswick, NJ: Transaction Publishers, 1986), 141–60.
12. See Note 10.
13. Energy Information Administration (EIA), Nuclear and Uranium: Weighted-Average Price of Uranium Purchased by Owners and Operators of U.S. Civilian Nuclear Power

Reactors, 1994–2011. *U.S. Energy Information Administration*, May 2012, http://www.eia.gov/uranium/marketing/html/summarytable1b.cfm. (Accessed February 21, 2013).

14. Regnier, Eva, "Oil and Energy Price Volatility," *Energy Economics* 29 (2012): 405–27.
15. Byrne, John, and Kurdgelashvilli, Lado, "The Role of Policy in PV Industry Growth: Past, Present and Future," In *Handbook of Photovoltaic Science and Engineering, Second Edition*, ed. Antonio Luque, and Steven Hegedus (Hoboken, New Jersey: John Wiley & Sons, Ltd.), 2011.
16. Lovins, Amory B., et al, *Winning the Oil Endgame—Innovation for Profits, Jobs, and Security* (Boulder, Colorado : Rocky Mountain Institute (RMI), 2004).
17. Ji, Qiang and Fan, Ying, "How Does Oil Price Volatility Affect Non-Energy Commodity Markets?," *Applied Energy* 89 (2012): 273–280.
18. Nazlioglu, Saban, Erdem, Cumhur, and Soytas, Ugur, *Volatility Spillover between oil and agricultural markets*, *Energy Economics* (2012): Accepted Manuscript: in Press.
19. Barbose, Galen L., et al., *The Future of Utility Customer Funded Energy Efficiency Programs in the United States: Projected Spending and Savings to 2025* (Berkeley, CA: Lawrence Berkeley National Laboratory (LBNL), 2013), LBNL-5803E.
20. Laitner, John A. "Skip", et al. *The Long-Term Energy Efficiency Potential: What the Evidence Suggests* (Washington, DC : American Council for an Energy-Efficient Economy (ACEEE), 2012.) ACEEE report E121.
21. Choi Granade, H., et al., *Unlocking Energy Efficiency in the U.S. Economy*. Retrieved from http://www.mckinsey.com/client_service/electric_power_and_natural_gas/latest_thinking/unlocking_energy_efficiency_in_the_us_economy on August 16: McKinsey & Company, 2009. Research report.
22. Griffith, Brent, et al., *Assessment of the Technical Potential for Achieving Net Zero-Energy Buildings in the Commercial Sector* (Golden, Colorado: National Renewable Energy Laboratory (NREL), 2007). NREL/TP-550-41957.
23. Organization for Economic Cooperation and Development (OECD), *OECD Green Growth Studies: Energy* (Paris, France : Organization for Economic Cooperation and Development (OECD); International Energy Agency (IEA), 2011).
24. See Note 14.
25. Chen, Shaohua, and Ravallion, Martin, *How Have the World's Poor Fared Since the Early 1990s?* (World Bank Working Paper, 2004).
26. Chen, Shaohua, and Ravallion, Martin. *The Developing World is Poorer than we Thought, But No Less Successful in the Fight Against Poverty* (Washington, D.C: The World Bank—Development Research Group, 2008), WPS4703.
27. Johansson, Thomas B., "The Imperatives of Energy for Sustainable Development," In *The Law of Energy for Sustainable Development*, ed. Adrian J. Bradbrook et al. (New York: Cambridge University Press, 2005), 46–52.
28. International Energy Agency (IEA), *Energy Poverty: How to Make Energy Access Universal?* (Paris, France: International Energy Agency (IEA); United Nations Development Program (UNDP); United Nations Industrial Development Organization (UNIDO), 2010).
29. United Nations (UN), Department of Economic and Social Affairs, Population Division, *World Urbanization Prospects—the 2011 Revision* (New York: United Nations (UN), 2012).
30. Bin, Shui and Jun, Li, *Building Energy Efficiency Policies in China—Status Report* (Paris, France: Global Buildings Performance Netwwork; American Council for an Energy-Efficient Economy (ACEEE), 2012).

31. Freud, Peter, and Martin, George, "Driving South: The Globalization of Auto Consumption and its Social Organization of Space," *Capital, Nature, Socialism,* 11 (2000): 51–71.
32. Wildavsky, Aaron B., *Searching for Safety.* (New Brunswick, NJ: Transaction Publishers, 1988).
33. Beck, Ulrich. "Politics of Risk Society." In *Debating the Earth: The Environmental Politics Reader*, ed. J. S. Dryzek, and D. Schlosberg (Oxford: Oxford University Press, 1998), 587–595.
34. Mumford, Lewis, "Authoritarian and Democratic Technics," *Technology and Culture* 5. (1964): 1–8.
35. See Note 14.
36. Mickwitz, Per, et al., *Climate Policy Integration, Coherence, and Governance* (Helsinki: Partnership for Eureopean Environmental Research (PEER) Report No. 2, 2009).
37. Nunan, Fiona, Campbell, Adrian, and Foster, Emma, *Environmental Mainstreaming: the Organisational Challenges of Policy Integration* (Public Administration and Development, 2012), 262–77.
38. European Commission, *Communication from the Commission to the European Parliament, the Council, the European Economic and Social Committee and the Committee of the Regions—Energy Roadmap 2050.* (Brussels, Belgium: European Commission, 2011). COM, 885 final.
39. European Commission, *Communication from the Commission to the European Parliament, the Council, the European Economic and Social Committee and the Committee of the Regions—Roadmap for Moving to a Competitive Low-Carbon Economy in 2050.* (Brussels: European Commission, 2011). COM, 112 final.
40. See Note. 14.
41. See Note. 14.

Contributors

Allen Barnett (Professor, University of New South Wales, Australia)

Dr. Barnett is Professor of Advanced Photovoltaics at the school of Photovoltaics and Renewable Energy Engineering at the University of New South Wales in Sydney. In 1976, Dr. Barnett joined the University of Delaware (UD; Newark, Delaware, USA) as Director of the Institute of Energy Conversion and Professor of Electrical Engineering. In 1993, he left UD to devote his time to AstroPower, Inc. but returned to UD in 2003 in several capacities: Executive Director of the Solar Power Program; Research Professor at the Department of Electrical and Computer Engineering; and Senior Policy Fellow at the Center for Energy and Environmental Policy. Dr. Barnett received his MS and BS in Electrical Engineering from the University of Ilinois and his Ph.D. in Electrical Engineering from Carnegie-Mellon University. He is a fellow at the Institute of Electrical and Electronic Engineers (IEEE) and was recognized for his work with the IEEE William R. Cherry Award and the Karl W. Böer Solar Energy Medal of Merit.

Roger Bezdek (President, Management Information Services Inc. [MISI])

Dr. Bezdek has thirty years' experience in research and management in the energy, utility, environmental, and regulatory areas and has served in private industry, academia, and the US Federal Government. Founder and president of the Washington, DC–based economic and energy research firm MISI, Dr. Bezdek has served as a consultant to the White House, Federal and state government agencies, and various corporations and research organizations such as the National Academies of Science, the National Science Foundation, NASA, DOE, DOD, EPA, IBM, and the Electric Power Research Institute. Co-author of *Peaking of World Oil Production: Impacts, Mitigation, & Risk Management* (known as the Hirsch Report) and the follow-up *Economic Impacts of Liqued Fuel Mitigation Options*, he is currently serving as a member of the joint U.S. National Academies of Science/Chinese Academy of Sciences Committee on Energy Futures and Air Pollution in Urban China and the United States.

John Byrne (Director Center for Energy and Environmental Policy and Distinguished Professor, University of Delaware, USA)

Dr. Byrne is Director of the Center for Energy and Environmental Policy (CEEP) and Distinguished Professor of Energy and Climate Policy at the University of Delaware. As contributor to Working Group III of the Intergovernmental Panel on Climate Change (IPCC) since 1992, Dr. Byrne shares the 2007 Nobel Peace Prize with the IPCC's authors and review editors. Dr. Byrne is the co-founder and co-executive director of the Joint Institute for a Sustainable Energy and Environmental Future, an innovative research and policy advocacy organization headquartered in South Korea, and is a founding member of the *International Solar Cities Initiative*, a pioneering program to assist cities around the world in building sustainable futures. Dr. Byrne is co-editor-in-chief of the invitation-only journal *Energy and Environment*, a new WIRE reference work series published by Wiley & Sons. Since 1983, he has been the editor of the annual book series *Energy and Environmental Policy,* published by Transaction Books.

Miriam Horn (Director Smart Grid Initiative, Environmental Defense Fund)

As Director of the Smart Grid Initiative at the Environmental Defense Fund (EDF), Ms. Horn oversees work across the energy value chain to ensure maximum environmental benefits from the deployment of new electricity infrastructure. She works with key stakeholders to establish specific environmental performance criteria for smart grid deployment and to develop regulatory reforms and new electric sector business models in order to transform conventional utilities into agents of change. In collaboration with EDF president Fred Krupp, Ms. Horn co-authored the New York Times bestseller *Earth: The Sequel, The Race to Reinvent and Stop Global Warming.* Ms. Horn contributed to the forthcoming book *Smart Grids: Infrastructure, Technology and Solutions.* She holds a bachelor's degree from Harvard University and studied Environmental Science at Columbia University and is on the Advisory Board for Gridweek, the Galvin Electricity Initiative's Perfect Power Seal of Approval and the Coalition tasked with implementing FERC's National Action Plan for Demand Response.

Gregory Hutchings (Ph.D. Candidate, University of Delaware, USA)

Mr. Hutchings is a Ph.D. candidate in Chemical Engineering at the University of Delaware. He completed his undergraduate study in chemical engineering at the University of Florida and is pursuing his Ph.D. under the direction of Dr. Feng Jiao. His current research interests lie in developing advanced cathode materials for the lithium–oxygen battery system. He co-authored several peer-reviewed papers such as a paper titled *Nanostructured Alkaline-Cation-Containing δ-MnO2 for Photocatalytic Water Oxidation* in Advanced Functional Materials.

Feng Jiao (Assistant Professor, University of Delaware, USA)

Dr. Jiao holds a BS degree in chemistry from Fudan University (China) and completed his Ph.D. studies on nanomaterials for energy storage and conversion at the University of St. Andrews (United Kingdom) before moving to Lawrence Berkeley National Laboratory as a visiting scholar. After two years in Berkeley working on solar fuel technology, Dr. Jiao joined the Chemical and Biomolecular Engineering department of the University of Delaware as an assistant professor in 2010. Dr. Jiao has received several major awards that recognize his contribution to the field, such as Student Research Award (battery division) from Electrochemical Society, Graduate Research Award from the Material Research Society, and Doctoral New Investigator Award from American Chemical Society Petroleum Research Fund. He serves as a journal reviewer on several major journals, and his research activities include synthesis of nanomaterials and their potential applications in energy storage and solar fuel production technologies.

Daniel M. Kammen (Chief Technical Specialist, the World Bank)

Dr. Kammen serves as the World Bank Group's Chief Technical Specialist for Renewable Energy and Energy Efficiency. Dr. Kammen was Class of 1935 Distinguished Professor of Energy at the University of California, Berkeley, and served on parallel appointments in the Energy and Resources Group, the Goldman School of Public Policy, and the department of Nuclear Engineering. Dr. Kammen is also the founding director of the Renewable and Appropriate Energy Laboratory (RAEL), Co-Director of the Berkeley Institute of the Environment, and Director of the Transportation Sustainability Research Center. Author and/or co-author of twelve books, Dr. Kammen has written more than 240 peer-reviewed journal publications and testified over forty times to U.S. state and federal congressional briefings, and has provided. He is a frequent contributor to or commentator in international news media, including *Newsweek, Time, The New York Times,* and *The Financial Times.* Dr. Kammen has appeared on *60 Minutes, Nova, Frontline,* and hosted the six-part Discovery Channel series *Ecopolis.*

Michael T. Klare (Director Five College Program and Professor, University of Massachusetts, USA)

Dr. Klare is the Director of, and professor at, the Five College Program in Peace and World Security Studies, a joint appointment at Amherst, Hampshire, Mount Holyoke, and Smith Colleges and the University of Massachusetts. Dr. Klare has written on world security affairs, the arms trade, and global resource politics. Recent books include *Resource Wars* (2001), *Blood and Oil* (2005), and *Rising Powers, Shrinking Planet* (2008). "Blood and Oil" was made into a documentary by the Media Education Foundation of Northampton, Massachusetts. Dr. Klare has also written for many publications, including *Current History, Foreign Affairs, The Nation, Newsweek,* and *Scientific American.*

He also serves as defense correspondent of *The Nation* and is a contributing editor of *Current History*. In addition to his academic and writing pursuits, Dr. Klare is active in disarmament, environmental, and human rights advocacy work. He serves on the board of the Arms Control Association and the National Priorities Project.

John A. "Skip" Laitner (Director of Economic and Social Analysis, American Council for an Energy-Efficient Economy [ACEEE])

Now Director of Economic and Social Analysis for the American Council for an Energy-Efficient Economy (ACEEE), Skip previously served as Senior Economist for Technology Policy for the US EPA but left in 2006 to focus his research on developing a more robust technology and behavioral characterization of energy efficiency resources. In 1998, Skip received the EPA's Gold Medal and in 2003 the US Combined Heat and Power Association acknowledged his contributions to the policy development of the CHP industry with an award. His 2004 paper *How Far Energy Efficiency?* catalyzed new research into the proper characterization of efficiency as a long-term resource. Author of over 260 journal articles, book chapters and reports, Skip has forty years of involvement in the energy, environmental, and economic policy arenas. He holds a master's degree in Resource Economics from Antioch University and among his latest publications is a book he co-authored with colleague Karen Erhardt-Martinez titled *People-Centered Initiatives for Increasing Energy Savings*.

Peter D. Lund (Full Professor, Aalto University, Finland)

Working on new energy technologies since 1979, Dr. Lund's primary interest is on future energy solutions including advanced technologies but also energy-sustainable communities. In 1984, Dr. Lund completed his Ph.D. degree in Engineering Physics from Helsinki University of Technology and received supplementary education from the London Business School in 1989. In 1988, Dr. Lund served as research director for the National Program on New Energy Technologies at the National Technology Agency (TEKES) and, from 2002–2006, chaired the Advisory Group Energy of the European Commission. Since 2008, Dr. Lund has co-chaired multiple European Commission Call evaluations and is a member of panels such as the European Platform of Universities in Energy Research (EPUE). He holds international advisory board positions in China, Saudi-Arabia, and Spain and is Editor-in-Chief for WIRE's Energy and Environment and is Editor of the International Journal of Energy Research. He received the 2004 Finnish Nature Conservation Society's Prize and the 2008 Fortum Prize and published over 450 research papers.

Leon Mach (Ph.D. candidate, University of Delaware)

Mr. Mach is a Ph.D. candidate at the Center for Energy and Environmental Policy (CEEP) at the University of Delaware. He is an experienced and passionate conservation educator with field experience in coastal restoration, international relations, and sustainable surf tourism. Focusing on issues related to governance, environment, technology, and tourism, Mr. Mach has combined his research interests with opportunities to lecture in CEEP's undergraduate program. Fully proficient in Spanish, he has also studied at the University of Wollongong, Australia; American University, DC, and the United Nations Mandated University for Peace in Costa Rica. Mr. Mach has published his work in journals such as The Journal of Power Sources.

Peter Newman (Director Curtin University Sustainability Policy and Professor, Curtin University, Australia)

Dr. Newman is Professor of Sustainability at Curtin University and Director of Curtin University Sustainability Policy (CUSP) Institute. He serves on the Board of Infrastructure Australia and is a Lead Author on Transport for the IPCC. In 2001–2003, Peter directed the production of western Australia's Sustainability Strategy in the Department of the Premier and Cabinet, the first state sustainability strategy in the world. In 2004–5 he was a Sustainability Commissioner in Sydney advising the government on planning issues. In 2006/7, he was a Fulbright Senior Scholar at the University of Virginia Charlottesville. Dr. Newman invented the term "automobile dependence" to describe how we have created cities that revolve around the automobile. Ever since he attended Stanford University during the first oil crisis thirty years ago, Dr. Newman has been cautioning cities to prepare for peak oil. Recent books that demonstrate his world include *Technologies for Climate Change Mitigation: Transport* for the UN Environment Program, *Resilient Cities: Responding to Peak Oil and Climate Change* and *Green Urbanism Down Under* for Island Press.

John H. Perkins (Faculty Emeritus, Evergreen State College, USA)

Dr. Perkins has worked in the energy and environmental field for forty years. In the early 1970s, Dr. Perkins worked at the National Research Council in Washington, DC, on a project to understand the issues involved with pesticides. In 1974, he was a founding member of the faculty at the Western College of Miami University in Ohio and worked to create an interdisciplinary undergraduate liberal arts program. Dr. Perkins started as Senior Academic Dean and member of the faculty at Evergreen State College in 1980 and led the founding of the graduate program on the environment. He served as the Director of this program from 1999–2005 and continued to teach until 2009. Editor of NAEP's *Environmental Practice* journal from 1995–2008, Dr. Perkins has also published over forty articles and other materials. In 1997, he authored *Geopolitics and the Green Revolution*. Since retirement, Dr. Perkins has worked on a history of nuclear power in the United States and aspects of the history of biofuels. Dr. Perkins also works with the National Council for Science and the Environment in Washington, DC, on forming a project to improve energy education for undergraduates.

Ajay K. Prasad (Professor, University of Delaware, USA)

Dr. Prasad holds the appointment of Engineering Alumni Distinguished Professor of Mechanical Engineering at the University of Delaware. He received his B.Tech. in Mechanical Engineering from IIT Bombay, and his PhD from Stanford University. After a post-doctoral fellowship at the University of Illinois at Urbana-Champaign, he joined University of Delaware faculty in 1992 as Professor of Mechanical Engineering. Dr. Prasad has held research professor and visiting scientist positions in the United States, the Netherlands, and India. Dr. Prasad's research interests lie in the area of clean energy including fuel cells, Li-ion batteries, wind and ocean current energy, and vehicle-to-grid technology. Dr. Prasad is the founding director of the Center for Fuel Cell Research at UD and, as Director, Dr. Prasad facilitates research collaborations amongst the approximately twenty UD faculty members working in this area as well as companies involved in fuel cells and hydrogen infrastructure activities. Dr. Prasad also directs the UD Fuel Cell Bus Program, which aims to develop and demonstrate fuel cell powered transit vehicles and refueling stations in the state of Delaware.

Susan Roaf (Professor, Herriot-Watt University, U.K.)

Dr. Roaf is Professor of Architectural Engineering at Herriot-Watt University and ex-Oxford City Councilor. She is an award-winning teacher, designer, and author. She works in Climate Change and Solar Energy Research and is on the international Board of the International Solar City Initiative (ISCI) and on the Board of Dundee Solar City. She is a member of the EPSRC College, Fellow of the Royal Society of Arts and the Schumacher Society and Freeman of the City of London. She is author of books on benchmarks for sustainable buildings, adapting buildings for climate change and eco house design. She is currently working in the field of Adaption to Climate Change as Herriot-Watt Director of the Edinburgh Climate Change Centre and Co-Director of the Adaptation Knowledge Exchange program for the Centre of Excellence in Climate Change for the Scottish Government.

Jon Rynn (CIUS Fellow, City University of New York [CUNY] Institute for Urban Systems, USA)

Dr. Rynn completed his Ph.D. in political science from the City University of New York Graduate Center and is the author of the book *Manufacturing Green Prosperity: The Power to Rebuild the American Middle Class* from Praeger Books. Dr. Rynn is currently a CIUS Fellow at the CUNY Institute for Urban Systems after being an adjunct professor of political science at Baruch College and a research assistant for the late Professor Seymour Melman of Columbia University for many years. His online work can be found on the blog of the Eleanor and Franklin Roosevelt Institute, NewDeal20.org, and at EconomicReconstruction.org.

Kathleen Saul (Ph.D. candidate, University of Delaware, USA)

Ms. Saul graduated from the University of Notre Dame in 1981/82 with a BA in French and a BS in chemical engineering. After several years working for Procter and Gamble, Ms. Saul pursued her MA in Management at the Wharton School of Business. She became deeply passionate for environmental issues after relocating to the Pacific Northwest, where she coordinated numerous tree planting/habitat restoration projects. Ms. Saul returned to school yet again to pursue Master's in Environmental Studies at the Evergreen State College, focusing on nuclear power plant construction in the U.S. After graduation, she served on the faculty at the Evergreen State College for a year and a half, teaching in the Evening and Weekend College and MES program. Ms. Saul now combines her academic and professional background in her pursuit for a Ph.D. at the Center for Energy and Environmental Policy at the University of Delaware.

Lee Schipper (Senior Researcher Engineer, Precourt Energy Efficiency Center, Stanford, USA)

Dr. Schipper, a senior researcher engineer at the Precourt Energy Efficiency Center at Stanford, passed away on August 16, 2011. Dr. Schipper joined the Center in September, 2008 to work on policy studies of efficient energy use in transport systems and became a highly valued member of the team as both researcher and educator. Awarded with a Ph.D. in astrophysics from Berkeley, Dr. Schipper devoted his career to earthly problems of transport, energy, and the environment. Dr. Schipper has had a long and rewarding career and was, among others, Senior Project Scientist at Global Metropolitan Studies (UC Berkeley); Director of Research for EMBARQ (the World Resources Institute [WRI] Center for Sustainable Transport), which he helped found in 2002; Staff Senior Scientist at the Lawrence Berkeley National Laboratory; and Visiting Scientist for the IEA in Paris. Dr. Schipper has authored over hundred technical papers and a number of books on energy economics and transportation around the world.

Elizabeth Brooke Stein (Senior Attorney, Climate and Energy Program, Environmental Defense Fund, USA)

Ms. Stein is a senior attorney at EDF's US Climate and Energy Program, focusing on energy efficiency in large commercial buildings and the development of a low-carbon energy system. She has focused on energy in buildings in New York City, including the City's adoption and implementation of the Greener, Greater Buildings laws. Ms. Stein has worked with the City to help create a market for energy efficiency through innovations relating to energy efficiency finance such as on-bill repayment and EDF's Investor Confidence Project. She has also engaged in state, regional, and federal proceedings to advocate for the promulgation of technology and practices that improve the flexibility and performance of the electric system to enable the full deployment of low-carbon resources, including demand response and intermittent renewable resources.

She is currently working to ensure post hurricane Sandy transformation of New York's electric system towards resiliency and sustainability. Ms. Stein holds an AB from Harvard and a JD from New York University.

Peter C. Syrett AIA, LEED AP BD+C (Architect, Perkins+Will, USA)

Mr. Syrett's vision for and experience in sustainability and living buildings have made him one of Perkins+Will's champions for green architecture and construction. Mr. Syrett leads the sustainability efforts of the New York office and leads teams in viewing the larger ecological picture, one that looks beyond LEED, overseeing projects from brainstorm to detail. He led the firm's effort in 2009 to develop the Precautionary List: the architectural industry's first major list of substances to avoid using in building design. Mr. Syrett also leads Perkins+Will's Sustainable Advisory Services in the northeast region, helping clients to strategize investments and returns to achieve a more sustainable future. His philosophy on design is the creation of a unique conceptual vocabulary that embodies a client's mission in space, material, form, and character. He lectures regularly on green institutional design and is a recognized expert in the field.

Job Taminiau (Ph.D. candidate, University of Delaware, USA)

Mr. Taminiau is a Ph.D. candidate at the Center for Energy and Environmental Policy at the University of Delaware. He holds a bachelor's degree in biology and a master's degree in energy and environmental Science, both from the University of Groningen in the Netherlands. Mr. Taminiau received first place in the climate policy thesis award from CE Delft for his Master thesis. Working as a researcher at the Joint Implementation Network, he gained experience in the energy and environmental policy field and contributed to a book on new and innovative solutions to climate change. Interested in international climate change policy making and, more specifically, decision-making structures, he focuses on the social origin and consequences of the issue of climate change from a political ecology perspective.

Young-Doo Wang (Associate Director Center for Energy and Environmental Policy and Professor, University of Delaware, USA)

Dr. Wang is the Associate Director of the Center for Energy and Environmental Policy (CEEP) in the College of Engineering at the University of Delaware. He is also Professor and Director of the Energy and Environmental Policy Program (ENEP). Dr. Wang currently serves as an Associate Editor of *WIRE: Energy and Environment*, as a member of the Editorial Board of the *Korean Journal of Policy Development*, and also as a member of the External Research and Technical Advisory Board (ERTAB) to review Nevada's NSF EPSCoR Research Infrastructure Improvement Grant. In 2009, he co-created the first undergraduate degree in the United States focused around energy and environmental policy.

Dr. Wang's recent books include *Energy Revolution: 21st Century Energy and Environmental Strategy* published by the Maeil Business Newspaper (a prestigious publisher in Korea) and *Water Conservation-Oriented Rates: Strategies to extend Supply, Promote Equity, and Meet Minimum Flow Levels* published by the American Water Works Association.

Xiaoting Wang (Solar Analyst, Bloomberg New Energy Finance, China)

Dr. Xiaoting Wang is a Solar Analyst at Bloomberg New Energy Finance in China. Prior to her current appointment, Dr. Wang joined the Center of Energy and Environmental Policy, University of Delaware in November 2011 as a post-doctoral researcher. Dr. Wang received her B.S. in Optical Engineering from Zhejiang University in China and her Ph.D. in Electrical Engineering from University of Delaware. During her Ph.D., Dr. Wang worked on a variety of programs such as Very High Efficiency Solar Cell (VHESC) project funded by DARPA. Next to work on new photovoltaic systems, recognized by NREL to have a record efficiency of 36.7 percent, Dr. Wang also systematically quantified the efficiency losses and analyzed the real energy production in a traditional high concentration photovoltaic module adopting monolithic multi-junction solar cells with a vertical spectrum splitting approach. Dr. Wang had fifteen publications during the Ph.D. study, and is preparing for four papers and one book chapter.

Robert M. Wendling (Vice President, Management Information Services, Inc., USA)

Mr. Wendling is Vice President of Management Information Services, Inc. and has thirty years of experience in consulting and management. Serving in private industry and the Federal government, his consulting background includes economic analysis of the renewables industry and markets, energy and environmental legislation, assessment of oil markets and their impact on the economy, and economic and employment forecasting. Mr. Wendling was Director of the Commerce's *STAT-USA* office and also served as the lead U.S. representative for Asian Pacific Economic Cooperative (APEC) tariff and trade data negotiations. Mr. Wendling has served in industry as Corporate CEO and President and as Corporate Vice President and senior positions in the U.S. Department of Commerce and the Department of Energy. He authored over seventy-five reports and professional publications on energy and environmental topics, lectures frequently, and co-authored *The Impending World Energy Mess.* He holds undergraduate and graduate degrees in Economics from Indiana and George Washington Universities.

Bryan Yonemoto (Ph.D. Candidate, University of Delaware)

Mr. Yonemoto graduated from Tulane University in 2010 with a BSE in Chemical Engineering. He is currently a Ph.D. chemical engineering candidate supervised by Prof. Feng Jiao at the University of Delaware. His research focuses on the development of nanoporous lithium-ion-battery cathodes in order to improve cathode performance. The technique that is at the core of his research is called ionothermal synthesis. Mr. Yonemoto co-authored the peer-reviewed article A General Synthetic Method for MPO_4 *(M=Co, Fe, Mn) Frameworks Using Deep-eutectic Solvents.*

Index